Food Quality, Safety and Technology

Food Quality, Safety and Technology

Editor: Cindy Featherstone

R CALLISTO REFERENCE

www.callistoreference.com

Callisto Reference,
118-35 Queens Blvd., Suite 400,
Forest Hills, NY 11375, USA

Visit us on the World Wide Web at:
www.callistoreference.com

ISBN: 978-1-64116-108-4 (Hardback)

Cataloging-in-Publication Data

Food quality, safety and technology / edited by Cindy Featherstone.
    p. cm.
Includes bibliographical references and index.
ISBN 978-1-64116-108-4
1. Food--Quality. 2. Food--Safety measures. 3. Food industry and trade--Appropriate technology.
I. Featherstone, Cindy.
TP372.5 .F66 2019
664--dc23

Table of Contents

Preface

This book aims to highlight the current researches and provides a platform to further the scope of innovations in this area. This book is a product of the combined efforts of many researchers and scientists, after going through thorough studies and analysis from different parts of the world. The objective of this book is to provide the readers with the latest information of the field.

Food quality refers to the accepted standards of taste, flavor, appearance, dietary and nutritional requirements as well as safety for consumption by humans. The scientific discipline of food safety is concerned with the handling, preparation and storage of food in a manner that is conducive to ensuring its safety. It involves a close inspection of food additives and pesticide residues. The allied field of food technology is a branch of food science concerned with the production methods of food. Food quality is an important aspect of food manufacturing. Besides nutritional requirements, sanitation is an essential consideration of food quality. This book covers the studies that are being conducted in these domains. It also elucidates the modern principles and concepts crucial to their development. It consists of contributions made by international experts working in the field of food science. This book will immensely benefit students, food scientists, nutritionists and engineers.

I would like to express my sincere thanks to the authors for their dedicated efforts in the completion of this book. I acknowledge the efforts of the publisher for providing constant support. Lastly, I would like to thank my family for their support in all academic endeavors.

Editor

Quality assessment and consumer acceptability of bread from wheat and fermented banana flour

Abiodun Omowonuola Adebayo-Oyetoro, Oladeinde Olatunde Ogundipe & Kehinde Nojeemdeen Adeeko

Department of Food Technology, Yaba College of Technology, P.M.B 2011, Yaba, Lagos, Nigeria

Keywords
Acceptability, banana, fermented, quality, wheat

Correspondence
Abiodun Omowonuola Adebayo-Oyetoro, Department of Food Technology, Yaba College of Technology, P.M.B 2011, Yaba, Lagos, Nigeria.

E-mail: wonunext@yahoo.com

Funding Information
No funding information provided.

Abstract

Bread was produced from wheat flour and fermented unripe banana using the straight dough method. Matured unripe banana was peeled, sliced, steam blanched, dried and milled, and sieved to obtain flour. The flour was mixed with water and made into slurry and allowed to stand for 24 h after which it was divided into several portions and blended with wheat flour in different ratios. Proximate and mineral compositions as well as functional, pasting, and sensory characteristics of the samples were determined. The results of proximate analysis showed that crude fiber ranged between 1.95% and 3.19%, carbohydrate was between 49.70% and 52.98% and protein was 6.92% and 10.25%, respectively, while iron was between 27.07 mg/100 g and 29.30 mg/100 g. Swelling capacity of the experimental samples showed a significant difference from that of control. Peak viscosity ranged between 97.00RVU and 153.63RVU for experimental samples compared with 392.35RVU obtained for the control. Most of the sensory properties for the experimental samples were significantly different from the control. This study showed that bread with better quality and acceptability can be produced from wheat–unripe banana blends.

Introduction

Bread is the loaf that results from the baking of dough which is obtained from a mixture of flour, salt, sugar, yeast, and water. Other ingredients like fat, milk, milk solids, egg, anti-oxidants, etc. may be added. Bread is an important food whose consumption is steady and increasing in Nigeria. It is, however, relatively expensive because it is made from wheat flour which has to be imported (Edema et al. 2004). Bread is an important staple food both in the developed and developing world (Abdelghafor et al. 2011). In India, bread has become one of the most widely consumed nonindigenous food (Das et al. 2012), whereas in Nigeria, bread has become the second most widely consumed nonindigenous food after rice (Shittu et al. 2007). Efforts have been made to promote the use of composite flour, in which flours can be made from locally grown crops replace a portion of wheat for use in bread, thereby, decreasing the demand for imported wheat

and producing protein enriched bread (Giami et al. 2005). The predominance of wheat flour for baking aerated breads is due to the properties of its elastic gluten protein which helps in producing a relatively large loaf volume with a regular, finely vesiculated crumb structure.

Banana (*Musa* spp) constitutes a rich energy source with carbohydrate accounting for 22–32% of the fruit weight. It is rich in vitamins A, B$_6$, and C as well as minerals particularly potassium, magnesium, phosphorus, and folate (Chandler 1995; Honfo et al. 2007a,b). Banana is a major staple crop for millions in developing countries. Meanwhile, a lot of postharvest losses result from the large quantities of this crop produced annually. It is therefore imperative to introduce diversification to its use thereby reducing this wastage. Preparation of banana flour from unripe flour has been reported by some researchers (Rodriguez-Ambriz et al. 2008). Acceptable bread from wheat–plantain composite flour using up to 80:20 w/w ratios of wheat: mature green plantain flour has been reported (Mepba et al. 2007). The aim of this study is to produce and evaluate the quality of bread prepared from blends of wheat and banana flour.

Materials and Methods

Materials

Hard wheat flour, bakers' yeast, granulated sugar, table salt, baking fat, and vegetable fat were purchased from Ojuwoye market, Mushin, Lagos. The unripe banana bunch was purchased from the fruit market, Ketu. Production of bread was carried out in the laboratory of the department of Food Technology, Yaba College of Technology, Yaba, Lagos, Nigeria.

Sample preparation

The banana slurry was produced by the modified method of Oloyede et al. (2013). A bunch of unripe banana whose peel is green and whose pulp is not soft was washed, peeled, and cut into round slices of 10 mm thickness. The slices obtained were steam blanched for 10 min, dried at 60°C for 24 h, milled, sieved, and mixed with water (10 g flour/3 mL water) before fermentation for 24 h. The fermented slurry was now mixed with wheat flour in ratio 90:10, 80:20, and 70:30 and other ingredients such as fat, yeast, and salt to make bread (Fig. 1).

Formulation of blends

Composite flour samples containing wheat and unripe banana flour were formulated, using ratio 100:0, 90:10, 80:20, and 70:30 and coded CON, KAN, TAD, and OAV, respectively.

Recipe for bread

Flour —————100%
Water—————65%
Fat—————3%
Sugar—————6%
Salt and yeast—————2–2.5% (Raymond, 2001 and Saus, 2008).

Fifty-five percentage water was used in this study because the banana was in the slurry form instead of the dried powder.

Production of bread using composite flour

The straight dough method of bread making as described by Badifu and Akaa (2001) was used (Fig. 2).

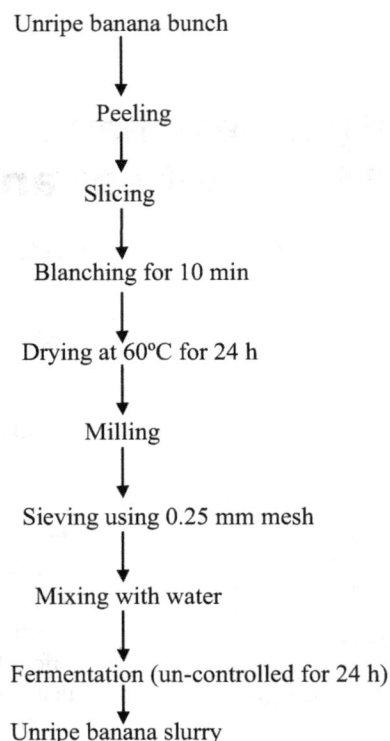

Figure 1. Flowchart for the production of unripe banana slurry. Source: Modified method of Oloyede et al. (2013).

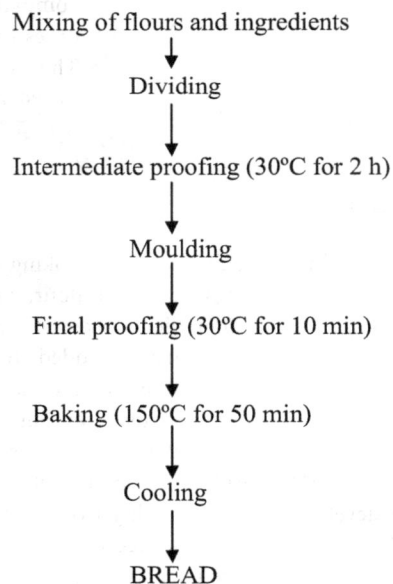

Figure 2. Flowchart for straight dough method of bread making. Source: Badifu and Akaa (2001).

Proximate composition

Proximate analyses of the bread samples were carried out using the AOAC (2005) for protein, crude fiber, fat

Table 1. Proximate composition of bread samples.

Sample	Moisture (%)	Ash (%)	Fat (%)	Crude fiber (%)	Crude protein (%)	Carbohydrate
CON	28.94 ± 7.32[c]	0.93 ± 0.04[b]	1.81 ± 0.00[a]	0.05 ± 0.01[d]	10.25 ± 0.13[a]	52.98 ± 0.01[a]
KAN	34.71 ± 0.27[b]	1.11 ± 0.01[a]	1.76 ± 0.00[a]	1.95 ± 0.52[c]	9.41 ± 0.01[b]	51.09 ± 0.00[a]
TAD	35.26 ± 0.00[b]	1.49 ± 0.05[a]	1.72 ± 0.04[a]	2.44 ± 0.62[b]	8.30 ± 0.14[b]	50.85 ± 0.05[a]
OAV	36.95 ± 0.21[a]	1.59 ± 0.09[a]	1.66 ± 0.03[a]	3.19 ± 0.13[a]	6.92 ± 0.02[c]	49.70 ± 0.15[b]

Sample CON: 100% wheat flour, KAN: 90% wheat and 10% unripe banana flour, TAD: 80% wheat and 20% unripe banana flour, OAV: 70% wheat and 30% unripe banana flour.
Mean values with the same letter within the same column are not significantly (*P* > 0.05) different.

content, ash content and moisture content. Carbohydrate content was determined by difference.

Functional properties

Bulk density of the flour was determined using the method of Onwuka (2005) while the swelling capacity was determined according to the method described by Oladele and Aina (2007).

Pasting properties

The pasting properties of samples were assessed in an RVA-4 (Rapid Visco Analyzer), using the RVA general pasting method.

Sensory evaluation

The sensory analysis was carried out using the scoring test as described by Akinjayeju (2009). The sensory attributes of the bread samples including crust color, crumb color, texture, taste, and aroma were evaluated about 1 h after baking by a semitrained 15-member panel that are familiar with bread.

Statistical analysis

All analysis was conducted in duplicate and the data were all subjected to ANOVA (analysis of variance) as described by Akinjayeju (2009) while the mean will be separated by DMRT (Duncan Mean Range Test).

Results and Discussion

Table 1 shows the proximate composition of wheat–plantain composite bread. Moisture content ranged between 28.94% and 36.95% with sample CON having the least and sample OAV having the highest, this was in agreement with the findings of Olaoye and Onilude (2011). It was observed that as the substitution level increases the moisture content also increases. Highest amount of ash is in sample OAV with 1.59 ± 0.09% and the least

Table 2. Mineral composition of wheat/plantain flour blend.

Samples	Iron (Fe) mg/100 g	Zinc (Zn) mg/100 g
CON	27.07 ± 1.17[b]	29.19 ± 1.31[b]
KAN	29.00 ± 0.17[a]	30.24 ± 0.08[a]
TAD	29.39 ± 0.18[a]	30.61 ± 0.38[a]
OAV	30.09 ± 0.22[a]	31.36 ± 0.16[a]

Sample CON: 100% wheat flour, KAN: 90% wheat and 10% unripe banana flour, TAD: 80% wheat and 20% unripe banana flour, OAV: 70% wheat and 30% unripe banana flour.
Mean values with the different letters within the same column are not significantly different at (*P* > 0.05).

was in sample KAN 1.11 ± 0.01%. This was in agreement with the findings of (Mongi et al. 2011). It was also observed that the crude protein was very high in CON with a value of 10.25 ± 0.01%, followed by KAN, TAD, and OAV with 9.41 ± 0.01%, 8.30 ± 0.14%, and 6.92 ± 0.02%, respectively. Sample CON has the highest value of carbohydrate with mean score of 52.98 ± 0.01. It was observed that the carbohydrate content was reducing with increase in the substitution level. This is in agreement with Olaoye and Onilude (2011) and contrary to the findings of (Mepba et al. 2007). The results showed significant difference between the proximate compositions of the samples.

Table 2 shows the mineral composition of flour samples. Minerals are essential nutrients that are needed in the body to facilitate proper functioning of certain organs (Amoakoah et al. 2015). Some minerals are needed in smaller quantity (micro) while others are needed in larger quantity (macro). This study analyzed iron and zinc as macro and micro nutrients, respectively. The result showed that the concentration of iron (Fe) in sample OAV is higher than other two samples with a mean value of 30.09 mg/100. Meanwhile the least concentration of iron (Fe) is found in the experimental sample KAN with value of 29.00 mg/100. However, the result also showed that the amount of iron in sample CON (27.07 mg/100) is lower than the other samples which were KAN (29.00 mg/100), TAD (29.39 mg/100), and OAV (30.09 mg/100), respectively. Furthermore, the table showed

Table 3. Functional properties of wheat/plantain flour blend.

Samples	Bulk density (%)	Swelling capacity (%)
CON	0.75 ± 0.17[a]	40 ± 0.00[c]
KAN	0.78 ± 0.01[a]	40 ± 0.00[c]
TAD	0.79 ± 0.03[a]	53 ± 1.29[b]
OAV	0.72 ± 0.14[b]	64 ± 2.71[a]

Sample CON: 100% wheat flour, KAN: 90% wheat and 10% unripe banana flour, TAD: 80% wheat and 20% unripe banana flour, OAV: 70% wheat and 30% unripe banana flour. Mean values with the different letters within the same column are not significantly different at ($P > 0.05$).

that the highest amount of Zinc (Zn) was found in sample OAV with a mean value of 31.36 mg/100, while sample KAN has the lowest value of 30.24 mg/100. It was observed that there is a significant difference between the experimental samples and the control samples in terms of iron and zinc content.

Table 3 shows the functional properties of wheat–plantain flour blends. The bulk density ranged from 0.72 to 0.78 g/cm^3 with sample OAV having the highest value while sample TAD has the lowest amount. This means the higher the bulk density, the denser the flour (Eke-Ejiofor and Kiin-Kabari 2012), suggesting that sample TAD is denser than other substituted samples. There is no significant difference between the control sample and samples KAN AND TAD. Swelling capacity is regarded as the quality criterion in some good formulations such as bakery products (Osungbaro et al. 2010). The sample OAV exhibited highest the swelling power 64% while sample KAN exhibited the lowest swelling power with 40%. Fermentation of the unripe banana flour and substitution level was observed to influence progressive increase in the swelling capacity. Fermentation and sun drying had been observed to clearly play a role in obtaining starch with high swelling power and desirable organoleptic properties. This has been found to facilitate production of high-quality cassava-wheat composite flours of which demand exist in bread making and various confectionery industries (Duffour et al. 1994). Sample OAV was significantly different from other samples in terms of bulk density while a significant difference also existed

between samples CON and KAN, and other samples (TAD and OAV), respectively.

The pasting properties of the flour samples are shown in table 4. The pasting properties of starch are used in assessing the suitability of its application as functional ingredient in food and other industrial products (Oluwalana and Oluwamukomi 2011). The most important pasting characteristics is its amylographic viscosity (Sandhu et al. 2007). The pasting temperature of the flour samples ranges between 76.60°C and 87.57°C and the control sample is 87.2°C. The pasting temperature is a measure of the minimum temperature required to cook a giving food sample (Sandhu et al. 2007). The peak time is the measure of the cooking time (Adebowale et al. 2005). Both KAN and CON have the highest peak time of 5.97 min and 5.97 min, respectively, followed by TAD with 5.87 min while OAV has the lowest value of 5.70 min. Peak viscosity is often correlated with the final product quality. It also provides an indication of the viscous load likely to be encountered during mixing (Maziya-Dixon et al. 2004). Higher swelling capacity is indicative of higher peak viscosity while higher solubility as a result of starch degradation or dextrinization results in reduced paste viscosity (Shittu et al. 2001). The hold period sometimes called shear thinning, holding strength, hot paste viscosity, or trough due to the accompanied breakdown in viscosity is a period when the sample was subjected to a period of constant temperature (usually 95°C) and mechanical shear stress. It is the minimum viscosity value in the constant temperature phase of the RVA profile and measures the ability of paste to withstand breakdown during cooling (Newport, 1998). Sample KAN has the highest trough value 115.8RVU followed by TAD with a value of 110.6RVU and the least was OAV with 101.2RVU, while the control sample has a trough value of 224.75 RVU. This period is often associated with a breakdown in viscosity (Ragaee et al. 2006). It is an indication of breakdown or stability of the starch gel during cooking. The lower the value the more stable is the starch gel. The breakdown is regarded as a measure of the degree of disintegration of granules or paste stability (Newport Scientific 1998). The breakdown viscosities ranged between 52.45RVU and 81.2RVU and the control

Table 4. Pasting properties of samples.

Samples	Peak (RVU)	Trough 1 (RVU)	Breakdown viscosity (RVU)	Final viscosity (RVU)	Setback (RVU)	Time (min)	Pasting temp. (°C)
KAN	197.00 ± 2.53	115.80 ± 18.38	81.25 ± 0.91	227.95 ± 1.85	112.15 ± 3.23	5.97 ± 0.05	87.57 ± 0.60
TAD	183.95 ± 1.85	110.60 ± 11.31	73.35 ± 3.53	215.50 ± 2.28	104.90 ± 1.97	5.87 ± 0.09	87.20 ± 0.07
OAV	153.65 ± 1.77	101.20 ± 5.65	52.45 ± 2.12	191.85 ± 1.61	90.65 ± 0.95	5.70 ± 0.05	86.83 ± 0.53
CON	392.35 ± 2.15	224.75 ± 12.36	167.60 ± 1.79	384.20 ± 1.98	159.45 ± 0.63	5.97 ± 0.33	76.60 ± 0.85

Sample CON: 100% wheat flour, KAN: 90% wheat and 10% unripe banana flour, TAD: 80% wheat and 20% unripe banana flour, OAV: 70% wheat and 30% unripe banana flour.

Table 5. Sensory analysis of bread samples.

Samples	Crumb color	Crumb texture	Crust color	Taste	Aroma	Overall acceptability
CON	5.0[a]	5.5[a]	6.2[a]	5.8[a]	5.2[a]	5.5[a]
KAN	4.5[b]	5.0[a]	6.0[a]	5.5[a]	5.0[a]	5.0[a]
TAD	4.0[b]	4.0[b]	3.5[b]	4.0[b]	3.5[a]	3.5[b]
OAV	3.0[c]	2.5[c]	3.5[b]	2.5[c]	3.0[a]	2.5[c]

KAN: 90% wheat and 10% unripe banana flour, TAD: 80% wheat and 20% unripe banana flour, OAV: 70% wheat and 30% unripe banana flour. Mean values with the different letters within the same column are not significantly different at ($P > 0.05$).

sample CON had the highest break down viscosity value of 224.75RVU. The viscosity after cooling to 50°C represents the setback or viscosity of cooked paste. It is a stage where retrogradation or reordering of starch molecules occurs. Higher setback values are synonymous to reduced dough digestibility (Shittu et al. 2001), while lower setback during cooling of the paste indicates lower tendency for retrogradation (Izonfuo and Omuaru 1988). The final viscosities of the samples were KAN (227.95RVU), TAD (215.5RVU), OAV (191.85RVU), and 384.20RVU for CON. The setback value for the control sample CON was 159.45RVU, while KAN, TAD, and OAV have 227.95RVU, 215.5RVU, and 191.85RVU, respectively. The setback viscosity indicates the tendency of the dough to undergo retrogradation, a phenomenon that causes dough to become firmer and increasingly resistant to enzyme attack (Ihekoronye and Ngoddy 1985). This has a serious implication on the digestibility of the dough when consumed. Higher setback values are synonymous to reduced dough digestibility (Shittu et al. 2001), while lower setback during the cooling of the paste indicates lower tendency for retrogradation (Sandhu et al. 2007).

The Table 5 below shows the result of sensory analysis. The result showed that sample KAN is the most acceptable in terms of crumb and crust color with mean score of 4.5 and 6.0 while sample OAV is the least preferred in terms of crumb and crust color, respectively. As the level of substitution increases, the acceptance of crumb color decreases. With reference to taste and aroma, sample KAN is the most preferred with an average mean scores of 5.5 and 5.0, respectively followed by sample TAD with mean scores of 4.0 and 3.5 while sample OAV is the least accepted with mean scores of 2.5 and 3.0, respectively. Sample KAN was the most preferred in terms of overall acceptability with an average mean score of 5.0 while sample OAV is the least accepted with a score of 2.5. This is as a result of substitution of wheat flour with unripe banana flour. These results are similar to the findings of Mongi et al. (2011) and Mepba et al. (2007). It was also observed that a significant difference occurs between the control sample and most of the experimental samples except KAN in most of the sensory parameters measured.

Conclusion

Addition of unripe banana was found to increase the crude fiber, ash, iron, and zinc content of the bread samples. Meanwhile, the products were acceptable by the sensory panelists although a significant difference was observed between the control and the experimental samples used in this study.

Conflict of Interest

Authors hereby declare that no conflict of interest exists.

References

Abdelghafor, R. F., A. I. Mustafa, A. M. H. Ibrahim, and P. G. Krishnan. 2011. Quality of bread from composite flour of sorghum and hard white winter wheat. Adv. J. Food Sci. Technol. 3:9–15.

Adebowale, Y. A., I. A. Adeyemi, and A. A. Oshodi. 2005. Functional and physico- chemical properties of flour of six Mucuna Species. Afr. J. Biotechnol. 4:1461–1468.

Akinjayeju, O. 2009. Quality control for the food industry: a statistical approach. Concept Publications, Lagos.

Amoakoah, T. L., I. D. Kottoh, I. K. Asare, W. Torby-Tetteh, E. S. Buckman, and A. Adu-Gyamfi. 2015. Physicochemical and elemental analyses of bananas composite flour for infants. Br. J. Appl. Sci. Technol. 6:277–284.

AOAC. 2005. Official Methods of Analysis of AOAc International. 18th ed. (AOAC-925.10), (AOAC-2003.05), (AOAC-923.03), (AOAC-960.52) (Nx6.25). AOAC International, Maryland, USA.

Badifu, G. I. O., and S. Akaa. 2001. Evaluation of shea fat performance as a shortening in breadmaking. J. Food Sci. Technol. 28:59–68.

Chandler, S. 1995. The nutritional value of banana. Pp. 77–89 in S. R. Gowen, ed. Banan and plantain. Chapman and Hall, London.

Das, L., U. Raychaudhri, and R. Chakraborty. 2012. Effects of baking conditions on the physical properties of herb bread using RSM. Int. J. Food Agri. Vet. Sci. 2:106–114.

Dufour, D., S. Larsonneur, S. Alarcon, C. Brabet, and G. Chuzel. 1996. Improving the bread-making potential of cassava sour starch. Pp. 133–142 *in* D. Dufour, G. M. O'Brian and R. Best, ed. Proceedings of meeting on cassava flour and starch. Progress in research and development. CIRAD/CIAT, CA.

Edema, M. O., L. O. Sanni, and A. Sanni. 2004. Evaluation of Plantain flour blends for plantain bread production in Nigeria. Afr. J. Biotechnol. 4:911–918.

Eke-Ejiofor, J., and D. B. Kiin-Kabari. 2012. Effects of substitution on the functional properties of flour, proximate and sensory properties of wheat/plantain composite bread. Int. J. Agri. Sci. 2:1–5.

Giami, S. Y., M. N. Adindu, A. D. Hurt, and E. O. Denenu. 2005. Effect of heat processing on in-vitro protein digestibility and some chemical properties of African breadfruits. Plant Foods Hum. Nutr. 56:117–126.

Honfo, F. G., A. P. P. Kayode, O. Coulibaly, and A. Tenkouano. 2007a. Relative contribution of banana and plantain products to the nutritional requirement for iron, zinc and vitamin A of infants and mothers in Cameroon. Fruits 62:267–277.

Honfo, F. G., K. Hell, O. Coulibaly, and A. Tenkouano. 2007b. Micronutrient value and contribution of plantain-derived foods to daily intakes of iron, zinc and β-carotene in Sountern Nigeria. Info-Musa 16:2–6.

Ihekoronye, A. I., and P. O. Ngoddy. 1985. Integrated food science and technology for the tropics. Macmillan Publishers Limited, London and Basingstoke.

Izonfuo, W.-A. L., and V. O. T. Omaru. 1988. Effect of ripening on the chemical composition of plantain peels and pulps. J. Food Agri. 45:333–336.

Maziya-Dixon, B., A. G. Dixon, and A. Adebowale. 2004. Targeting different end uses of cassava: genotypic variations for cyanogenic potentials and pasting properties. A paper presented at ISTRC-AB Symposium, 31 October – 5 November 2004, Whitesands Hotel, Mombassa, Kenya.

Mepba, H. D., L. Eboh, and S. U. Nwaojigwa. 2007. Chemical composition, functional and baking properties of wheat-plantain composite flours. Afr. J. Food Agri. Nutr. Dev. 7:1–22.

Mongi, R. J., B. K. Ndabikunze, B. E. Chove, P. Mamiro, C. C. Ruhembe, and J. G. Ntwenya. 2011. Proximate composition, bread characteristics and sensory evaluation of cocoyam-wheat composite breads. Afr. J. Food Agric. Nutr. Dev. 11:1–14.

Newport Scientific. 1998. Applications manual for the Rapid Visco Analyzer using thermocline for windows. Newport Scientific Pty Ltd., Warriewood, NSW 2012, Australia. Pp. 2–26.

Oladele, A. K., and J. O. Aina. 2007. Chemical composition and functional properties of flour from two varieties of tiger nut (*Cyperus esculentus*). Afr. J. Biotechnol. 6:2473–2476.

Olaoye, O. A., and A. A. Onilude. 2011. Microbiological, proximate analysis and sensory evaluation of baked products from blends of wheat-breadfruit flours. Afr. J. Food Agric. Nutr. Dev. 8:1–12.

Oloyede, O. O., O. B. Ocheme, and L. M. Nurudeen. 2013. Physical, sensory and microbiological properties of wheat-fermented unripe plantain flour. Niger. Food J. 31:123–129.

Oluwalana, I. B., and M. O. Oluwamukomi. 2011. Proximate composition, rheological and sensory qualities of plantain (*Musa parasidiaca*) flour blanched under three temperature regimes. Afr. J. Food Sci. 5:769–774.

Onwuka, G. I. 2005. Food analysis and instrumentation (theory and practice). 1st Ed. Napthali Prints, Surulere, Lagos, Nigeria. Pp. 140–160.

Osungbaro, T. O., D. Jimoh, and E. Osundeyi. 2010. Functional and pasting properties of composite cassava-sorghum flour meals. Agri. Biol. J. Nutr. Am. 1:715–720.

Ragaee, S., E. M. Abdel-Aal, and M. Noaman. 2006. Antioxidant activity and nutrient composition of selected cereals for food use. Food Chem. 98:32–38.

Raymond, C. 2001. The taste of bread. Gaithersburg Md: Aspen Publishers Philadephia, U.S.A. Pp. 31.

Rodriguez-Ambriz, S. L., J. J. Isla-Hernandez, E. AgamaAcevedo, J. Jovar, and L. A. Bello-Perez. 2008. Characterisation of fibre-rich powder prepared by liquefaction of unripe banana flour. Food Chem. 107:1515–1521.

Sandhu, K. S., N. Singh, and N. S. Malhi. 2007. Some properties of corn grains and their flour I: physicochemical, functional and chapatti-making properties of flour. Food Chem. 101:938–946.

Saus, M. 2008. Advanced bread and pastry: A Professional Approach. DEMAR CENGAGE Learning. Clifton Park, New York, U.S.A. Pp. 52.

Shittu, T. A., O. O. Lasekan, L. O. Sanni, and M. O. Oladosu. 2001. The effect of drying methods on the functional and sensory characteristics of pukuru-a fermented cassava product. ASSET Int. J. 1:9–16.

Shittu, T. A., A. O. Raji, and L. O. Sanni. 2007. Bread from composite cassava-wheat flour: I. Effect of baking time and temperature on some physical properties of bread loaf. Food Res. Int. 40:280–290.

2

Effect of conventional milling on the nutritional value and antioxidant capacity of wheat types common in Ethiopia and a recovery attempt with bran supplementation in bread

Genet Gebremedhin Heshe[1], Gulelat Desse Haki[2], Ashagrie Zewdu Woldegiorgis[1] & Habtamu Fekadu Gemede[1,3]

[1]Center for Food Science and Nutrition, College of Natural Sciences, Addis Ababa University, P.O. BOX 1176, Addis Ababa, Ethiopia
[2]Department of Food Science and Technology, Botswana College of Agriculture, Private Bag 0027, Gaborone, Botswana
[3]Department of Food Technology and Process Engineering, Wollega University, P.O. Box: 395, Nekemte, Ethiopia

Keywords
Antioxidant, bran, nutritional, refined milling, wheat, white flour

Correspondence
Habtamu Fekadu Gemede, Center for Food Science and Nutrition, College of Natural Sciences, Addis Ababa University, P.O. BOX 1176, Addis Ababa, Ethiopia.

E-mail: fekadu_habtamu@yahoo.com

Funding Information
No funding information provided.

Abstract

The effect of wheat flour refined milling on nutritional and antioxidant quality of hard and soft grown in Ethiopia was evaluated. Bread was prepared with the supplementation of the white wheat flour with different levels (0%, 10%, 20%, and 25%) of wheat bran. Whole (100% extraction) and white wheat (68% extraction) flours were analyzed for proximates, minerals, and antioxidants. Results indicated that at a low extraction rate (68%), the protein, fat, fiber, ash, iron, zinc, phosphorous, and antioxidant contents of the samples significantly ($P < 0.05$) decreased by milling. The TPC (total phenolic content) of the white wheat flours, which ranged from 3.34 to 3.49 mg GAE (gallic acid equivalent)/g, was significantly ($P < 0.005$) lower than those of the whole wheat flours, whose TPC ranged from 7.66 to 8.20 GAE/g). At 50 mg/mL, the DPPH (2-diphenyl-1-picrylhydrazyl) scavenging effect of the wheat extracts decreased in the order of soft whole, hard whole, soft white, and hard white wheat flour, which was 90.39, 89.89, 75.80, and 57.57%, respectively. Moreover, the proximate and mineral contents of the bran-supplemented breads increased significantly ($P < 0.05$) with the bran level of the bread, and the highest values (protein, 12.0 g/100 g; fat, 2.6 g/100 g; fiber, 2.5 g/100 g; ash, 3.3 g/100 g; iron, 4.8 mg/100 g and zinc, 2.33 mg/100 g) were found in 25% bran supplemented bread. The sensory evaluation of bread showed that all the supplementation levels had a mean score above 4 for all preferences on a 7- point hedonic scale. The results indicated that refined milling at 68% extraction significantly reduces the nutritional and antioxidant activity of the wheat flours. Bread of good nutritional and sensory qualities can be produced from 10% and 20% bran supplementations.

Introduction

Wheat has accompanied humans since remote times (as far back as 3000–4000 BC) in their evolution and development, evolving itself (in part by nature and in part by manipulation) from its primitive forms (emmer wheat) into the presently cultivated species (Curtis et al. 2002). Wheat crop is widely adapted to a variety of environments and is cultivated in tropical, subtropical, and tem-perate areas (Hussain et al. 2010). It is widely consumed by humans in over 100 countries that are primary producers and in other countries where wheat cannot be grown (Shewry 2009). It also occupies 27% of the total cereal production worldwide (Curtis et al. 2002). It is thus, an important agricultural commodity, which is consumed in large amount all over the world among all grains.

Ethiopia is the largest wheat producer in sub-Saharan

Africa (MOA, 2011). Nationally, wheat ranks fourth in total area coverage (1,389,215.00 ha). It is also third in productivity (after maize and sorghum) among cereals (CSA, 2005). It is one of the most important crops grown and consumed in Ethiopia both in terms of total production (2.85 million MT in 2010/11) (CSA, 2011) and the proportion of total calories consumed in the country (19.6% of calories consumed) (Rashid et al. 2010).

Wheat possesses several health benefits, especially when utilized as a whole-grain product. According to Kumar et al. (2011), wheat provides protection against diseases such as constipation, ischaemic, heart disease, diverticulum, appendicitis, diabetes, and obesity. These benefits are attributed in part to the presence of different compounds such as dietary fibers, phytochemicals, proteins, vitamins, and minerals (Ragaee et al. 2012). Whole-wheat grain consists of bran, germ, and endosperm. When conventionally milled, only carbohydrate- rich endosperm is retained. This results in a big loss of many nutritionally valuable biochemical compounds such as dietary fiber, vitamins, minerals, and antioxidant compounds, which play an important role in reducing CVD (cardiovascular disease) (Mellen et al. 2008). When white flour is produced, many important nutrients and fiber are removed because these components are mainly located in bran and germ (Iuliana et al. 2012). Wheat bran is rich in protein (~14%), carbohydrates (~27%), minerals (~5%), and fat (~6%) (Anwarul et al. 2002). In addition, wheat bran is the main by-product of conventional flour milling. Wheat bran is a most important fiber source, which is inexpensive and available. It is a good source of not only dietary fiber. The loss of vitamins and minerals in the refined wheat flour has led to widespread prevalence of constipation and other digestive disturbances and nutritional disorders (Kumar et al. 2011).

Milling is the critical process affecting the concentrations of nutrients in wheat-derived food products. The outer parts of the kernel, especially the aleurone layer and the germ are richer in minerals. Conventional milling reduces nutritional content of flour and concentrates them in the milling residues (Cubadda et al. 2009). White flour with a milling extraction rate 68% mean up to 32% of the original grain is not in the flour. Whole grain flour includes all parts of the seed and is 100% milling extraction rate. Milling of wheat into highly refined flours not only precludes considerable amounts of nutrients from human consumption, but the remaining flours have a much poorer nutritive value than flours made from whole wheat.

Over the past 20 years, wheat production and consumption have both increased in Ethiopia (Bergh et al. 2012). In Ethiopia, 28% of consumers purchase wheat flours and flour products, and about 22 million people use wheat flours (FDRE, 2011). However, no published information is available regarding the effects of conventional milling refining on nutritional and antioxidant capacity of wheat that are commonly grown in Ethiopia, though wheat is widely distributed and consumed. The nutritional value and antioxidant properties of wheat grain are significantly influenced by soil type and richness, growing temperatures, moisture levels, other climatic differences, and genotype (Adom et al. 2003). It is therefore, very important to understand the nutritional value of Ethiopian wheat and evaluate the effects of conventional milling. In addition, it is necessary to find a way to improve the nutrient quality of wheat products without compromising palatability.

Materials and Methods

Samples

Hard wheat (Kubsa) and soft wheat (ET-13) samples were obtained from the Kebron food complex (Oromia region) and Wedera farmers cooperative (Debrebrhan), respectively, in Ethiopia. Bran sample was obtained from the Universal Food Complex (Addis Ababa, Ethiopia) (Fig. 1).

Milling of wheat

The amount of water required for tempering was calculated according to AACC (2000). One kilogram of each sample from both the soft and hard wheat were cleaned and tempered separately to 14% moisture level and kept for 6 and 24 h, respectively, at ambient temperature in a closed plastic jar. After tempering, wheat samples were milled at the extraction rates of 68% and 100% by using. The milling of the flour was conducted at Kokeb Flour and Pasta Factory and extraction rate was calculated according to Slavin et al. (2000).

Formulation of bread

Flour blends were prepared by mixing wheat flour with wheat bran in the proportions of 100: 0, 90:10, 80:20, and 75:25 (wheat flour to bran) using homogenizer and 100% white wheat flour was used as the control. The formulation was made based on the preliminary test (unpublished). The four flour samples were packaged in black low-density polyethylene bags and stored in plastic containers at room temperature from where samples were taken for bread production.

Bread manufacture

Bread was prepared with the formulated flours in 2.3, and the dough was prepared based on the method de-

Hard wheat   Bran   Soft wheat

Figure 1. Hard wheat bran soft wheat.

scribed by Hertzberg and Francois (2007) with some modifications; Each formulated flour (400 g) was mixed (Linkrich-B15; with sugar (8 g), salt (4 g), oil (8 g), and yeast (1.6 g; baker' s yeast) for 30 min and then water was added to the dough for the desired consistency. The dough was weighed and divided into three equal portions for replications. These were placed in baking pans and left for 1 h. Then they were transferred into an oven preheated to about 180–250°C and allowed to bake for 20 min. The baked products were left to cool.

Nutritional analysis of wheat flour, bran and bread

All samples were analyzed for moisture, crude protein, crude fat, and total ash by standard methods (AOAC, 2000).

Determination of mineral contents

Iron and zinc were determined according to the standard method of AOAC (2000) using flame Atomic Absorption Spectrophotometer. Ash was obtained from dry ashing of the samples. The ash was wetted completely with 5 mL of 6N HCl, and dried on a low temperature on hot plate. A 7 mL of 3N HCl was added to the dried ash and heated on the hot plate until the solution just boiled. The ash solution was cooled to room temperature in a hood and was filtered using filter paper (Whatman 45). A 5 mL of 3N HCl was added into each crucible dishes and was heated until a solution boiled then cooled and filtered into the flask. The crucible dishes are again washed three times with deionized water, the washing was filtered into a flask. Then, the solution was cooled and diluted to 50 mL with deionized water. A blank was prepared

by taking the same procedure as the sample.

Sample extraction

Samples were extracted based on the procedures as outlined by Woldegiorgis et al. (2014). The powdered wheat samples were homogenized and weighed (10 g) before extraction by stirring with 100 mL of methanol at 250 C at 150 rpm for 24 h using an incubator shaker (ZHWY-103) and then filtered through Whatman No. 1 filter paper. The residue was then extracted with two additional 100 mL portions of methanol as described above. The combined methanolic extracts were evaporated at 400°C to dryness using a rotary evaporator and redissolved in methanol at a concentration of 50 mg/mL and stored at 40°C for further use.

Determination of free radical scavenging activity

The hydrogen atoms or electrons donation ability of the corresponding extracts and some pure compounds were measured from the bleaching of purple colored methanol solution of DPPH (Gursoy et al. 2010). Antioxidant activity of the methanol extracts was determined by DPPH radical scavenging method as described by Woldegiorgis et al. (2014). A 0.004% solution of DPPH radical solution in methanol was prepared and then 2 mL of DPPH solution was mixed with 1 mL of various concentrations (0.1–50 mg/mL) of the extracts in methanol. Finally, the samples were incubated for 30 min in the dark at room temperature. Scavenging capacity was read spectrophotometrically by monitoring the decrease in absorbance at 517 nm. Ascorbic acid was used as a standard and mixture without extract was used as the control. Inhibition of free radical DPPH in percent (I %) was then calculated.

Total phenolics determination

Phenolic compounds concentration in the wheat was estimated with Folin–Ciocalteu reagent according to the Singleton & Rossi method (1965) as described by Woldegiorgis et al. (2014). One milliliter of sample (5000 μg) was mixed with 1 mL of Folin and Ciocalteu's phenol reagent. After 3 min, 1 mL of saturated sodium carbonate (20%) solution was added to the mixture and adjusted to 10 mL with distilled water. The reaction was kept in the dark for 90 min, after which the absorbance was read at 725 nm. Gallic acid was used to construct the standard curve (5–80 μg/mL). The results were mean values + standard error of mean and expressed as mg of GAEs (gallic acid equivalents/g of extract).

Sensory evaluation of bread

Sensory evaluation was conducted for the freshly baked breads by 30 semitrained panelists consisting of male and female students, aged from 23 to 43 years old, from the Food Science and Nutrition Center of the Addis Ababa University. The samples were presented randomly in identical containers, coded with three digit numbers. The sensory test was conducted using a seven- point hedonic scale, where 1 = dislike very much, 2 = dislike moderately, 3 = dislike slightly, 4 = neither like nor dislike, 5 = like slightly, 6 = like moderately and 7 = like very much. The sensory attributes evaluated were taste, odor, color, texture, and overall acceptability. Samples were considered as acceptable when their average score for the overall acceptability was >4 (neither like nor dislike) (Lazaridou et al. 2007).

Statistical analysis

The data were subjected to ANOVA (analysis of variance) and Duncan's multiple range tests were used for mean separation at $P < 0.05$. Linear regression analysis was used to calculate IC50 value. Pearson correlation between DPPH scavenging (%) and TPC (total phenolic content) was considered at $P < 0.05$.

Results and Discussion

Proximate composition of wheat, wheat bran, and bread

The mean value for moisture, crude protein, crude fat, total ash and crude fiber of wheat bran, wheat flour (hard and soft), white wheat flour (hard and soft), and bran supplemented bread are presented in Tables 1–3. The mean values for moisture contents of different whole wheat and white wheat flours are presented in Table 1. It ranged from 10.5% to 12.3%. The highest moisture level, 12.3%, was found in hard white wheat flour. The moisture content varies significantly between whole and white flour and between hard and soft white flour ($P < 0.05$).The increment on moisture content of both soft and hard white wheat as compared to the whole wheat flour could be due to the addition of water during the tampering process to facilitate milling of wheat, which resulted in retaining more water in refined wheat flour than whole wheat flour.

There is also a significant difference ($P < 0.05$) on total ash contents of all flour samples (Table 1). The highest ash content (1.6%) was found in hard whole wheat flour whereas the soft white flour showed the lowest (0. 4%) ash content. The results were comparable to Azizi et al. (2006) values obtained from different extraction rate of wheat flour, which ranged 1.51% to 0.54% ash content at 93% and 70% extraction rate, respectively.

The result for crude fat content is shown in Table 1 and the values showed significant difference ($P < 0.05$) between whole wheat flours and white wheat flours. The fat content decreased in white wheat flour. The highest fat content, 1.83%, was found in whole wheat flour (100% extraction rate); whereas, the lowest, 1.32%, was found in white wheat flour (low extraction rate). The high percentage of fat in whole wheat flour is because wheat germ is ground along with endosperm during milling (Farooq et al. 2001).

The results of this study also indicated that the protein contents for all flours varied significantly ($P < 0.05$). The protein contents decreased with in both hard and soft

Table 1. Proximate composition of wheat.

Parameters	Wheat samples			
	HWF	HWWF (refined)	SWF	SWWF (refined)
Moisture (%)	10.75 ± 0.38[a]	12.30 ± 0.09[c]	10.48 ± 0.10[a]	11.60 ± 0.23[b]
Ash (%)	1.62 ± 0.03[d]	0.65 ± 0.01[b]	1.41 ± 0.07[c]	0.38 ± 0.07a
Fat (%)	1.82 ± 0.04[b]	1.43 ± 0.18[a]	1.78 ± 0.10[b]	1.32 ± 0.11[a]
Protein (%)	14.40 ± 0.30[d]	11.91 ± 0.087[c]	9.11 ± 0.12[b]	7.13 ± 0.06a
Fiber (%)	2.6 ± 0.08[b]	0.42 ± 0.06[a]	2.5 ± 0.08[b]	0.36 ± 0.07[a]

HWF, Hard whole wheat flour; HWWF, Hard white wheat flour (refined); SWF, Soft whole wheat flour; SWWF, Soft white wheat flour (refined). Data are average of triplicate ± SE. Mean value with different superscript in the same rows are significantly different ($P < 0.05$).

Table 2. Proximate composition of wheat bran.

Parameter	Composition (g/100 g)
Protein	15.26 ± 0.35
Fat	3.12 ± 0.7
Fiber	9.97 ± 0.27
Ash	4.5 ± 0.16

white wheat flour; at the same time, there was significant difference between hard and soft whole wheat flour. The highest protein content found on hard whole wheat flour was 14.40%; whereas the protein content of soft whole wheat flour was 9.11%. The result of this study was in line with the value of hard and soft whole wheat reported by Blakeney et al. (2009).

The mean value for fiber content of hard whole wheat, hard white wheat, soft whole wheat, and soft white wheat flour are 2.6, 0.42, 2.5 and 0.36 g/100 g, respectively (Table 1). The result of this study showed that there is significant difference ($P < 0.05$) between whole wheat flour and white wheat flour. Both hard and soft whole wheat flour exhibited high crude fiber content (2.6 and 2.5%). The white wheat flour showed less fiber content because wheat bran was removed during milling process, which decreased the amount of fiber in flour. The result of this study is in agreement with Azizi et al. (2006), who reported crude fiber in the range of 0.30–2.24% white wheat and whole flour, respectively.

The proximate composition of wheat bran samples are given in Table 2. Wheat bran was found to contain highest amounts of crude protein, fat, fiber, and ash with mean values of 15.26%, 3.12%, 9.97%, and 4.45%, respectively (Table 2). The objective of milling is to separate the bran and germ from the starchy endosperm so that the endosperm can be ground into flour. The aleurone layer, which is rich in protein, minerals, and vitamins, usually breaks away with the outer layer of the bran in the milling process, thus, contributing significantly to the nutritional quality of the bran fraction (Posner 2000).

Proximate composition of different bran supplemented bread and control were also analyzed for proximate

composition. The mean values for moisture contents of the bread samples are presented in Table 3, which ranged from 30.92% to 32.83%. The highest moisture level, 32.83%, was found in 25% bran supplemented bread. The moisture content of the control bread decreased from the three samples bread significantly ($P < 0.05$).

The statistical analysis for crude protein is presented in Table 3. The mean value for protein content of all the study samples of bread ranged from 9.42 for control to 12.04 for WBB (75:25). The protein contents for three of the breads (0%, 10%, 20% bran supplemented bread) varied significantly ($P < 0.05$). However, there was no significant difference between 20% and 25% bran supplemented bread. The result of protein contents are in agreement with the findings of Butt et al. (2004) who reported an increase in contents of protein with an increase in bran proportion.

The result of this study indicates that crude fat showed significant difference ($P < 0.05$) among all breads. The fat content increased with an increase in bran level. The highest fat content, 2.61%, is found in 25% bran supplemented bread, whereas the lowest, 1.56%, was found in the control bread. The increase in fat content is because of the germ which is grounded along with bran and endosperm during milling, results in bread with higher fat content than the control bread (Farooq et al. 2001).

The mean values of crude fiber contents of different bread samples are given in Table 3. The statistical analysis showed significant ($P < 0.05$) effect on the quantity of crude fiber. The crude fiber contents ranged from 0.38% to 3.27%. The 25% bran supplemented bread exhibited the highest crude fiber (3.27%), whereas the control bread contained the lowest crude fiber (0.38%). The crude fiber increased with an increase in bran supplementation rate. The control showed less fiber contents because the bread was made from refined bread with no addition of bran.

Ash is the mineral residue remaining after a sample has been completely oxidized in a manner such that all organic volatile material is driven off, while preventing any mineral from being lost (Posner 2000). Ash varied significantly among all the bread (Table 3). The statistical analysis showed significant ($P < 0.05$) effect on total ash contents. The results indicate that ash content ranged

Table 3. Proximate composition of Bran supplemented bread and control (Dry weight basis). WFB-(white wheat flour bread) – control, WF:BR 90:10, 10% bran supplemented bread, WF:BR 80:20 – 20% bran supplemented bread, WF:BR 75:25 – 25% bran supplemented bread.

Samples	Moisture %	Protein %	Fat %	Ash %	Fiber %
Control (WFB)	30.92 ± 0.38[a]	9.42 ± .22[a]	1.56 ± .06[a]	1.38 ± 0.10[a]	0.38 ± 0.02[a]
WF:BR (90:10)	33.14 ± 0.26[b]	10.70 ± 0.55[b]	2.14 ± 0.032[b]	2.04 ± 0.32[b]	2.13 ± 0.05[b]
WF:BR (80:20)	33.24 ± 0.69[b]	11.66 ± 0.89[c]	2.36 ± 0.04[c]	2.29 ± 0.04[c]	3.11 ± 0.06[c]
WF:BR (75:25)	32.83 ± 0.58[b]	12.04 ± 0.84[c]	2.61 ± 0.06[d]	2.48 ± 0.02[d]	3.27 ± 0.008[d]

Data are average of triplicate ± SE. Mean value with different superscript in the same column are significantly different ($P < 0.05$).

from 1.38% to 2.48%. The highest ash content (2.48%) was found in 25% bran supplemented bread, whereas the control showed the lowest (1.38%) ash content. The addition of 10% to 25% wheat bran to the bread increased the ash content.

Mineral content of wheat flour and bread

The mineral content of the whole and white flour samples are shown in Table 4. According to the results of this study, the iron content level in hard and soft whole wheat flour is significantly different ($P < 0.05$) from hard and soft white wheat flour. The iron content of whole wheat flour ranged from 2.95 to 4.15 mg/100 g whereas the iron content of the white wheat flour ranged from 2.51 to 3.35 mg/100 g.

Dewettinck et al. (2008) reported that the iron content of the whole wheat was (1–5 mg/100 g), which is in agreement with the result of this study. However, the level of iron in the white flour decreases significantly. The milling process removes many important nutrients when white flour is produced. The bran and the germ are relatively rich in minerals and the milled products contain less of these than the original grain. As a result of milling, the palatability is increased, but the nutritional value of the products is decreased (Hoseney 1992).

Zinc content varied significantly among all whole wheat and white wheat flour (Table 4). The zinc content of the whole wheat ranged from 3.59 to 2.47 mg/100 g. However, the white flour contained zinc content in the range of 0.58 to 1.39 mg/100 g. The highest Zn content 3.59 mg/100 g was found in hard whole wheat flour. This study showed that the low rate of extraction of wheat reduce the zinc content of the wheat significantly ($P < 0.05$). The hard whole wheat flour zinc content was 3.59 mg/100 g whereas milling reduced the zinc content to 1.39 mg/100 g. Whereas, in case of soft wheat, the decrease was from whole wheat to white wheat flour 2.47 to 0.58 mg/100 g of zinc, respectively. According to Lopez

et al. (2003), 80% of the total amounts of minerals are concentrated in the aleurone layer of pericarp (bran), which was removed during milling process while only 20% minerals are present in endosperm.

The mean total content of phosphorous on hard whole and soft whole wheat flour was 337.99 and 313.98 mg/100 g, respectively, however, there was significant decrease in phosphorous content ($P < 0.05$) on the refined milled product of hard and soft white wheat flour which was 144.69 and 77.03 mg/100 g, respectively. Wheat is one of the cereals which is classified as rich sources of phosphorous. The result of the whole wheat flour was in line with the finding of Dewettinck et al. (2008), which reported the presence of phosphorous from 200 to 1200 mg/100 g.

The replacement effect of different levels of wheat bran on the mineral content of bread is shown in Table 5. All the mean value for iron content varied among the bread samples. Results showed that the mineral progressively increased when levels of bran were increased. Control contained 1.98 mg/100 g iron and at the level of 25% bran replacement, the value increased to 4.84 mg/100 g of iron. The result of this study was in agreement with the study of Butt et al. (2004).

The replacement effect of different levels of wheat bran on the zinc content of bread is shown in Table 5. Results showed that zinc progressively increased when levels of bran were increased. The result ranged from 0.93 to 2.33 mg/100 g of zinc. The value of zinc at the bran supplementation of between 20 and 25% was not significantly different ($P > 0.05$).

Micronutrient malnutrition greatly increases mortality and morbidity rates, diminishes cognitive abilities of children, lowers labor productivity, and reduces the quality of life for all those affected. Deficiency of micronutrients, such as iron and zinc, is critical and major problem. It could be concluded that the addition of bran improves the nutritional quality of bread and could be a means of providing adult their daily requirements of iron and zinc. Although supplementation of bran improves the

Table 4. Mineral composition of whole and refined wheat flour.

Wheat samples	Mineral content mg/100 g		
	Iron	Zinc	Phosphorous
HWF	4.15 ± 0.12[c]	3.59 ± 0.063[d]	337.99 ± 0.56[d]
HWWF	2.51 ± 0.16[a]	1.39 ± 0.036[b]	144.69 ± 0.61[b]
SWF	3.35 ± 0.17[b]	2.47 ± 0.04[c]	313.98 ± 1[c]
SWWF	2.95 ± 0.26[a]	0.58 ± 0.01[a]	77.03 ± 0.51[a]

HWF, hard whole wheat flour; HWWF, Hard white wheat flour; SWF, soft whole wheat flour; SWWF, Soft white wheat flour. Mean value with different superscript in the same column are significantly different ($P < 0.05$).

Table 5. Mineral analysis of bread. WFB-(white wheat flour bread) – control, WF:BR 90:10, 10% bran supplemented bread, WF:BR 80:20 – 20% bran supplemented bread, WF:BR 75:25 – 25% bran supplemented bread.

Samples	Iron mg/100 g	Zinc mg/100 g
Control (WFB)	1.98 ± 0.056[a]	0.93 ± 0.064[a]
WF:BR (90:10)	2.34 ± 0.159[b]	1.697 ± 0.108[b]
WF:BR (80:20)	2.95 ± 0.0753[c]	2.28 ± 0.131[c]
WF:BR (75:25)	4.83 ± 0.074[d]	2.33 ± 0.066[c]

Data are average of triplicate ± SE. Mean value with different superscript in the same column are significantly different ($P < 0.05$).

Table 6. Percent yield and total phenolics content.

Sample	Yield % (g/100 g)	Total phenolics (mg GAE/g)
HWF (whole)	8.2	7.66 ± 0.70[b]
HWWF(refined)	7.58	3.49 ± 0.86[a]
SWWF (refined)	7.7	3.34 ± 0.14[a]
SWF(whole)	7.3	8.20 ± 0.35[b]

HWF, hard whole wheat flour; HWWF, Hard white wheat flour (refined); SWF, Soft whole wheat flour; SWWF, Soft white wheat flour (refined). Data are average of triplicate ± SE. Values in the same column with different superscript are statistically significant ($P < 0.05$).

mineral content of the bread due to increase in phytic acid content, the availability of these minerals may be reduced. Therefore, such issues would need further evaluation.

Antioxidant capacity of wheat

The percentage yields of extracts were 7.58% w/w (Hard refined wheat), 8.2% w/w (Hard whole flour wheat), and 7.7% w/w (Soft refined wheat flour) and 7.3% w/w (soft whole flour wheat).

It has been recognized that the TPC of plant extracts is associated with their antioxidant activities due to their redox properties, which allow them to act as reducing agents, hydrogen donors, and singlet oxygen quenchers. TPC was expressed as milligrams of GAE (Gallic acid equivalent) per gram (mg/g) of dry flour samples. As shown in Table 6, the TPC in whole wheat was highest. The TPC of white wheat flours (refined), which ranged from 3.34 to 3.49 mg GAE/g which were significantly lower ($P < 0.005$) than those of whole wheat flours (range 7.66–8.20 mg GAE/g).

However, the mean content did not vary much between whole hard and soft wheat type. Also, there was no significant variation between soft and hard white wheat flour. The difference in the TPC between whole and white wheat flour could be due to the process of milling. Research found antioxidants in wheat concentrated mostly in the aleurone layer of bran with some in the pericarp, nucellar envelope, and germ (Fulcher and Duke 2002; Žilic et al. 2012).

The ability of wheat extracts to quench reactive species by hydrogen donation was measured through the DPPH radical scavenging activity test. The antioxidants can react with DPPH, a violet colored stable free radical, converting it into a yellow colored α,α-diphenil-β- picrylhydrazine. The discoloration of the reaction mixture can be quantified by measuring the absorbance at 517 nm, which indicates the radical scavenging ability of the antioxidant. The antioxidant capacity of whole and refined wheat was measured as the DPPH• scavenging activity.

The DPPH radical scavenging effects of wheat methanol extracts was shown in Figure 2. As the concentration of sample increased, the percent inhibition of DPPH radical also increased (Haung et al. 2005). At the concentration of 50 mg/mL, the scavenging effect of ascorbic acid, and wheat extracts, on the DPPH radical scavenging decreased in the order of L- ascorbic acid > soft whole > hard whole > soft white > hard white wheat flour, which were 92.53, 90.39, 89.89, 75.80, 57.57%, respectively. Therefore, the percentage of DPPH radical scavenging capacity of soft whole and hard whole wheat extracts are comparable with commercial antioxidants, L- ascorbic acid at concentration of 50 mg/mL. This suggested that whole wheat contain compounds that can donate electron/hydrogen easily and stabilizes free radicals.

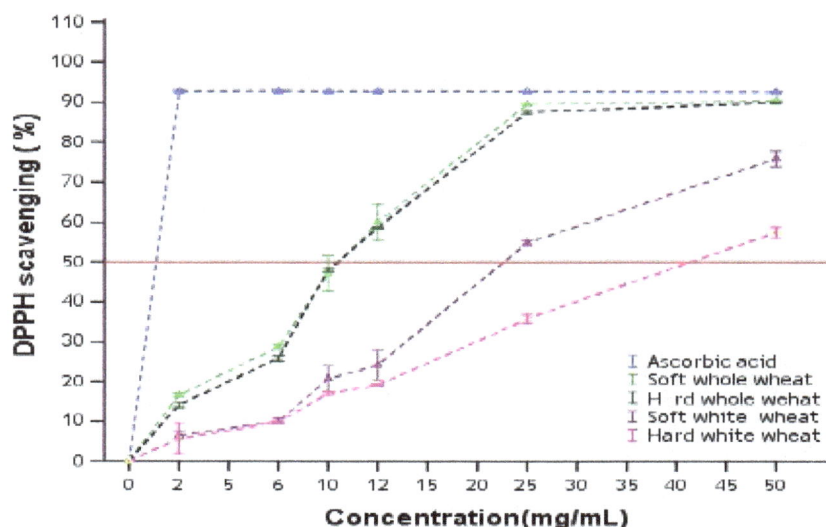

Figure 2. DPPH radical scavenging activity (%) of wheat extracts and standard (Values are average of triplicate measurements [mean ± SEM]).

The IC50 values of all the extracts were calculated from plotted graph of percentage scavenging activity against concentration of the extracts (Fig. 2). The lower the IC50 value, the higher is the scavenging potential. The IC50 values ranged from 10.56 mg/mL for whole wheat extracts to 41.25 mg/mL for hard white wheat extracts. Strongest scavenging activity (lower IC50 values) was recorded for whole hard and soft wheat extracts, which appeared more than four times stronger than that of hard white flour and two times stronger than that of soft white wheat extracts. IC50 value of ascorbic acid, a well-known antioxidant, was relatively more pronounced than that of the extracts (Fig. 2). The results of this study demonstrate that the antioxidant content of wheat has been affected by the refined extraction/milling process. According to Fikreyesus et al. (2013), the DPPH (IC50) for whole wheat flour is 15.56, which is different form the result obtained in this research. It is known that the antioxidant properties of wheat grain are significantly influenced by the genotype and environmental conditions (Adom et al. 2003).To the best of our knowledge, there are no or few studies conducted on antioxidant content of Ethiopian wheat and particularly on comparison of antioxidant content on the whole and white wheat flour.

A relationship between phenolic content and antioxidant activity was extensively investigated, and both positive and negative correlations were reported. Bakchiche et al. (2013), Petra et al. (2012) and many other research groups stated that there was a positive correlation. However, a few evidences of no significant correlation were reported (Mohammad et al. 2008). In this study, the dependence of DPPH scavenging activity (%) in relation to the TPC was also evaluated. The TPC correlated significantly with DPPH scavenging activity ($R^2 = 0.637$, $P < 0.05$). Thus, the phenolics from the wheat extracts showed a good hydrogen-donating capacity, as well as high reactivity to free radicals, leading to the stabilization and termination of the radical chain reactions.

Sensory analysis of the bread

The sensory attributes of bread made from bran (Fig. 3) using different ratio were evaluated using 7-point hedonic scale at Addis Ababa University; Center for Food Science

Figure 3. Wheat bran supplemented bread.

and Nutrition by semitrained panelists of first year M.Sc program students of Food Science and Nutrition stream and the mean scores of evaluated sensory attributes were presented in Table 7.

Taste is an important parameter when evaluating sensory attribute of food. The product without acceptable good test is likely to be unacceptable by consumers. The observed mean score of taste in experiential bran supplemented bread ranged from 4.43 to 5.93 (Table 7). Control (100% white wheat flour bread) had the highest mean sore in taste (5.93) followed by 10% bran supplemented bread (5.93). The 100% white wheat flour (control) bread had significant difference ($P < 0.05$) with 10% bran (5.26), 20% bran bread (4.80), and 25% bran bread (4.43).

The control bread had significant difference ($P < 5$) from three of the bran supplemented bread (10%, 20%, 25%). The 10% bran supplemented bread scores >5 indicating that it is moderately likable by panelists, and 20% and 25% bran supplemented bread were rated as neither like nor dislike by the panelists (scored as 4.80 and 4.43, respectively). This is due to the addition of the bran to the bread.

Table 7. Sensory characteristics of wheat bran supplemented Bread. WFB-(white wheat flour bread) – control, WF:BR 90:10, 10% bran supplemented bread, WF:BR 80:20 – 20% bran supplemented bread, WF:BR 75:25 – 25% bran supplemented bread.

Sample	Test	Odor	Color	Texture	Overall acceptability
Control (WFB)	5.93 ± 0.14[c]	5.53 ± 0.21[b]	5.93 ± 0.18[c]	5.70 ± 0.19[b]	5.93 ± 0.11[c]
WF:BR (90:10)	5.26 ± 0.16[b]	5.43 ± 0.18[b]	5.46 ± 0.14[c]	5.30 ± 0.13[b]	5.43 ± 0.15[b]
WF:BR (80:20)	4.80 ± 0.24[b]	5.33 ± 0.23[b]	4.76 ± 0.22[b]	4.60 ± 0.20[a]	4.96 ± 0.16[b]
WF:BR (75:25)	4.43 ± 0.30[a]	4.76 ± 0.22[a]	4.13 ± 0.24[a]	4.36 ± 0.23[a]	4.50 ± 0.22[a]

Data are average of triplicate ± SE. Mean value with different superscript in the same column are significantly different ($P < 0.05$).

The mean score of odor of bread ranged from 4.76 to 5.53 (Table 7). Most of the samples were similar in odor while only 25% supplemented bread significantly ($P < 0.05$) decreased. Three of the samples were liked moderately while 25% bran supplemented bread was rated as neither like nor dislike. As is shown in Table 7 color of bread had low score as a result of increasing the level of wheat bran. The color of control and 10% supplemented bread were similar in appearance while 20% and 25% bran supplemented bread were decreased (4.76 and 4.13) significantly ($P < 0.05$). The results indicated that no significant difference ($P > 0.05$) was observed by panelists between the control and 10% supplemented bread. The texture of the control and 10% bran supplemented bread were relatively most preferred (liked moderately) by the panelists. While the bread prepared from increasing level of bran supplement from 25% were scored as 4.5.

Generally, among the bread products, the control was highly acceptable by the panelists, with a score of 5.93. Next to this, 10% and 20% bran supplemented bread scored, 5.43 and 4.96, respectively. These two are liked moderately. The result obtained for 25% bran supplemented bread was significantly different from the previous two. The latter was neither like nor disliked by the panelists and it scored 4.5. In relation to this, Lazaridou et al. (2007) reported that those samples were considered as acceptable, which their average score for the overall acceptability were greater than 4 which mean neither like nor dislike. Thus, whole wheat flour with high extraction rate (100%) needed to be given high emphasis by consumers because of its nutritional and antioxidant capacity of the product.

Conclusions

Wheat and wheat products are important staple foods that are commonly consumed in Ethiopia. Consumption of whole grains as part of the diet is recommended for health reasons because they are good source of minerals, fibers, protein, and antioxidants. There are no studies on the effect of refining on the nutritional content and antioxidant capacity of wheat grown in Ethiopia. This study showed that wheat extraction/refining at the lower rate have significantly reduced the proximate composition as well as the antioxidant content of wheat in both hard and soft wheat samples.

Addition of wheat bran to white wheat flour improves the nutritional value of the bread. Based on obtained results, the incorporation of wheat bran in the ratio of 10–20% showed better sensory acceptability though the proximate composition and mineral content increased at 25% bran supplemented bread. This indicates that bread of good nutritional and sensory qualities could be produced from 10% and 20% bran supplementation. The result of this study also indicated that wheat bran as a good source of minerals and fibers and can be used to supplement bread.

Acknowledgments

Authors acknowledge the Addis Ababa University for covering the research costs.

Conflict of Interest

None declared.

References

AACC. 2000. Approved methods of American Association of Cereal Chemists. 15th ed. American Association of Cereal Chemists, Arlington, VA.

Adom, K., M. Sorrells, and R. Liu. 2003. Phytochemical profiles and antioxidant activity of wheat varieties. J. Agric. Food Chem. 51:7825–7834.

Anwarul, H., U. Shams, and A. Anwarul. 2002. The effect of aqueous extracted wheat bran on the baking quality of biscuit. Int. J. Food Sci. Technol. 37:453–462.

AOAC. 2000. Official methods of analysis. 17th ed. The Association of Official Analytical Chemists, Arlington, VA.

Azizi, M. H., S. Sayeddin, and S. Payghambardoost. 2006. Effect of flour extraction rate on flour composition, dough rheology characteristics and quality of flat breads. J. Agric. Sci. Technol. 8:323–330.

Bakchiche, B., A. Gherib, A. Smail, G. Custodia, and M. Grac. 2013. Antioxidant activities of eight Algerian plant extracts and two essential oils. Ind. Crops Prod. 46:85–96.

Bergh, K., A. Chew, M. Gugerty, and C. Anderson. 2012. Wheat Value Chain: Ethiopia. University of Washington: Evans School Policy Analysis and Research (EPAR). No 204.

Blakeney, A., R. Cracknell, G. Crosbie, S. Jefferies, D. Miskelly, L. O'Brien, et al. 2009. Understanding wheat quality: a basic introduction to Australian wheat quality. Wheat Quality Objectives Group, Kingston, Australia.

Butt, B., M. Ihsanullah, F. M. Anjum, A. Aziz, and M. Atif. 2004. Development of minerals enriched brown flour by utilizing wheat milling by-products. Intern. J. Food Saf. 3:15–20.

CSA. 2005. Report on area and production of major crops. Statistical Bulletin. Addis Ababa, Ethiopia.

CSA. 2011. Agricultural sample survey report-production. Ethiopia Central Statistics Agency, Addis Ababa.

Cubadda, F., F. Aureli, A. Raggi, and M. Carcea. 2009. Effect of milling, pasta making and cooking on minerals in durum wheat. J. Cereal Sci. 49:92–97.

Curtis, B., S. Rajaram, and H. Macpherson, eds. 2002. Bread wheat improvement and production. Food and Agriculture Organization of the United Nations, Rome.

Dewettinck, K., F. Van Bockstaele, B. Kuhne, D. Van de Walle, T. Courtens, and X. Gellynck. 2008. Nutritional value of bread: influence of processing, food interaction and consumer perception. Review. J. Cereal Sci. 48:243–257.

Farooq, Z., S. Rehman, and M. Q. Bilal. 2001. Suitability of wheat varieties/lines for the production of leavened flat bread (Naan). J. Res. Sci. 12:171–179.

FDRE, Ministry of Health. 2011. Assessment of feasibility and potential benefits of food fortification in Ethiopia. Addis Ababa.

Fikreyesus, S., H. Vasantha, and T. Astatke. 2013. Antioxidant capacity, total phenolics and nutritional content in selected Ethiopian staple food ingredients. Int. J. Food Sci. Nutr. 64:915–920.

Fulcher, R., and T. Duke. 2002. Whole-grain structure and organization: implications for nutritionists and processors. Pp. 9–45 in L. Marquart, L. Slavin and R. G. Fulcher, eds. Whole-grain foods in health and disease. American Association of Cereal Chemists, St. Paul, MN.

Gursoy, N., C. Sarikurkeu, B. Tepe, and M. Solak. 2010. Evaluation of antioxidant activities of 3 edible mushrooms: ramaria flava (Schaef.: Fr.) Quel., *Rhizopogon roseolus* (Corda) T.M. Fries., and *Russula delica* Fr. Food Sci. Biotechnol. 19:691–696.

Haung, D., B. Ou, and R. L. Prior. 2005. The chemistry behind antioxidant capacity assays. J. Agric. Food Chem. 53:1841–1856.

Hertzberg, J., and Z. Francois. 2007. Artisan bread in five minutes a day: the discovery that revolutionizes home baking. St. Martin's Press, New York.

Hoseney, R. C.. 1992. Principles of cereal science and technology. AACC, Paul, MN.

Hussain, A., H. Larsson, R. Kuktaite, and E. Johansson. 2010. Mineral composition of organically grown wheat genotypes: contribution to daily minerals intake. Int. J. Environ. Res. Public Health 7:3442–3456.

Iuliana, B., S. Georgeta, S. Violeta, and A. Iuliana. 2012. Effect of the addition of wheat bran stream on dough rheology and bread quality. Ann. Univ. Dunarea de Jos of Galati Fascicle VI – Food Technol. 36:39–52.

Kumar, P., R. Yadava, B. Gollen, S. Kumar, R. Verma, and S. Yadav. 2011. Nutritional contents and me dicinal properties of wheat: a Review. Life Sci. Med. Res. 22:1–10.

Lazaridou, A., D. Duta, M. Papageorgiou, C. Belc, and C. Biliaderis. 2007. Effects of hydrocolloids on dough rheology and bread quality parameters in gluten-free formulations. J. Food Eng. 79:1033–1047.

Lopez, H. W., V. Krespine, A. Lemairet, C. Coudray, C. Coudray, A. Messager, et al. 2003. Wheat variety has a major influence on mineral bioavailability: studies in rats. J. Cereal Sci. 37:257–266.

Mellen, P., T. Walsh, and D. Herrington. 2008. Whole grain intake and cardiovascular disease: a meta-analysis. Nutr. Metab. Cardiovasc. Dis. 18:283–290.

MOA. 2011. Animal and Plant Health Regulatory Directorate Crop Variety Register. No.4. Addis Ababa, Ethiopia.

Mohammad, A., P. Fereshteh, and R. Ahmad. 2008. Iron chelating activity, phenol and flavonoid content of some medicinal plants from Iran. Afr. J. Biotechnol. 7:3188–3192.

Petra, T., C. Barbara, P. Nata˘sa, and A. Helena. 2012. Studies of the correlation between antioxidant properties and the total phenolic content of different oil cake extracts. Ind. Crops Prod. 39:210–217.

Posner, E. S. 2000. Wheat. Pp. 1–29 in K. Kulp and J. G. Ponte Jr, eds. Handbook of cereal science and technology. 2nd ed. Marcel Dekker Inc., New York, NY.

Ragaee, S., I. Guzar, E.-S. M. Abdel-Aal, and K. Seetharaman. 2012. Bioactive components and antioxidant capacity of Ontario hard and soft wheat varieties. Can. J. Plant Sci. 92:19–30.

Rashid, S., K. Getnet, and S. Lemma. 2010. Maize value chain potential in Ethiopia. IFPRI, Addis Ababa, Ethiopia.

Shewry, P. 2009. Wheat. J. Exp. Bot. 60:1537–1553.

Singleton, V. L., and J. A. Rossi. 1965. Colorimetry of total phenolics with phospho molybdic phosphotungstic acid reagents. Am. J. Enol. Vitic. 37:144–158.

Slavin, J. L., D. Jacobs, and L. Marquart. 2000. Grain processing and nutrition. Crit. Rev. Food Sci. Nutr. 40:309–326.

Woldegiorgis, A., D. Abate, G. D. Haki, and G. R. Ziegler. 2014. Antioxidant property of edible mushrooms collected from Ethiopia. Food Chem. 157:30–36.

Žilic, S., A. Serpen, G. Akillioglu, M. Jankovic, and V. Gokmen. 2012. Distributions of phenolic compounds, yellow pigments and oxidative enzymes in wheat grains and their relation to antioxidant capacity of bran and debranned flour. J. Cereal Sci. 56:652–658.

3

Effect of γ radiation processing on fungal growth and quality characteristcs of millet grains

Nagat S. Mahmoud[1], Sahar H. Awad[1], Rayan M. A. Madani[1], Fahmi A. Osman[1], Khalid Elmamoun[2] & Amro B. Hassan[1]

[1]Environment and Natural Resource and Desertification Research Institute (ENDRI), National Center for Research, PO Box 6096, Khartoum, Sudan
[2]Sudanese Atomic Energy Commission (SAEC), Khartoum, Sudan

Keywords
Antinutritional factors, fungal growth, germination, millet, protein solubility, radiation

Correspondence
Amro B. Hassan, Environment and Natural Resource and Desertification Research Institute (ENDRI), National Center for Research, PO Box 6096, Khartoum, Sudan.

E-mail: amrobabiker@yahoo.com

Funding Information
No funding information provided.

Abstract

The aim of this study was to evaluate the effect of gamma radiation processing of millet grains on fungal incidence, germination, free fatty acids content, protein solubility, digestible protein, and antinutritional factors (tannin and phytic acid). The grains were exposed to gamma radiation at doses 0.25, 0.5, 0.75, 1.0, and 2.0 kGy. Obtained results revealed that radiation of millet grains at a dose level higher than 0.5 kGy caused significant ($P < 0.05$) reduction on the percentage of fungal incidence and the free fatty acid of the seeds, while, no significant change in the germination capacity was observed of the grains after radiation. Additionally, the radiation process caused significant ($P < 0.05$) reduction on both tannins and phytic acid content and gradual increment on in vitro protein digestibility of the grains. On the other hand, the treatments significantly ($P < 0.05$) increased the protein solubility of the grains. Obtained results indicate that gamma irradiation might improve the quality characteristics of millet grains, and can be used as a postharvest method for disinfestations and decontamination of millet grains.

Introduction

Pearl millet (*Pennisetum gluucum* L.) is considered to be the staple food for most people in Asia and Africa. It is considered as a good source of needed elements (Abdalla et al. 1998). However, it contains high amounts of antinutrients such as tannin and phytic acids which reduce its nutritional value (Abdel Rahaman et al. 2007). During postharvest storage, millet is susceptible to attack by a variety of insects and microorganisms. This infestation cause physical losses and reduces the nutritional value of grains which leads to the loss of the economic value of stored grain. Moreover, infestation with the insects result in contamination with dead insect's bodies and their products, as well as fungal growth that favor the spread of *Aspergillus flavus*, a mold which produces aflatoxin (Rees 2004).

Generally, chemical fumigants are used to disinfest grains (Arthur 1996), however, continuous application of these pesticides have a negative impact either on the environment or human health (Cherry et al. 2005). Therefore, the industry has been forced to explore nonchemical alternatives. One possible alternative is the application of gamma irradiation. Radiation processing is considered to be a safe alternative to chemical methods, enhance quality and nutritional characteristics of stored products as well as maintain its shelf-life. It has shown great promise in accomplishing disinfestations and decontamination of food and agricultural products (Loaharanu 1994; Fombang et al. 2005). Besides disinfection criteria, gamma radiations enhance the nutritional value of grains and improves the functional properties of its flours (Rahma and Mostafa 1988; Dario and Salgado 1994; Dogbevi et al. 2000). Furthermore, it has been reported that gamma radiation causes a significant reduction in antinutrients and enhances the nutritional quality of grains (Hassan et al. 2013; Osman et al. 2014). Although, application of gamma radiation has many advantages over other physical methods, however, the application of this technology in the industry is limited. Therefore, further research is needed to improve their efficiency to reach the reasonable usage stage and in controlling stored-grain pests as well as to improve the nutritive value of stored products.

Thus, in the present study, gamma radiation was applied as a preserving method to investigate its efficiency in controlling fungal growth and enhancement of quality characteristics of millet grains.

Material and Methods

Sample preparation

Millet grains were cleaned manually and freed from broken seeds and impurities, and then stored in plastic bags at 4°C during the study.

Radiation treatments

Radiation processing was done at Kaila irradiation processing unit, Sudanese Atomic Energy Corporation (SAEC). About 250 g of millet grains packed in polyethylene bags were irradiated, using a γ- ray ^{60}Co radiator. The seeds were evenly exposed to radiation doses 0.25, 0.5, 0.75, 1.0, and 2.0 kGy with a dose rate of 33 Gy/min. Unirradiated seeds (0 kGy) served as control.

Fungal culture and incidence

The fungal incidence and the colony formation unit per gram (cfu/g) of treated and untreated samples was determined after platted on double strength Sabaroud Dextrose Agar and incubated at 25°C for 5 days according to standard methods (AOAC 1995).

Determination of germination

The germination of grains was determined according to the international Seed Testing Association (ISTA 2006). Twenty five seeds were platted on filter paper in a Petri dish and saturated with distilled water. The plates were incubated at 25 ± 2°C for 7 days.

Determination of free fatty acid content

Free fatty acid of millet was determined according to Aibara et al. (1986) cited by Zhao et al. (2007) with slight modification. About 25 mL of ethanol was added

to 5 g of millet flour. After shaking, the mixture was filtrated and additional 25 mL ethanol was added. The filtrate was titrated with 0.1 N KOH, using phenolphthalein (3%) as indicator. Flour acidity was calculated as mg KOH required neutralizing free fatty acid from one gram grain on dry matter basis.

Determination of tannins and Phytic acid content

Tannins content of grains was estimated according to Price et al. (1978). Phytic acid content was determined by the method described by Wheeler and Ferrel (1971).

Crude and digestible protein determination

The crude protein was determined following the Kjeldahl method described by AOAC (1995). The digestible protein was determined by the procedure of Maliwal (1983) cited from Monjula and John (1991).

$$\text{Protein digestibility } \% = \frac{\text{digestible protein}}{\text{total protein}} \times 100$$

Protein solubility

Soluble protein solubility was determined in millet grains after extracted by water, using the method described by Hagenmaier (1972).

Statistical analysis

All data were the average of triplicates. Data were analyzed using one-way analysis of variance (ANOVA). Significant differences were calculated ($P < 0.05$), using least significant difference (LSD).

Results and Discussion

Effect of radiation process on the fungi incidence and germination rate

Table 1 presents the percentage of fungal incidence and colony formation (cfu/g) in raw and treated millet grains. Before radiation, the fungal incidence was found to be 100% in raw grains. Radiation of seeds up to 0.50 kGy caused no significant reduction in fungal incidence, however, radiation of grains at higher doses 0.75, 1.0 and 2.0 kGy sharply decreased the fungal incidence to 21.3, 18.7, and 5.3%, respectively. Similarly, the effectiveness of gamma radiation in reducing the formation of fungi was observed. Prior to radiation, the colony formation was found to be 5.3×10^4 cfu/g, where as it was decreased to 2.1×10^4, 2.1×10^4, 4×10^2, 3×10^2 and 3×10 cfu/g

Table 1. Effect of gamma irradiation on fungal growth (%) and colony formation (cfu/g) in pearl millet. Error bars indicate the standard deviation ($n = 3$). Values not sharing a common superscript are significantly ($P < 0.05$) different.

Gamma dose (kGy)	Fungal incidence (%)	Colony formation (cfu/g)
0.0	$100 \pm (0.000)^a$	5.3×10^4
0.25	$100 \pm (0.000)^a$	2.1×10^4
0.50	$100 \pm (0.000)^a$	2.1×10^4
0.75	$21.3 \pm (9.238)^b$	4.0×10^2
1.0	$18.7 \pm (2.309)^b$	3.0×10^2
2.0	$5.3 \pm (2.309)^c$	3.0×10^1

after radiation treatment at doses of 0.25, 0.50, 0.75, 1.0, and 2.0 kGy, respectively (Table 1). Similar observation on walnut kernel was reported by Al-Bachir (2004), who found that application of gamma radiation at doses of 0.5, 1.5, and 2 kGy reduced the fungal load on walnut kernels. Furthermore, it was reported that doses of 1.5 and 3.5 kGy reduced the number of fungi in many raw fruits and vegetables (Aziz and Moussa 2004). Hilmy et al. (1995) concluded that radiation process peanuts with doses up to 1 kGy inhibit the incidence of mycelium and toxins secretion. Reduction in the fungal incidence rate in millet grains after radiation might be due to high sensitivity of the fungus and mold to gamma radiation, since the radiation process causes direct and indirect damage to the DNA (Refai et al. 1996; McNamara et al. 2003).

Germination test is comparatively assessing the quality losses of grain after treatments. It is directly associated with various characteristics of grain quality (Beckett and Morton 2003). As shown in Figure 1, no significant change in seed germination rate after treatments was observed. Maximum decrease in germination rate of millet (90.7%) was observed with 2 kGy treatment. Obtained results were in accordance with El-Naggar and Mikhaiel (2011), who found that the germination capacity of wheat grains was not changed after radiation at dose up to 1 kGy. Moreover, the results of Melki and Marouani (2009) showed that

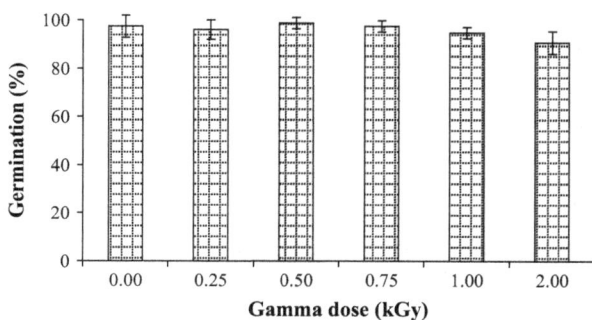

Figure 2. Effect of gamma irradiation on free fatty acids (FFA) in pearl millet. Error bars indicate the standard deviation ($n = 3$). Values not sharing a common superscript are significantly ($P < 0.05$) different.

there was no significant change in germination capacity in wheat after radiation.

Effect of radiation process on free fatty acids (FFA) content

Figure 2 presents the free fatty acids (FFA) content in mg/g in millet grains for the control and radiated samples. The FFA content of millet flour was found to be 217.7 mg/100 g prior to the radiation treatment. After radiation at dose levels of 0.75, 1.0, and 2.0 kGy, it is clearly observed that the FFA content of millet grains significantly ($P < 0.05$) reduced to 190.9, 190.6, and 190.5 mg/100 g, respectively. Significant reduction on FFA content might be due to lipase activity reduction in treated grains, which result in dropping the FFA formation. Pankaj et al. (2013) demonstrated that the radiation treatment significantly reduced the lipase activity in wheat germ. Therefore, the results indicated that gamma radiation is an effective method for stabilization and to extend the shelf life of grains, since free fatty acids content is an index of the rancidity and contributes to the development of off-flavor and off- odors in oil during storage.

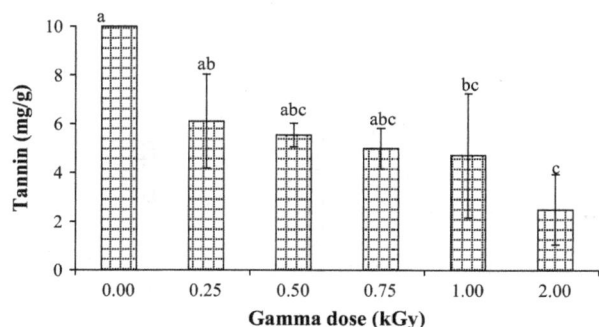

Figure 3. Effect of gamma irradiation on tannin content in pearl millet. Error bars indicate the standard deviation ($n = 3$). Values not sharing a common superscript are significantly ($P < 0.05$) different.

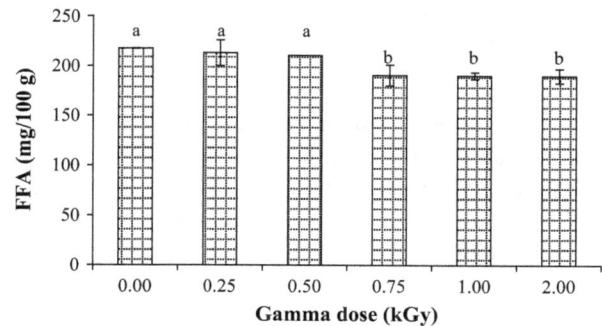

Figure 1. Effect of gamma irradiation on germination rate in pearl millet. Error bars indicate the standard deviation ($n = 3$).

Effect of radiation process on tannin and phytic acid content

As illustrated in Figure 3, the tannin content of millet grains was found to be 9.99 mg/g prior to radiation process. Tannin content of the examined grains presented a dose-dependent decrease. It was clearly observed that radiation significantly ($P < 0.05$) reduced tannin content of millet grains. The reduction in tannin content was found to be 38.9, 44.4, 50, and 52.8, and 74.9% when millet grains were irradiated at dose 0.25, 0.50, 0.75, 1.0, and 2.0 kGy, respectively, compared to control one. These findings are in agreement with those stated by several researchers. Hassan et al. (2009) concluded that radiation process significantly reduced tannins content of sorghum and maize grains. Similar observation was reported by Pinn (1992) who stated that radiation of white beans at dose levels 2, 4, 6, 8, 10, 15, and 20 kGy followed by cooking significantly reduced tannins content. Moreover, El-Niely (2007) found that the tannin content of legume seeds decreased after radiation treatments. Decrease in tannin content might be result of chemical degradation by the action of free radicals formed by the radiation.

On the other hand, before radiation treatment, the phytic acid of millet grains was found to be 1.27 mg/g (Fig. 4). After radiation, phytic acid of millet grains significantly ($P < 0.05$) decreased. The level of reduction increased with an increase in radiation dose. Decreases in phytic acid were 3.9, 29.9, 42.5, 43.3, and 52.8%, at dose 0.25, 0.50, 0.75, 1.0, and 2.0 kGy, respectively. This reduction might be the result of the action of free radicals, since they are able to cleave to the phytate ring (De Boland et al. 1975). Obtained results were in agreement with El-Niely (2007) who stated that radiation after processing significantly ($P \leq 0.05$) decreased the level of phytic acid of legumes and cereal grains.

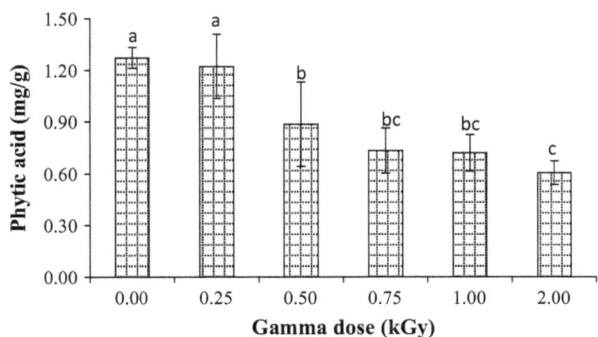

Figure 5. Effect of gamma irradiation on in vitro protein digestibility (%) in pearl millet. Error bars indicate the standard deviation ($n = 3$).

Effect of radiation process on in vitro protein digestibility (IVPD) and protein solubility

Figure 5 summarizes the in vitro protein digestibility (IVPD) of millet before and after radiation. The IVPD of untreated seeds was found to be 32.8%. Radiation process of the seeds caused a minor increment of the IVPD and was increased as the dose was increased. Increment in protein digestibility might be due to the reduction in the antinutrients particularly tannin content of grains as reported by Hassan et al. (2009). Since disulphide and hydrogen bonds are involved in stabilizing protein structure, their breaking can result in loss of conformational or structural integrity that exposed additional peptide bonds, thus enhancing proteolysis. Irradiation can cause change in their protein structure that enhances denaturation of the protein and hence improve its digestibility (Koppelman et al. 2005).

Protein solubility is doubtless the most important function among the functional properties of proteins. Data in Figure 6 showed that the protein solubility of raw

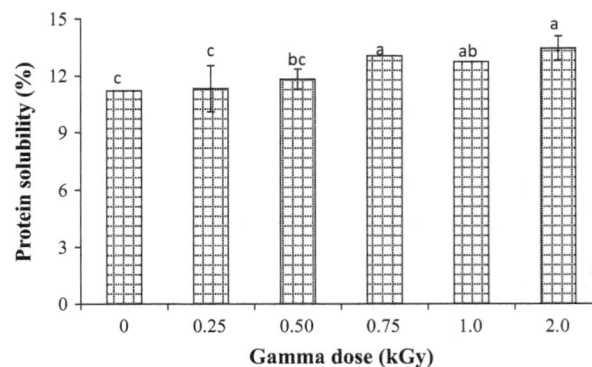

Figure 4. Effect of gamma irradiation on phytic acid content in pearl millet. Error bars indicate the standard deviation ($n = 3$). Values not sharing a common superscript are significantly ($P < 0.05$) different.

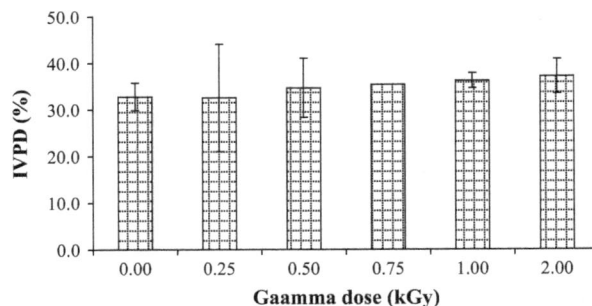

Figure 6. Effect of gamma irradiation on protein solubility in pearl millet. Error bars indicate the standard deviation ($n = 3$). Values not sharing a common superscript are significantly ($P < 0.05$) different.

millet was found to be 11.20%. The protein solubility of millet was increased significantly ($P < 0.05$) to 11.32, 11.82, 13.04, 12.74, and 13.44% after radiation at 0.25, 0.50, 0.75, 1.0, and 2.0 kGy, respectively. Increase in protein solubility after radiation is likely due to the high proteolytic activity during radiation, which may lead to hydrolysis of the stored proteins.

Conclusion

The obtained results revealed that gamma irradiation processing of millet grains up to 2 kGy significantly reduced the fungal incidence and free fatty acids content of the grains. On the other hand, it caused a decrease in the amount of antinutrients namely, tannin and phytic acid and gradually increase the in vitro protein digestibility and protein solubility of the grains. According to these results, therefore, gamma radiation can be applied as a safe postharvest method in order to extend the shelf life of millet grains.

Conflict of Interest

None declared.

References

Abdalla, A. A., A. H. El Tinay, B. E. Mohamed, and A. H. Abdalla. 1998. Proximate composition, starch, phytate and mineral contents of 10 pearl millet genotypes. Food Chem. 63:243–246.

Abdel Rahaman, S. M., H. B. ElMaki, W. H. Idris, A. B. Hassan, E. E. Babiker, and A. H. El Tinay. 2007. Antinutritional factors content and hydrochloric acid extractability of minerals in pearl millet cultivars as affected by germination. Int. J. Food Sci. Nutr. 58:6–17.

Aibara, S., I. A. Ismail, H. Yamashita, and H. Ohta. 1986. Changes in rice bran lipids and fatty acids during storage. Agric. Biol. Chem. 50:665–673.

Al-Bachir, M. 2004. Effect of gamma irradiation on fungal load, chemical and sensory characeristics of walnut (Juglans regial). J. Stored Prod. Res. 40:355–362.

AOAC. 1995. Official methods of analysis of Association of Official Analytical Chemists, 16th ed. Association of Official Analytical Chemists, Washington, DC.

Arthur, F. H. 1996. Grain protectants: current status and prospects for the future. J. Stored Prod. Res. 32:293–302.

Aziz, N. H., and L. A. A. Moussa. 2004. Reduction of fungi and mycotoxins formation in seeds by gamma-radiation. J. Food Safety 24:109–127.

Beckett, S. J., and R. Morton. 2003. Mortality of Rhyzopertha dominica (F.) at grain temperatures ranging from 50°C to 60°C obtained at different rates of heating in a spouted bed. J. Stored Prod. Res. 39:313–332.

Cherry, A. J., P. Abalo, and K. Hell. 2005. A laboratory assessment of the potential of different strains of the entomopathogenic fungi Beauveria bassiana (Balsamo) Vuillemin and Metarhizium anisopliae (Metschnikoff) to control Callosobruchus maculatus (F.) (Coleoptera: Bruchidae) in stored cowpea. J. Stored Prod. Res. 41:295–309.

Dario, A. C., and J. M. Salgado. 1994. Effect of thermal treatments on the chemical and biological value of irradiated and non-irradiated cowpea bean (Vigna unguiculata L.Walp.) flours. Plant Food Hum. Nutr. 46:181–186.

De Boland, A. R., G. B. Garner, and B. L. O'Dell. 1975. Identification and properties of 'phytate' in cereal grains and oil-seed products. J. Agric. Food Chem. 23:1186–1189.

Dogbevi, M. K., C. Vachon, and M. Lacroix. 2000. Effect of gamma irradiation on the microbiological quality and on the functional properties of proteins in dry red kidney beans (Phaseolus vulgaris). Radiat. Phys. Chem. 57:265–268.

El-Naggar, S. M., and A. A. Mikhaiel. 2011. Disinfestation of stored wheat grain and flour using gamma rays and microwave heating. J. Stored Prod. Res. 47:191–196.

El-Niely, H. F. G. 2007. Effect of radiation processing on antinutrients, in vitro protein digestibility and protein efficiency ratio bioassay of legume seeds. Radiat. Phys. Chem. 76:1050–1057.

Fombang, E. N., J. R. N. Taylor, C. M. F Mbofung, and A. Minnaar. 2005. Use of γ irradiation to alleviate the poor protein digestibility of sorghum porridge. Food Chem. 91:695–703.

Hagenmaier, R. 1972. Water binding of some purified oil seed proteins. J. Food Sci. 37:965–966.

Hassan, A. B., G. A. M. Osman, M. A. Rushdi, M. M. Eltayeb, and E. E. Diab. 2009. Effect of gamma irradiation on the nutritional quality of maize cultivars (Zea mays) and sorghum (Sorghum bicolor) grains. Pak. J. Nutr. 8:167–171.

Hassan, A. B., E. E. Diab, N. S. Mahmoud, R. A. A. Elagib, M. A. H. Rushdi, and G. A. M. Osman. 2013. Effect of radiation processing on in vitro protein digestibility and availability of calcium, phosphorus and iron of groundnut. Radiat. Phys. Chem. 91:200–203.

Hilmy, N., R. Chosdu, and A. Matsuyama. 1995. The Effect of Humidity after gamma-irradiation on aflatoxin B production of A. flavus in ground nutmeg and peanut. Radiat. Phys. Chem. 46:705–711.

International seed testing Assoiation, ISTA. 2006. International rules for seed testing. Seed science and technology. ISTA, Basserdorf, Switzerland.

Koppelman, S., W. F. Nieuwenh, M. Gaspari, L. M. J. Knippels, A. H. Pennincs, E. F. Knol, et al. 2005. Reversible denaturation of Brazil nut 2S albumins and implications and destabilization on digesion by pepsin. J. Agric. Food Chem. 53:123–131.

Loaharanu, P. 1994. Food irradiation in developing courtiers: a practical alternative. IAEA Bulletin. 36:30–35.

Maliwal, B. P. 1983. In vitro method to assess the nutritive value of leaf concentrate. J. Agric. Food Chem. 31:315–319.

McNamara, N. P., H. I. J. Black, N. A. Beresford, and N. R. Parekh. 2003. Effects of acute gamma irradiation on chemical, physical and biological properties of soils. Appl. Soil Ecol. 24:117–132.

Melki, M., and A. Marouani. 2009. Effects of gamma rays irradiation on seed germination and incidence of hard wheat. Environ. Chem. Lett. 8:307–331.

Monjula, S., and E. John. 1991. Biochemical changes and in vitro protein digestibility of endosperm of germinating *Dolichos lablab*. J. Sci. Food Agric. 55:229–233.

Osman, A. M., A. B. Hassan, G. A. M. Osman, N. Salih, M. A. H. Rushdi, E. E. Diab, et al. 2014. Effects of gamma irradiation and/or cooking on nutritional quality of faba bean (*Vicia faba* L.) cultivars seeds. J. Food Sci. Technol. 51:1554–1560.

Pankaj, K. J., V. B. Kudachikar, and K. Sourav. 2013. Lipase inactivation in wheat germ by gamma irradiation. Radiat. Phys. Chem. 86:136–139.

Pinn, A. B. R. O. 1992. Efeitos das radiações gama sobre adisponibilidade do ferro em feijões (Phaseolus vulgaris) (129 p).São Paulo: Dissertação (Mestrado). Faculdade de CiênciasFarmacêuticas Universidade de São Paulo.

Price, M. L., S. Van Socoyoc, and L. G. Butter. 1978. A critical evaluation of the vanillin reaction as an assay for tannin in sorghum grain. J. Agric. Food Chem. 26:1214–1218.

Rahma, E. H., and M. M. Mostafa. 1988. Functional properties of peanut flour as affected by different heat treatments. J. Food Sci. Technol. 25:11–15.

Rees, D. 2004. Insects of stored products. CSIRO publishing, Collingwood, VIC, Australia.

Refai, M. K., N. H. Aziz, F. El-Far, and A. A. Hassan. 1996. Detection of ochratoxin produced by A. ochraceus in feedstuffs and its control by gamma radiation. Appl. Radiat. Isot. 47:617–621.

Wheeler, E. I., and R. E. Ferrel. 1971. Methods for phytic acid determination in wheat and wheat fractions. Cereal Chem. 48:312–320.

Zhao, S., S. Xiong, C. Qiu, and Y. Xu. 2007. Effect of microwaves on rice quality. J. Stored Prod. Res. 43:496–502.

4

Glycemic responses to maize flour stiff porridges prepared using local recipes in Malawi

Vincent Mlotha, Agnes Mbachi Mwangwela, William Kasapila, Edwin W.P. Siyame & Kingsley Masamba

Faculty of Food and Human Sciences, Lilongwe University of Agriculture and Natural Resources (LUANAR)P.O. Box 219, Lilongwe, Malawi

Keywords

Consumer behavior, diabetes mellitus, diet, fermentation, *ugali*

Correspondence

Agnes Mbachi Mwangwela, Department ofFood Science and Technology, LUANAR, P.O. Box 219, Lilongwe, Malawi.

E-mail: agnesmwangwela@yahoo.com

Funding Information

The United States Agency for International Development (USAID).

Abstract

Glycemic index is defined as the incremental area under the blood glucose response curve of a 50 g carbohydrate portion of a test food expressed as a percent of the response to the same amount of carbohydrate from a standard food taken by the same subject. This study investigated glycemic index of maize stiff porridges consumed as staple food in Malawi and a large majority of other countries in sub-Saharan Africa to identify areas for improvement in consumer diets. Stiff porridges were prepared using flour from whole maize, maize grits, and fermented maize grits. The porridges were served to 11 healthy volunteers for 3 weeks, with two serving sessions a week. Glucose was served as a reference food during weekly serving sessions. Results from descriptive analysis revealed that glycemic responses varied across subjects and porridge types. Porridge prepared from fermented maize grits had moderate glycemic index of 65.49 and was comparable in nutrient composition and sensory characteristics with the other test porridges. Glycemic indices of the porridges prepared from whole maize flour and grits were high at 94.06 and 109.64, respectively, attributed to the effect of traditional maize flour processing, preparation, and cooking methods used. The study also calculated glyaemic load of the porridges and drew recommendations to inform diet planning and modifications for healthy and diabetic individuals.

Introduction

There is a growing interest in glycemic index (GI) of carbohydrates-rich foods to help consumers make healthy food choices within specific food groups. Glycemic index is defined as the incremental area under the blood glucose response curve of a 50 g carbohydrate portion of a test food expressed as a percent of the response to the same amount of carbohydrate from a standard food taken by the same subject (FAO and WHO, 1998). It is used to classify carbohydrate foods based on their blood glucose raising potential. Low-GI foods are those that are digested and absorbed slowly, resulting in low fluctuations in blood sugar levels (Brand-Miller et al. 2014). Good examples of such foods are whole grain cereals, whole kernel bread, beans, and fruits. High-GI foods, by virtue of their rapid digestion and absorption, produce marked fluctuations in blood sugar levels and they include white bread and highly processed grains, cereals, and potatoes (Brand-Miller et al. 2014).

Various authors have studied in vitro digestion models that mimic in vivo situation to examine diets and their effect on postprandial glycemia (Kelly et al. 2004). Many interventions and epidemiological studies have also investigated the short-term biological and health effects of foods, meals, and diets of varying GI to understand implications on health (Meyer et al. 2000; Foster-Powell et al. 2002; Benton et al. 2003; Rizkalla et al. 2004; Fatema et al. 2013). At present, there is global research-based evidence that reductions in daily glycemic load (GL) may lead to a reduced risk for developing noncommunicable diseases (NCDs), such as type 2 diabetes, cancer, and coronary heart disease.

In 1998, Food and Agriculture Organization (FAO) and World Health Organization (WHO) recommended low-GI diets (GI of 0–55) as a viable way to prevent and address the burden of NCDs. In several countries (Australia, France, Sweden, Canada, and South Africa), the use of the GI concept has been integrated in dietary guidelines given by health professionals, and an increasing number of food companies market low-GI products (Brouns et al. 2005). In line with these developments, a large number of academic and commercial laboratories have undertaken measurements of glycemic indices of food for both research and commercial application purposes (Brouns et al. 2005).

Glycemic indices of food have also been examined and reported in various studies conducted elsewhere in Africa (Foster-Powell et al. 2002; Mahgoub et al. 2013). Unfortunately, reported GI values vary widely across studies and cannot be used to guide diets for the region. More research is needed in a wider variety of countries to get population specific data necessary for guiding nutritionists, healthcare practitioners, and consumers.

In sub-Saharan Africa, it is estimated that 12.1 million people are living with diabetes and this figure is projected to increase to 23.9 million by 2030 (Msyamboza et al. 2014). Malawi is one of the countries affected by the NCD epidemic in the region. For example, results from a 2009 WHO STEPS survey showed that in every 100 Malawian adults six have diabetes, 33 have hypertension and five are obese (Msyamboza et al. 2014). A large majority of patients in the country suffer from type 2 diabetes. Type 2 diabetes occurs when blood glucose levels are higher (>108 mg/dL) than normal (68–108 mg/dL), and can be prevented through health food choices, physical activity, and weigh loss. Diabetes type 1 is caused by genetic factors, such as shortage of insulin and decreased ability to use insulin, a hormone that allows glucose (sugar) to enter cells and be converted to energy. Blood sugar levels rise and lead to the disease in this regard.

In tandem with the global strategy to address NCDs, the Malawi government through the Ministry of Health has included dietary management of cardiovascular diseases and diabetes among priorities of essential health package for 2011–2016 (Government of Malawi, 2012). Unfortunately, glycemic index data for local carbohydrate-rich foods are nonexistent and nutrition labeling, which also help consumers make informed food choices, is voluntary and not well understood, when available on food packaging, by the majority of consumers.

Nsima is a local dish made from either maize flour and eaten widely in Malawi and across Africa where it has different local names, including *Ugali* in eastern Africa and *fufu* in western Africa. It is usually eaten together with side dishes locally known as relish—legumes, meat, fish, and vegetables—and various chilli and tomato sauces. Figure 1 shows stages in preparation of different flours from maize. The aim of this study was to determine the glycemic index of maize stiff porridges consumed locally as staple food in Malawi to help consumers make informed choices about low-GI foods.

Materials and Methods

Ethical approval for the study

The study protocol was reviewed and approved by the National Health Sciences Research Committee (NHSRC) under the approval number NHSRC 1130, having been approved by the Department of Food Science and Technology at Lilongwe University of Agriculture and Natural Resources. Medical examinations of the subjects, finger pricking, and blood glucose determination were done by three registered nurses at the University clinic. All subjects provided formal consent before participation in the study. On the other hand, researchers guaranteed confidentiality of information obtained and abided by professional ethical conduct such as neutrality, respect for respondent's dignity, culture, and data verification.

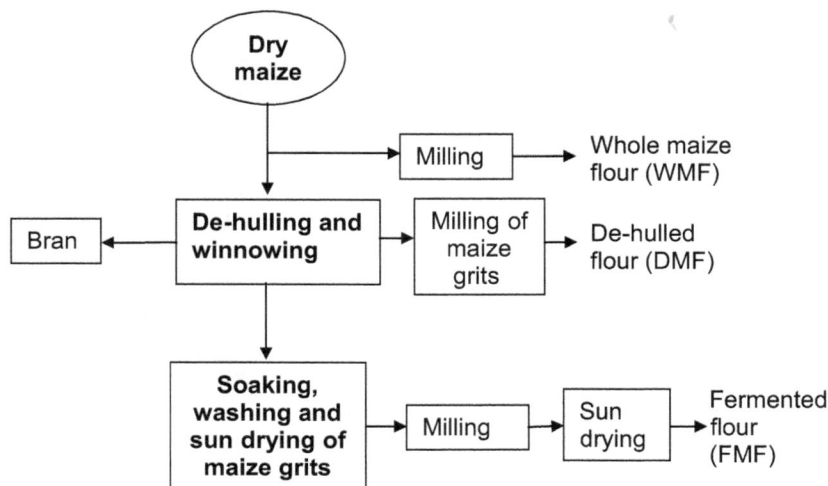

Figure 1. Processing of maize into different flours. Source: Matumba et al. (2009); Mpulula (2013).

Recruitment of research subjects

An announcement was made on the notice boards of the University for healthy volunteers, students, and staff of both sexes, to participate in this study. Thirteen subjects responded to the call and were screened for eligibility. Screening involved diagnosis of insulin sensitivity, glucose tolerance status, and diabetes to minimize outlier glycemic response to test foods. Weight and height were taken to determine levels of overweight and obesity. Subjects were also asked about their past and current medical history as part of medical examination for the study. A total of 12 subjects were deemed eligible and recruited. Participants were also informed about study objectives and commitment required from them—to avoid heavy meals, alcohol, smoking, and vigorous physical activity a day before and on the morning of the test. Other study protocols such as participant's availability at 8:00 h once a week for a period of 2 months and the requirement to consume study samples and have finger-prick tests were communicated. Researchers reiterated that participation is voluntary and that one could withdraw from the study at any time after having given consent to participate. One subject withdrew and the study was conducted on 11 participants.

Test foods for the study

The test food for this study was *nsima* prepared traditionally from three flours known locally as *mgaiwa*, *gramil* and white cornmeal (*woyera*). Thick porridges were prepared from maize flour and water. Porridges were stirred for 5 min to create a stiff paste by adding more flour. Portioning was done using a wooden spoon into serving sizes containing 50 g carbohydrate determined using food composition database and an analytical balance. These portions were 399.91 g, 292.71 g, and 252.77 g for *nsima* prepared from the above-mentioned flours, respectively. Two GI testing sessions took place every week for a period of 3 weeks. At each session of the week, subjects consumed a particular type of *nsima* and glucose, the reference food (Table 1). Nutrient composition of flours used was determined using Association of Official Analytical Chemists (AOAC) methods.

Preparation of standard food

Glucose powder (Royale, batch number 1001) was purchased from a local pharmacy and used as a standard food for this study. Fifty grams of glucose was measured using an analytical balance (Sartorius, L01201S, serial number 50511449) and diluted into 250 mL of tap water.

Table 1. GI testing sessions for the study.

Week	Day	Food material
1	Tuesday, 16 April 2013	Glucose
	Friday, 19 April 2013	Whole maize flour thick porridge
2	Monday, 22 April 2013	Glucose
	Thursday, 25 April 2013	Dehulled, degermed maize flour thick porridge
3	Monday, 29 April 2013	Glucose
	Thursday, 2 May 2013	Fermented dehulled and degermed maize grits thick porridge

Measurement of Glycemic Index

Many methods can be used to calculate GI of food. This study used incremental area under the curve (incremental AUC) method as recommended by FAO and WHO (1998). This method has been used for most calculations of GI (Foster-Powell et al. 2002; Brouns et al. 2005). Measured portions of test foods containing 50 g of available carbohydrate were served to subjects at 8:00 h, breakfast time. Capillary finger-prick blood samples were taken at 0 (fasting), 15, 30, 45, 60, 90, and 120 min after starting to eat the test meal using a safety lancet and glucose stick. Blood glucose was determined using glucometers (Betachek® G5, Blood glucose monitoring system, National Diagnostic Products, Sydney, NSW, Australia). These blood glucose values were used to calculate the incremental area under the curve (iAUC), a reflection of the total rise in blood glucose levels after eating the test food. Steps for calculation of glycemic index using iAUA are given in Table 2.

Data analysis

All analyses were done using the SPSS software for Windows (version 16) (SPSS Inc., 2006, Chicago, IL). The incremental areas under the curve (iAUC) were calculated by the standardized criteria (FAO and WHO, 1998), ignoring any area

Table 2. Steps in GI Calculation using iAUA.

1. Recruitment of healthy individuals
2. Defining the amount of the test food containing 50 g of glycemic carbohydrates
3. Determining the standard food to be used (white bread or glucose)
4. Preparation and serving of test/standard food
5. GI testing sessions and collection of capillary blood samples over a 2-hour period for analysis
6. Calculating individual GI ratings by dividing blood glucose responses for test food by reference food
7. Calculating the GI value for the test food as an average GI value for the 10 people
8. Classification of food

Table 3. Socio-demographic data for the study volunteers

	Mean ± SD
Socio-demographic characteristic	
Age (Years)	23.48 ± 3.52
Weight (kg)	59.06 ± 7.38
Height (cm)	164.89 ± 5.38
BMI (kg/m^2)	21.76 ± 3.06
Fasting blood glucose (mmol/L)	3.92 ± 0.53
Physical activity (hours)	
Resting (includes sleeping)	12.09 ± 5.19
Light exercise	8.18 ± 4.96
Moderate exercise	3.45 ± 2.38
Heavy exercise	0.27 ± 0.65

Values are means and standard deviation (SD) of 11 subjects.

below the baseline. The average iAUC for the three glucose tests was used as the reference value, and each subject's individual GI for each food was calculated. Glycemic values were calculated by dividing the iAUC for the test foods by the iAUC for the reference food and multiplying by 100. The average of the glycemic ratings from all eleven subjects was recorded as the GI for that food. Glycemic load of each food was calculated by multiplying the amount of available carbohydrate in a typical serving of the food and the GI of that food divided by 100 (Table 2).

Results

Characteristics of the study subjects

The study engaged 12 health subjects, seven male and five female, with an average age of 24.5 years, mean BMI of 21.76 kg/m^2 and normal average fasting blood glucose of 3.92 mmol/L. The majority of them spent most of their time resting and doing light physical exercises (Table 3).

Composition of stiff porridges

Table 4 shows the composition of the test foods for this study. Maize flours used to prepare these foods were

processed differently that contributed to variations in composition of the porridges. Stiff porridge made from whole maize flour contained more nutrients than the other two porridges, with exception of carbohydrate. For example, a 399.91 g serving portion of whole maize flour stiff porridge that provided 50 g glycemic carbohydrate contained 11.36 g of protein, 28.27 g of fat, and 2.3 g of crude fiber, figures that were higher than those for the other two porridges (Table 4). FAO (1992) states that maize portion that remains after removal of the hulls and germ through processing are chiefly composed of starch (Fig. 1).

Incremental area under the curve

The capillary blood glucose responses to reference and test foods from three separate GI tests of this study are summarized in Table 5. Glucose caused rapid and very high increase in the glycemic response of the volunteers. It picked to 6.78 mmol/L after 30 min of ingestion of the food and decreased steadily to 4.69 mmol/L at 120 min, a trend that was also observed for whole maize and grits flour porridges. As shown in Table 5, there were no significant differences in the fasting (at 0 min) blood glucose values for the 3 days of GI testing. However, after 15 min of post ingestion significant differences were observed in blood glucose responses between porridges prepared from grits and fermented flours.

GI of the test foods

As shown in Table 6, there were variations in GI responses across individuals and test foods. Observed differences ranged from 46.13–157.13 for whole maize flour porridge, 73.22–177 for porridge from grits flour, and 24.0–133.88 for fermented porridge. Individual GI responses for the three porridges were also different. Large differences between highest and lowest GI responses were observed for subjects number 2 (108), 11 (101.62), 3 (89.25), and 7 (85.35). Average values calculated showed that fermented porridge had the lowest GI of 65.49 followed by whole maize (94.06) and grits (109.64) flour porridges, respectively.

Table 4. Proximate composition of thick porridges prepared from different maize flours (g/100 g).

Parameter	Whole maize flour	Dehulled, degermed maize flour	Maize flour made from fermented dehulled and degermed maize grits	P-value
Moisture	76.54 ± 0.33[a]	78.54 ± 0.02[b]	76.46 ± 0.71[a]	0.002
Ash	0.48 ± 0.03[b]	0.26 ± 0.08[a]	0.22 ± 0.08[a]	0.007
Crude protein	2.84 ± 0.44[b]	1.63 ± 0.46[a]	1.64 ± 0.42[a]	<0.001
Crude fat	7.07 ± 0.75[b]	2.26 ± 0.17[a]	1.71 ± 0.96[a]	<0.001
Crude fiber	0.58 ± 0.08[b]	0.23 ± 0.08[a]	0.19 ± 0.05[a]	0.009
Carbohydrate	13.08	17.31	19.97	

Values are means of three determinations and means with different superscripts in the same row are significantly different.
Carbohydrate value is a calculation by difference (Total carbohydrate—Crude fiber).

Table 5. Blood glucose responses to thick porridges made from whole maize flour, dehulled, degermed maize flour, and maize flour made from fermented dehulled and degermed maize grits (mmol/L, $n = 11$)

Time (minutes)	Whole maize flour	Dehulled, degermed maize flour	Maize flour made from fermented dehulled and degermed maize grits	P-value
0	4.10 ± 0.56	4.10 ± 0.79	3.80 ± 0.61	0.477
15	4.89 ± 0.75[ab]	5.24 ± 0.80[b]	4.40 ± 0.65[a]	0.040
30	6.06 ± 0.63[ab]	6.54 ± 0.71[b]	5.47 ± 0.95[a]	0.012
45	6.35 ± 0.65	6.66 ± 1.16	5.66 ± 1.11	0.069
60	6.14 ± 0.72[b]	6.36 ± 0.76[b]	5.01 ± 0.78[a]	<0.001
90	5.51 ± 0.93[b]	5.56 ± 0.95[b]	4.54 ± 0.85[a]	0.020
120	5.37 ± 0.72[b]	5.36 ± 0.91[b]	4.38 ± 0.83[a]	0.011

Values are means of 11 subjects, means with different superscripts in the same row are significantly different ($P < 0.05$).

Discussion

Changes in lifestyle and diet have contributed to an increased prevalence of diabetes in many low- and middle-income countries, including Malawi, where the burden of infectious diseases, such as HIV/AIDS and TB, is already high. This dual burden of disease is a serious and growing challenge for health systems. The aim of this study was to investigate glycemic indices (GI) of staple stiff porridges, locally known as *nsima*, to identify areas for improvement in consumer diets. A rigorous review of the extant literature reveals that in Malawi only Chilenga (2012) and Mpulula (2013) have conducted scholarly research on processing methods of selected maize flour products and their impact on physiologic and sensory characteristics. Our study is the first of its kind in this regard and has important implications for nutritionists and healthcare practitioners.

Analysis of our descriptive data showed that GI responses varied across subjects and test foods. More so, 11 healthy subjects engaged were, at the very least, not aware of their blood glucose responses to staple foods they eat daily

neither did they know the GI concept nor understand how it relates to diabetes. The problem is aggravated by lack of resources, poor health infrastructure, and inadequate laboratory facilities typical of developing countries and, likely, people die from the disease without ever being diagnosed. Latest statistics show that six in every 100 people (5.6%) in Malawi are diabetes sufferers and 3.9% of the cases remain undiagnosed (Msyamboza et al. 2014).

In this study, fermented porridge has shown to have the lowest GI of 74.90 (Table 6). We recommend fermented flour to be promoted for the preparation of various maize-based foods meant for the diabetics in the country. This flour has also shown to have most preferred sensory characteristics for the majority of local consumers (Chilenga 2012). The other two types of *nsima* tested had higher GI values and the potential to raise blood glucose in the body. Table 6 shows that a 50 g portion of *nsima* made from whole maize flour had a GI of 106.72. Paradoxically, this was the test food that showed to have the highest amount of protein, fat and dietary fiber (Table 4). These results suggest that *nsima* made from whole maize flour may also

Table 6. Incremental area under the curve (iAUC) and glycemic indices of reference and test foods.

Subject	Glucose	Whole maize flour	Dehulled, degermed maize flour	Maize flour made from fermented dehulled and degermed maize grits (Differences)	P-value
1	62.56	99.38	73.22	54.75 (44.63)	
2	88.88	157.13	133.88	49.13 (108)	
3	130.13	69.38	118.88	29.63 (89.25)	
4	85.75	70.13	85.13	24.00 (61.13)	
5	85.95	73.50	77.22	58.00 (19.22)	
6	73.75	139.50	120.38	119.25 (20.25)	
7	120.63	90.00	129.00	43.65 (85.35)	
8	77.38	46.13	99.38	39.03 (60.35)	
9	125.25	109.13	82.50	93.75 (26.63)	
10	87.60	72.38	109.50	133.88 (61.5)	
11	86.72	108.00	177.00	75.38 (101.62)	
Mean iAUC	93.14 ± 22.16	94.06 ± 32.99[ab]	109.64 ± 30.86[b]	65.49 ± 36.18[a]	0.014
GI values		106.72 ± 47.83	121.97 ± 38.99	74.90 ± 46.22	0.055

Values for iAUC are means of 11 subjects and means with different superscripts in the same row are significantly different ($P < 0.05$).

be a good choice for the general population considering the problem of malnutrition in the country and since local side dishes used have low GI, such as legumes, vegetables, and sauces. In every 100 Malawian children less than 5 years of age 48 are stunted, 13 underweight and five wasted (National Statistical Office, 2010). Foods rich in both macro and micronutrients are needed to combat the problem.

Previous studies undertaken elsewhere in Africa show mixed results for GI of maize flour products. Foster-Powell et al. (2002) have summarized figures ranging from 71 to 109 for studies conducted in South Africa and Kenya. In Nigeria, values reported range from 26.6 to 54.83 (Fasanmade and Anyakudo 2007) and 86.8–92.3 in the study by Panlasigui et al. (2010). These variations can be explained at least in part by discrepancies in specific test foods, traditional food preparation methods used as well as differences in sample sizes used, which ranged from 8 to 50, and characteristics of study subjects (healthy vs type 2 diabetes mellitus patients).

We found that glycemic index and glycemic load patterns (calculated as GL = GI/100 × CHO grams per serving) were consistent. Glycemic load measures the degree of glycemic response and insulin demand produced by a specific amount of a specific food (FAO/WHO, 1998). While GI ranks carbohydrates based on their immediate blood glucose response, GL reflects both the quality and quantity of dietary carbohydrates. In addition to this, GL helps predict blood glucose response to specific amount of specific carbohydrate food. Glycemic loads for whole maize, grits and fermented grits flour porridges were 47.03, 54.82, and 32.75 respectively. We predict that porridges cooked using the current Malawian maize flours and recipes would continue to raise blood sugar levels in the long-term considering high-GL figures recorded (defined as ≥20).

There is no easy way to change people's diets and food habits developed through experiences over the life course (Furst et al. 1996; Rozin 1996; Clarke 1998; EUFIC, 2005). According to the Health Belief Model (HBM) proposed by Rosenstock (1966) and modified later by Becker (1974), people need some kind of cue, such as being personally threatened by a disease, to take action with respect to changing dietary behavior. Taken together, considering that *nsima* is the main meal in Malawi interventions that stress small diet changes and consumer education are likely to succeed. An example of how to achieve this change is replacing maize with another staple, such as cassava, rice, and potatoes, at one meal to start with and then slowly increasing the amount of these staples until they are fully incorporated into food habits.

The study is not without limitations. Our data have limitations similar to other studies that measured GI based on 50 g available carbohydrate of single food as a reference amount for GI testing. In real life situations, people eat larger portion sizes of food and in this regard *nsima* is always eaten with sauces, including meat, legumes, fish, and vegetables. Classification of GI responses using relative glycemic effect (RGE) method, which is based on total serving size, could help minimize the bias in this regard. Moreover, a day before the test each subject ought to consume a meal of choice and repeat that meal before each test. This meal pattern was difficult to follow because subjects were not under controlled conditions. Different types of food and nutrients (simple sugars, carbohydrates, and fats) have different chemical compositions and should vary in terms of the amount of energy released (amount of heat per gram of food burned) during the calorie test. Unfortunately, the test was not conducted in a bid to avoid making the study too broad for researchers.

Conclusion

In conclusion, notwithstanding the aforesaid limitations this study has implications for future research on diet and prevention of diabetes mellitus and related diseases in Malawi. We allude that diet cannot by itself suffice to curb the current burden of noncommunicable diseases in light of their complexity and healthcare infrastructure constraints in the country. Interventions need to be multidisplinary and wide reaching in approach, encompassing issues of consumer education, policy framework, and enabling environment for stakeholders to be effective in this regard.

Acknowledgments

The study was done in partial fulfillment of the requirements for the primary author to be awarded the degree of Master of Science in Food Science and Human Nutrition at LUANAR (Lilongwe University of Agriculture and Natural Resources). We are grateful to The United States Agency for International Development (USAID) for providing the financial support for the research to be carried out. We thank all the volunteers who participated in the study and acknowledge valuable contribution of Mr W. Kamboyi, Mrs Kalengamaliro, and other medical staff at the University's clinic.

Conflict of Interest

Authors have no conflict of interest to disclose.

References

AOAC. 2002. Official methods of analysis of association of official analytical chemist, 17th edn. Association of Official analytical Chemist, Arlington, VA, USA.

Becker, M. H. 1974. The health belief model and sick role behavior. Health Educat. Monog. 2:409–419.

Benton, D., M. P. Ruffin, T. Lassel, S. Nabb, M. Messaoudi, S. Vinoy, et al. 2003. The delivery rate of dietary carbohydrates affects cognitive performance in both rats and humans. Psychopharmacology (Berlin) 166:86–90.

Brand-Miller, J. M. D., K. M. Foster-Powell, and J. McMillan-Price. 2014. The low GI diet revolution: the definitive science-based weight loss plan. Da Capo Press, Boston, MA.

Brouns, F. I., K. N. Bjorck, A. L. Frayn, V. Gibbs, G. Slama Lang, and T. M. S. M. Wolever. 2005. Glycaemic index methodology. Nutr. Res. Rev. 18:145–171.

Chilenga, C. 2012. Effect of variety and processing methods on physiological characteristics of maize flour and sensory properties of maize nsima. Master of Science Thesis, University of Malawi, Bunda College of Agriculture, Lilongwe. Malawi.

Clarke, J. E. 1998. Taste and flavour: their importance in food choice and acceptance. Proceed. Nut. Soc. 57:639–643.

EUFIC [The European Food Information Council]. 2005. The determinants of food choice. http://www.eufic.org/article/en/expid/review-food-choice/ (accessed 20 February 2015).

Fasanmade, A. A., and M. M. C. Anyakudo. 2007. Glycaemic indices of selected Nigerian flour meal products in male type 2 diabetes subjects. Diabetologia Croatica 36:33–38.

Fatema, K., C. B. Mikkelsen, F. Rahman, N. Sumi, and L. Ali. 2013. Glycemic and insulinemic responses to isis cookies and Danish Traditional Cookies in Healthy Subjects. J. Food Nut. Disord. 2:1–5.

Food and Agriculture Organization of the United Nations(FAO). 1992. Maize in human nutrition, introduction. FAO; Food and Nutrition Series, 25. Http://www.fao.org/docrep/t0395e/T0395E00.htm (accessed 1 August 2013).

Food and Agriculture Organization of the United Nations (FAO)/World Health Organization (WHO). 1998. Expert consultation. Carbohydrates in human nutrition: report of a joint FAO/WHO Expert Consultation, Rome, 14–18 April, 1997. Rome: Food and Agriculture Organization, (FAO Food and Nutrition paper 66). http://www.who.int/nutrition/publications/nutrientrequirements/ scientific_update_carbohydrates/en/[Accessed on 12 August 2013].

Foster-Powell, K., S. H. Holt, and J. C. Brand-Miller. 2002. International table of glycaemic index and glycaemic load values. Am. J. Clin. Nutr. 76:5–56.

Furst, T., M. Connors, C. A. Bisogni, J. Sobal, and L. Winter Falk. 1996. Food choice: a conceptual model of the process. Appetite 26:247–266.

Government of Malawi. 2012. Health sector strategic plan 2011–2016. Likuni Press and Publishing House, Lilongwe, Malawi.

Kelly, S. A. M, G. Frost, V. Whittaker, and C. D. Summerbell. 2004. Low glycaemic index diets for coronary heart disease. Cochrane Database of Systematic Reviews. 4: 1–66.

Mahgoub, S. O., M. Sabone, and J. Jackson. 2013. Glycaemic index of selected staple carbohydrate-rich foods commonly consumed in Botswana. South Afr. J. Clin. Nut. 26: 182–187.

Matumba, L., M. Monjerezi, E. Chirwa, D. Lakudzala, and P. Mumba. 2009. Natural occurrence of AFB1 in maize and effect of traditional maize flour production on AFB1 reduction in Malawi. Afr. J. Food Sci. 3:413–425.

Meyer, K. A., L. H. Kushi, D. R. Jacobs, J. Slavin, T. A. Sellers, and A. R. Folsom. 2000. Carbohydrates, dietary fibre and incident type 2 diabetes in older women. Am. J. Clin. Nutr. 71:921–930.

Mpulula, O. 2013. Effect of home-based maize processing methods and variety on apparent iron and zinc bio-availability. M.Sc. thesis. University of Malawi, Bunda College of Agriculture. Lilongwe. Malawi.

Msyamboza, K. P., C. J. Mvula, and D. Kathyola. 2014. Prevalence and correlates of diabetes mellitus in Malawi: population-based national NCD STEPS survey. BMC End. Disord. 14:41.

National Statistical Office (NSO). 2010. Malawi demographic and health survey. NSO, Zomba, Malawi.

Panlasigui, L. N., C. L. T. Bayaga, E. B. Barrios, and K. L. Cochon. 2010. Glycaemic response to quality protein maize grits. J. Nut. Metab. Pp 1-6 http://dx.doi.org/10.1155/2010/697842.

Rizkalla, S. W., L. Taghrid, M. Laromiguiere, D. Huet, J. Boillot, A. Rigoir, et al. 2004. Improved plasma glucose control, whole-body glucose utilization, and lipid profile on low-glycemic index diet in type 2 diabetic men: a randomized controlled trial. Diabetes Care 27:1866–1872.

Rosenstock, I. M. 1966. Why people use health services. Milbank Memorial Fund Quarterly. 83:1–32.

Rozin, P. 1996. Sociocultural influences on human food selection. Pp. 233–263 in E. D. Capaldi, ed. Why we eat what we eat: the psychology of eating. American Psychological Association, Washington, DC.

Wolever, T. M., and C. Mehling. 2003. Long-term effect of varying the source or amount of dietary carbohydrate on postprandial plasma glucose, insulin, triacylglycerol, and free fatty acid concentrations in subjects with impaired glucose tolerance Am. J. Clin. Nutr: 77: 612–621.

5

Effect of environment and genotypes on the physicochemical quality of the grains of newly developed wheat inbred lines

Noha I. A. Mutwali[1], Abdelmoniem I. Mustafa[1], Yasir S. A. Gorafi[2] & Isam A. Mohamed Ahmed[1]

[1]Department of Food Science and Technology, Faculty of Agriculture, University of Khartoum, Shambat, 13314 Khartoum, Sudan
[2]Wheat Research Program, Agricultural Research Corporation, P.O. Box: 26, Wad Medani, Sudan

Keywords
Falling number, gluten quality, growing environment, wheat genotypes

Correspondence
Isam A. Mohamed Ahmed, Department of Food Science and Technology, Faculty of Agriculture, University of Khartoum, Shambat, 13314, Khartoum, Sudan.

E-mail: isamnawa@yahoo.com

Funding Information
No funding information provided.

Abstract

To meet the increased demand for wheat consumption, wheat cultivation in Sudan expanded southward to latitudes lower than 15°N, entering a new and warmer environment. Consequently, wheat breeders developed several wheat genotypes with high yields under these environmental conditions; however, the evaluation of the end-use quality of these genotypes is scarce. In this study, we assessed the end-use quality attributes of 20 wheat genotypes grown in three different environments in the Sudan (Wad Medani, Hudeiba, and Dongola). The results showed significant differences ($P \leq 0.01$) in all quality tests among environments, genotypes and genotypes Versus environments. The findings obtained, covered wide ranges of test weight (TW, 76.6–85.25 kg/hL), thousand kernel weight (TKW, 28.70–48.48 g), protein (PC, 9.96–14.06%), wet gluten (WG, 28.63–46.53%), gluten index (GI, 36.36–92.77%), water holding capacity (WHC, 168.42–219.32%), falling number (FN, 508.00–974.67 sec), and sedimentation value (SV, 19.00–40.00 mL). Analysis of the traits, genotypes, and traits versus genotypes showed varied correlations in the three growing environments. The genotype G3 grown in either one or all of the three environments exhibits worthy performance and stability for most of the tested quality traits. The crossing of this genotype with high yield genotypes could produce cultivars with sufficient quality and marketability.

Introduction

Wheat is an important and most widely cultivated food crop in the world. This crop played a central role in combating hunger and improving the global food security. Wheat is ranked second in total cereal production behind corn, with rice being the third (FAO, 2012). The grains of this plant provide about 20% of all calories and proteins consumed by people on the globe (Shiferaw et al. 2013). In recent years, demand for wheat has significantly increased as a result of the global population growth, and thus wheat production has a strategic role in food security and the world economy. As a result, horizontal expansion of wheat production has arisen in recent years by moving wheat into nontraditional areas formerly considered unacceptable for production. However, the global

warming introduced various abiotic stresses such as drought, temperature extremes, and salinity that adversely affect the yield and grains quality of wheat (Huseynova and Rustamova 2010). To meet the demands of future population's explosions and ensure grain production in these environments, cultivars must be developed and evaluated for their high yield and high quality. Thus, the objective of wheat breeders is to produce well-adapted and high-yielding varieties with finest end-use quality (Lopes et al. 2012; Li et al. 2013).

In Sudan, wheat is the second most essential cereal food and the main staple food for many peoples in both rural and urban areas. This crop is traditionally cultivated in the northern region of Sudan where the winter conditions are favorable for plant growth and grain yield. However, in the last decades wheat cultivation in Sudan expanded southward to latitudes lower than 15°N, entering a new and warmer environment and inhabiting most of the irrigated sectors in central and northern states (Elsheikh et al. 2015). The average annual production of wheat during the period 2009–2013 was 242,000 tons and is forecasted to rise to 320,000 tons in 2014 with 32% change (FAO, 2014). Nevertheless, the rate of wheat grain production in the Sudan is far below the consumption needs. High temperature and drought stresses, low nitrogen content, and lack of quality seeds of improved varieties are the main constraints limiting wheat production in Sudan (Ali et al. 2006; El Siddig et al. 2013). To overcome these limitations wheat breeders have developed several varieties and inbred lines with enhanced tolerance to most of these stresses (Elahmadi 1996; Ali et al. 2006), and with better grain yield and quality. Although the grain yield of these advanced wheat lines has been extensively studied by many researchers, reports on the end-use quality of these lines are rare (Ali et al. 2006). Therefore, the primary objective of this study was to examine the effect of growing environment on end-use quality characteristics of twenty wheat genotypes grown in three different environments (Wad Medani, Hudeiba, and Dongola) in the Sudan.

Materials and Methods

Plant materials and field trials

In this study, 20 wheat genotypes representing a broad range of yield and adaptability to the environment of the Sudan were used (Table 1). These genotypes were developed through extensive wheat breeding programs at the Agricultural Research Corporation (ARC), Gazira, Sudan. All materials were grown for two constitutive seasons (2003/2004 and 2004/2005) in three different environments (Dongola, Hudeiba, and Wad Medani) representing both the traditional and new wheat growing environments in the Sudan. The three growing environments are characterized by their different soil and environmental conditions with no precipitation during the whole crop cycle. Dongola is located in Northern State (Lat. 19° 08′ N, Long. 30° 40′ E, and Alt. 240 m) with a warm (average temperature 21.8°C) and dry winter. The soil of Dongola is classified as sandy clay loam with very low organic matters (>5%), high water permeability, and a pH of 8.0. Whereas, Hudeiba is located in the Nile State (Lat. 17° 34′ N, Long. 33° 56′ E, and Alt. 350 m) and has warm (average temperature 24.2°C) and dry winter. The soil of Hudeiba is classified as karusoil clay (contained 4% sand, 40% silt, and 56% clay) with little nitrogen (360 ppm), phosphorus (8 ppm), and organic carbon, and a pH of 8.1. While Wad Medani is located in central Sudan (Lat. 14° 24′ N, Long. 29°

Table 1. Genotypes used in the current study.

Genotypes code	Pedigree/Variety
G1	ELNeilain
G2	Debeira
G3	RGO/SERI/TRAP//Bow
G4	KAU2 * CHEN//BCN. CMB
G5	SON64/SRC – LR64A) G155
G6	427F4/2000-1
G7	PYT#23 (DWR39xCONDOR "S")14PxT
G8	KAUZ "S" 6 57C1-3-6.2-2-1-2
G9	TEVEE "S"/SHUHA "S"
G10	N5732/HER//CASKOR
G11	ELNEILAIN/SASARIBE
G12	CONDOR "S"/14PYT//DWR39
G13	VERONA/KAUZ//KAUZ
G14	ELNEILAIN/DEBEIRA
G15	OASIS/KUAZ//3 * BCN
G16	CONDOR "S"/BALADI//DEBEIRA
G17	DH5
G18	DH8
G19	IHSGE # 19
G20	IHSGE # 20

33′ E, and Alt. 407 m) having a slightly hot (average temperature 26.8°C) and dry winter. The soil type of Wad Medani is heavy cracking vertisoil (58–66% clay) with very low water permeability, organic carbon (0.35%), nitrogen (0.03%), phosphorus (4 ppm), and a pH 8.3

In the three growing environments, the experiments were structured in a randomized complete block design with three replications. After soil harrowing and leveling, the seeds were seeded manually in rows of 0.2 m apart in plots consisting of 4 rows of 5 m length at a seeding rate of 120 kg/ha. The seeds were treated with Gaucho (Imidacloprid 35% WP) at a rate of about 1 g/1 kg seed to control pests mainly termites and aphids. Phosphorus was applied by furrow placement prior to sowing at the rate of 43 kg P_2O_5/ha, while nitrogen, in the form of urea, was implemented before the second irrigation at a rate of 86 kg N/ha. Irrigation intervals were every 10–12 days, and weeding was carried out manually at least twice.

Samples preparation

The wheat grains were manually cleaned and then the grains thousand kernels and test weights were evaluated. The samples were tempered and milled into straight grade flour (72% extraction rate) using a Brabender Quadrumat Junior Mill (Brabender, GmbH & Co. KG, Duisburg, Germany). After that, the flour samples were placed in a separate plastic container and stored in a deep freezer until used for biochemical analysis. Three independent replicates of each sample were used for biochemical analysis.

Chemical composition

Moisture, ash, crude protein (N × 5.7), and fat content of the flour samples were measured according to the official standard method (AACC, 2000).

Gluten quantity and quality

Wet gluten (WG), dry gluten (DG), water holding capacity (WHC), and gluten index (GI) were determined according to the standard method 38–12.02 (AACC, 2000) using a Glutomatic 2200 systems and a Perten 2015 centrifuge (Perten Instrument, AB, Huddinge, Sweden).

Falling number

Falling number (FN) of wheat flours was determined using the falling number instruments following the Official Method 56–81.03 (AACC, 2000) and expressed on 14% moisture basis.

Sedimentation value

Zeleny sedimentation value of wheat flours was measured according to the standard method 56–60.01 (AACC, 2000) and expressed on 14% moisture basis.

Statistical analyses

For the grains of individual wheat genotypes grown in each environment, the data of three independent experiments were first separately analyzed, and then the results were combined to determine the interactive effects of genotypes and growing environments. The data were assessed by analysis of variance (Gomez and Gomez 1984) and Duncan's multiple range test (DMRT). Correlation coefficients among all quality traits were evaluated based on the means of all genotypes in the individual environment using Stat View software. Exploratory multivariate statistical analysis of the data was performed using HJ-biplot methods included in the MULTBIPLOT software (Vicente-Villardón 2010). The HJ-Biplot method allows the plotting of both genotypes and wheat quality traits with an optimum quality of representation and hence provides easy and fast information about the interrelation of the plotted data. To ascribe a set of individuals to a particular group, we performed hierarchical clustering analysis with the Euclidean distance using the principal components scores and the Ward's technique as the process of linkage. Significance was accepted at $P \leq 0.05$, $P \leq 0.01$, and $P \leq 0.001$.

Results and Discussion

Grain physical characteristics

Thousand kernels (TKW) and test weight (TW) of wheat genotypes grown in three different environments are shown in Table 2. The TKW and TW of all genotypes were significantly varied ($P \leq 0.01$) between the three environments. These results revealed that the variation in the environmental and soil conditions between the three environments could contribute to the differences in the wheat grain weight. In addition, the interaction between the wheat genotypes and the growing environment was also significant ($P \leq 0.01$) for both traits. The highest mean values of TKW and TW were observed for G5 at Hudeiba and G19 at Dongola while the lowest values were obtained for G12 at Wad Medani and G3 at Hudeiba, respectively. Regarding the environments, the highest mean values of TKW and TW for all genotypes were at Hudeiba and Dongola while the lowest values were at Wad Medani and Hudeiba, respectively. Throughout all the three environments, the mean TW was in the range of 78.02–82.83 kg/hL, which indicates that all wheat genotypes exhibit well-filled grains (Kaya and Akcura, 2014; Li et al. 2013). Overall, the results of the TKW and TW demonstrated that the environmental conditions (temperature and soil fertility), agronomic practices (irrigation and fertilization), and wheat genotypes could affect the grain physical characteristic and hence the flour yield and end-use quality. Previous reports showed that environmental conditions and fertilizers application had a significant impact on the TKW and TW of various wheat genotypes (Lopes et al. 2012; Mohammadi 2012; Li et al. 2013; Bouacha et al. 2014; Kaya and Akcura 2014). In addition, water deficit and elevated temperatures above average during grain filling reported to reduce the TKW for winter wheat (Erekul and Kohn 2006). Mohammadi (2012) concluded that wheat cultivars capable of maintaining high TKW under heat stress appeared to possess a greater tolerance for warm environments. While Lopes et al. (2012) suggested the use of TKW for selection of wheat genotypes under a warm environment of the Sudan.

The grain physical characteristics TKW and TW have no correlation with all other quality traits of all genotypes grown in Wad Medani, whereas they showed some correlations with the quality traits in other environments (Table 3). At Hudeiba, TKW showed a significant positive correlation (*$P < 0.05$, $r = 0.49$) with PC, while it showed highly significant negative correlation (**$P < 0.01$, $r = -0.61$) with GI. The TW, on the other hand, showed significantly negative (*$P < 0.05$) correlations with the SV at Hudeiba ($r = -0.44$) and with GI at Dongola ($r = -0.50$). These results suggest that the association

Effect of environment and genotypes on the physicochemical quality of the grains...

33

Table 2. Thousand kernel weight (g) and test weight (kg/ha) of 20 wheat genotypes grown in three environments.

Genotypes	1000 Kernel weight (TKW)				Test weight (TW)			
	Wad Medani	Hudeiba	Dongola	Mean	Wad Medani	Hudeiba	Dongola	Mean
G1	35.81^{tuv}	46.21^a	36.97^{ij}	39.66^{fg}	82.03^i	80.40^{nop}	84.54^b	82.32^b
G2	35.20^v	44.82^{cd}	36.76^{rs}	38.92^{hi}	82.53^{fg}	79.65^{rs}	82.45^{gh}	81.54^f
G3	36.38st	44.24^{de}	40.44^{hi}	40.35^{de}	77.33^v	76.60^w	80.14^{opq}	78.02^m
G4	35.52^{uv}	44.53^d	41.08^{gh}	40.37^d	81.45^{jk}	80.89^{lm}	84.40^b	82.24^b
G5	39.48^{jkl}	48.48^a	43.66^{ef}	43.87^a	82.50^g	80.89^{lm}	83.58^{cd}	82.32^b
G6	37.17^{pqr}	45.50^{bc}	38.98^{klm}	40.55^{cd}	81.43^k	79.63^{rs}	83.45^d	81.50^{fg}
G7	35.86^{tuv}	39.70^{ijk}	36.18^{stu}	37.25^j	82.11^{hi}	79.95^{qr}	79.55st	80.54^i
G8	31.22^y	36.54^{rst}	33.52^{xy}	33.76^l	82.08^{hi}	78.11^u	82.45^{gh}	80.88^h
G9	33.77^{xy}	43.81^{de}	39.71^{ijk}	39.09^h	81.55^{jk}	80.25^{op}	83.35^d	81.71^{ef}
G10	33.63^{xy}	44.76^c	39.38^{ijk}	39.25^{gh}	79.21^t	77.50^v	80.50^{no}	79.07^l
G11	37.88^{nop}	44.32^{de}	40.83^{gh}	41.01^{bc}	82.70^{efg}	81.50^{jk}	84.30^b	82.83^a
G12	28.70^z	40.75^h	34.30^{wx}	34.58^K	80.25^{op}	80.05^{pq}	84.35^b	81.55^f
G13	37.84^{opq}	44.64^{cd}	41.73^g	41.40^B	81.81^{ij}	79.54st	82.66^{fg}	81.34^g
G14	31.42^y	37.63^{opq}	34.70^w	34.58^K	78.11^u	81.40^k	79.55st	79.68^j
G15	30.09^z	38.88^{klm}	35.26^v	34.74^k	81.45^{jk}	80.25^{op}	83.85^c	81.85^{de}
G16	30.71^z	35.52^{uv}	33.17^{yz}	33.13^m	82.01^i	81.30^k	83.00^e	82.10^{bc}
G17	31.24^y	46.14^b	38.22^{mno}	38.54^i	82.75^{efg}	81.20^{kl}	82.90^{ef}	82.28^b
G18	33.42^y	42.82^f	39.20^{jkl}	38.48^i	79.31st	79.14^t	79.55st	79.33^k
G19	31.91^y	41.55^g	38.72^{lmn}	37.39^j	80.73^{mn}	79.95^{qr}	85.25^a	81.97^{cd}
G20	35.14^v	44.71^{cd}	39.96^{ij}	39.94^{ef}	80.50^{no}	79.55st	82.45^{gh}	80.83^h
Mean	34.12^c	42.78^a	38.14^b	38.34	81.09^b	79.88^c	82.61^a	81.19

Means followed by the same letter are not significantly different ($P \leq 0.01$) from each other, according to Duncan's multiple range test.

between grain physical characteristics (TKW and TW) and the flour quality traits mainly PC, GI, and SV depend primarily on the environmental conditions rather than the genetic makeup of the cultivar. Similarly, positive correlations between TKW and PC have been reported for wheat cultivars and landraces grown in the subhumid region following K-fertilizers treatment (Bouacha et al. 2014). By contrast, negative correlations between grain physical characteristics and flour PC and SV for other wheat genotypes have also been reported (Ozturk and Aydin 2004; Tahir et al. 2006).

Chemical composition

The results of moisture, ash, protein, and fat content are presented in Table 4. The moisture content (MC) was in the range of 11.38–13.13%, 10.85–12.48%, and 10.21–12.21% for the genotypes grown at Wad Medani, Hudeiba, and Dongola, respectively. Significant differences ($P \leq 0.01$) in MC among all environments, indicating that environments had influenced the flour moisture content. Although it has no correlation with other traits in Wad Medani and Hudeiba, MC demonstrated extremely negative correlation (***$P < 0.001$, $r = -0.67$) with FN and positive correlation (*$P < 0.05$, $r = 0.54$) with PC at Dongola (Table 3). The highest mean value for MC (12.31%) among all genotypes was observed in Wad Medani while lowest

value (11.61%) was scored in Dongola. Regardless of the environment, the highest MC was recorded for G5 that was significantly different ($P \leq 0.01$) from all other genotypes, while the lowest value was observed for G6. Regarding the interaction between wheat genotype and growing environment, the highest MC (13.13%) was obtained for G13 grown at Wad Medani, whereas the lowest value (10.21%) was noticed for G2 grown at Dongola. The results achieved agreed with values obtained by Makawi et al. (2013) who stated that the moisture content of Sudanese wheat cultivars ranged from 10.40 to 12.07%. The variation in MC of the different wheat genotypes may be due to the variations in environmental conditions between the three environments, the genotypes, and their interaction. Moisture content is mostly affected by relative humidity at harvest and during storage (Makawi et al. 2013).

Statistical analysis showed significant differences ($P \leq 0.01$) in AC among the growing environments, indicating that the environment had affected the flour ash content (Table 4). Throughout the three areas, the highest mean value (0.68%) of AC for all genotypes was obtained at Wad Medani while the lowest value (0.65%) was observed at Hudeiba. The AC showed significantly negative correlations (*$P < 0.05$) with SV at Wad Medani ($r = -0.53$) and Dongola ($r = -0.51$), with TW at Wad Medani ($r = -0.47$), and with WG and Dongola ($r = -0.49$)

Table 3. Correlation coefficient between the physicochemical quality parameters of wheat genotypes grown in three different environments (Wad Medani, Hudeiba, and Dongola).

	TKW	AC	MC	TW	GI	WHC	FN	SV	WG	DG	PC
Wad Medani											
AC	0.10										
MC	0.38	0.03									
TW	0.19	−0.17	0.21								
GI	−0.04	0.19	−0.34	−0.04							
WHC	0.18	−0.32	−0.20	−0.25	0.53*						
FN	0.17	0.25	0.19	−0.32	0.37	0.35					
SV	−0.16	−0.53*	0.06	−0.12	−0.25	0.11	−0.23				
WG	−0.21	−0.37	0.22	−0.06	−0.83***	−0.44	−0.39	0.55*			
DG	−0.21	−0.22	0.24	0.08	−0.86***	−0.68***	−0.45*	0.42	0.94***		
PC	−0.19	−0.37	0.21	−0.20	−0.43	0.02	0.03	0.69***	0.64***	0.50*	
FC	−0.36	0.19	−0.26	−0.43	0.25	0.11	−0.08	0.35	0.04	−0.02	0.07
Hudeiba											
AC	−0.27										
MC	0.01	−0.17									
TW	−0.05	−0.47*	0.29								
GI	−0.61**	−0.20	−0.02	0.02							
WHC	−0.22	−0.06	0.24	0.10	0.17						
FN	−0.02	0.07	−0.14	−0.03	−0.01	−0.05					
SV	−0.15	−0.15	−0.01	−0.44*	0.25	0.30	0.47*				
WG	0.41	0.02	0.02	−0.36	−0.47*	−0.39	0.48*	0.45*			
DG	0.39	0.04	−0.03	−0.35	−0.42	−0.64***	0.41	0.29	0.96***		
PC	0.49*	−0.02	−0.18	−0.32	−0.61**	−0.40	0.36	0.35	0.83***	0.80***	
FC	−0.20	0.09	−0.31	−0.17	0.24	0.10	0.67***	0.59**	0.26	0.21	0.26
Dongola											
AC	−0.05										
MC	0.15	−0.24									
TW	0.10	−0.02	−0.18								
GI	−0.41	0.02	−0.26	−0.50*							
WHC	0.42	−0.07	−0.20	−0.28	0.25						
FN	−0.39	0.21	−0.67***	−0.02	0.42	0.04					
SV	−0.17	−0.51*	0.25	−0.44	0.50*	0.15	−0.11				
WG	0.05	−0.49*	0.33	0.20	−0.48*	−0.28	−0.35	0.21			
DG	−0.09	−0.31	0.35	0.26	−0.51*	−0.70***	−0.31	0.05	0.87***		
PC	0.12	−0.10	0.54*	−0.11	−0.03	−0.16	−0.22	0.31	0.52*	0.48*	
FC	0.02	0.19	0.21	−0.14	−0.18	−0.17	0.07	−0.14	−0.16	0.02	0.14

TKW, Thousand-kernel weight; AC, Ash content; MC, Moisture content; TW, test weight; GI, Gluten index; WHC, Water holding capacity; FN, Falling number; SV, Sedimentation value; WG, Wet gluten; DG, Dry gluten; PC, Protein content; FC, Fat content.
Values in bold are significant at *$P < 0.05$; **$P < 0.01$; ***$P < 0.001$.

(Table 3). Among wheat genotypes grown in the three environments, G13 had the highest AC, whereas G9 had the lowest. The differences seen in the AC in the present study may be attributed to differences in wheat genotypes and environmental conditions (temperature and soil conditions) as well as fertilizers application (Makawi et al. 2013).

Wheat grain protein is of primary importance in determining the end use quality of the flour and variations in both protein content and composition could significantly affect the flour quality. The crude protein (PC) content was found to be in the range of 9.59–13.40%, 9.96–14.06%, and 11.40–13.87% for the growing environments Wad Medani, Hudeiba, and Dongola, respectively (Table 4).

The results revealed significant differences ($P \leq 0.01$) in the PC among the wheat genotypes and their interaction with the growing environments. These findings indicated that both the genotypes and the growing environment had influenced the flour protein content. Throughout the three growing environments, the highest mean value (13.43%) of PC was found for genotypes grown in Dongola while the lowest value (10.99%) was observed in those grown in Wad Medani. Regarding the interaction, the highest value was obtained for G4 in Hudeiba and Dongola while the lowest value was obtained for G16 at Wad Medani. This result agreed with the outcome of Elmobarak et al. (2004) who stated that wheat grown at Wad Medani gave lower grain protein content compared to that of

Table 4. Moisture, ash, protein, and fat content (%) of 20 local wheat genotypes grown in three environments.

Variety/Lines	Moisture (MC)				Ash (AC)				Protein (PC)				Fat (FC)			
	Wad Medani	Hudeiba	Dongola	Mean	Wad Medani	Hudeiba	Dongola	Mean	Wad Medani	Hudeiba	Dongola	Mean	Wad Medani	Hudeiba	Dongola	Mean
G1	12.79^b	11.44^z	11.60^{uvwxyz}	11.94^{defg}	0.69^{efghi}	0.61^{klmno}	0.61^{klmno}	0.64^{fgh}	10.58^u	12.31^{lmn}	12.12^{mno}	11.67^h	1.15^{lmn}	1.03^{pqr}	0.93^r	1.04^g
G2	12.63^c	11.31^z	10.21^z	11.38^j	0.67^{fghijk}	0.68^{fghij}	0.71^{defgh}	0.68^{cde}	9.97^w	13.03^{fg}	12.04^{no}	11.68^h	1.44^{de}	1.74^a	1.13^{mno}	1.43^a
G3	12.11^{jk}	11.71^{qrstuvwx}	12.15^{hijk}	11.99^{cd}	0.73^{cdef}	0.76^{bcd}	0.73^{cdef}	0.74^b	11.15^q	13.60^{bcd}	12.36^{klm}	12.37^c	1.38^{efg}	1.41^{def}	1.46^d	1.41^a
G4	12.07^{jklm}	11.65^{rstuvwx}	11.82^{nopqr}	11.85^{efg}	0.63^{jklmn}	0.57^o	0.79^b	0.66^{efg}	10.94^{qrs}	14.06^a	13.87^{ab}	12.96^b	0.99^r	1.35^{fgh}	1.22^{jkl}	1.19^e
G5	12.15^{hijk}	12.48^{cde}	12.18^{hij}	12.27^a	0.61^{jklmn}	0.61^{klmno}	0.69^{ghijk}	0.63^{gh}	9.99^w	13.24^{ef}	12.55^{jkl}	11.92^g	1.03^{pqr}	1.00^r	0.93^r	0.99^h
G6	12.27^{fghi}	11.22^z	10.27^z	11.25^k	0.63^{jklmn}	0.63^{jjklm}	0.72^{cdefg}	0.66^{efg}	10.06^w	11.05^q	11.96^o	11.02^i	0.96^r	1.01^{qr}	1.17^{klmn}	1.04^g
G7	12.37^{efg}	11.60^{uvwxyz}	12.06^{jklm}	12.01^{bc}	0.63^{jklmn}	0.77^{bc}	0.59^{nno}	0.66^{ef}	10.36^v	12.63^{ijk}	13.55^{cd}	12.18^{def}	1.05^{opqr}	1.23^{jk}	1.09^{nopq}	1.12^f
G8	12.61^c	11.08^z	11.81^{nopqrs}	11.83^{fg}	0.69^{fghi}	0.66^{ghijk}	0.69^{ghijkl}	0.67^{ef}	11.40^p	12.09^{mno}	12.79^{ghij}	12.09^{ef}	1.01^{qr}	1.45^{def}	1.40^{def}	1.28^d
G9	12.62^c	11.52^{yz}	12.19^{ghij}	12.11^b	0.47^p	0.50^p	0.59^{nno}	0.52^i	13.40^{de}	13.13^f	13.76^b	13.43^a	0.91^r	1.15^{lmn}	1.35^{fgh}	1.14^f
G10	12.33^{efgh}	11.53^{yz}	12.19^{ghij}	12.02^{bc}	0.59^{mno}	0.64^{ijklmn}	0.59^{nnno}	0.61^{hi}	10.66^{tu}	13.00^{fg}	13.42^{de}	12.36^c	1.44^{de}	1.14^r	1.27^{hij}	1.28^d
G11	12.61^c	11.55^{wxyz}	11.85^{nopq}	12.00^{cd}	0.75^{bcde}	0.56^o	0.71^{defgh}	0.67^{ef}	11.53^p	11.53^p	13.16^f	12.07^{efg}	1.26^{ij}	1.11^{mnop}	1.23^{jk}	1.20^e
G12	12.24^{fghij}	11.81^{nopqrs}	11.93^{lmno}	11.99^{cd}	0.62^{klmno}	0.74^{bcde}	0.72^{cdefg}	0.70^{cd}	12.25^{lmn}	12.63^{ijk}	13.68^{bc}	12.85^b	1.41^{def}	1.13^{mno}	1.10^{mnop}	1.21^e
G13	13.13^a	11.18^z	11.77^{opqrstu}	12.02^{bc}	0.85^a	0.78^{bc}	0.71^{defgh}	0.78^a	10.77^{stu}	13.19^f	13.03^{fg}	12.33^{cd}	1.13^{mno}	1.29^{ghi}	1.46^d	1.29^d
G14	11.91^{mnop}	11.97^{jklmn}	11.63^{stuvwwy}	11.83^{fg}	0.81^a	0.63^{jjklmn}	0.78^{bc}	0.74^b	10.94^{qrs}	9.96^w	13.51^{cd}	11.47^h	1.32^{ghi}	1.29^{ghi}	1.00^r	1.20^e
G15	12.56^{cd}	10.85^z	11.65^{stuvwx}	11.68^h	0.61^{klmno}	0.71^{defg}	0.70^{defgh}	0.68^{def}	11.04^{qr}	12.97^{fgh}	13.05^{fg}	12.35^c	1.35^{fgh}	1.46^d	1.46^d	1.42^a
G16	11.78^{opqrst}	12.41^{def}	11.57^{vwxyz}	11.92^{cdef}	0.67^{fghij}	0.61^{klmno}	0.71^{defgh}	0.66^{ef}	9.59^x	11.19^q	11.40^p	10.72^i	1.23^{jk}	1.24^{jk}	1.18^{klm}	1.21^e
G17	11.38^z	11.47^z	11.64^{rstuvwy}	11.49^i	0.70^{defgh}	0.49^p	0.65^{hijklm}	0.61^{hi}	10.88st	13.08^{fg}	12.73^{hij}	12.23^{cde}	1.42^{def}	1.54^c	0.94^r	1.30^d
G18	12.22^{ghij}	11.61^{tuvwwxy}	11.54^{xyz}	11.79^g	0.72^{cdefg}	0.78^{bc}	0.75^{bcde}	0.75^b	11.35^p	12.39^{klm}	12.36^{klm}	12.03^{fg}	1.64^b	1.29^{hij}	1.23^{jk}	1.38^b
G19	12.10^{jkl}	11.74^{pqrstu}	11.23^z	11.69^h	0.70^{efgh}	0.78^{bc}	0.67^{fghijk}	0.71^{bc}	10.11^{vw}	11.13^q	12.15^{lmno}	11.13^i	1.64^b	1.26^{ij}	1.05^{opqr}	1.32^{cd}
G20	12.37^{efg}	12.05^{jklmn}	11.25^z	11.89^{defg}	0.58^{no}	0.59^{mno}	0.60^{lmno}	0.59^i	12.81^{ghi}	12.55^{jkl}	13.11^f	12.82^b	1.65^b	1.40^{def}	1.01^{qr}	1.35^{bc}
Mean	12.31^a	11.61^b	11.63^b	11.85	0.67^c	0.65^c	0.68^a	0.67	10.99^c	12.44^b	12.83^a	12.08	1.27^a	1.27^a	1.18^b	1.24

Means followed by the same letter are not significantly different ($P \leq 0.01$) from each other, according to Duncan's multiple range test.

North Sudan. The variation in PC in the current study may be due to variation in environmental conditions such as heat, drought, and soil fertility (Elmobarak et al. 2004), as well as genotypes. Tolbert (2004) found out that increasing nitrogen fertilizer increased the protein content of flour and the arrival time of dough. Many experiments and practical experience of wheat researchers show that the protein content of the grains and flours is greatly depend on agronomical practices, genotypes, soil N content, heat, and drought stresses (Morris et al. 2004; Tahir et al. 2006; Li et al. 2013; Bouacha et al. 2014; Kaya and Akcura 2014). In the current study, PC showed varied degrees of positive correlations with both WG and DG at the three growing environments (Table 3). It showed highly (***$P < 0.001$) positive correlation with WG at Wad Medani ($r = 0.64$) and Hudeiba ($r = 0.83$) and with DG at Hudeiba ($r = 0.80$) as well as a positive (*$P < 0.05$) correlation with WG at Dongola ($r = 0.52$) and DG at Wad Medani ($r = 0.50$) and Dongola ($r = 0.48$). Although, these results suggest the dependence of these quality traits on the genotypes rather than the growing environment. However, the crop management practices could have some impacts on these characters. Similar to our findings, previous reports showed the definite interrelation between PC, WG, and DG (Ozturk and Aydin 2004; Kaur et al. 2013; Kaya and Akcura 2014).

There were significant differences ($P \leq 0.01$) in fat content (FC) within the three environments and wheat genotypes (Table 4). Regarding the growing area, the highest mean value of FC was obtained for Hudeiba and Dongola while the lowest value was obtained for Wad Medani. Among genotypes, G2 showed that the highest FC while G5 showed the lowest value. Overall all, MC, AC, PC, and FC of the wheat genotypes of the current study depended greatly on the genotypes, the growing environment and the interaction between genotypes and environments.

Gluten quantity and quality

Mean values of wet gluten (WG) of wheat genotypes grown in the three environments were significantly varied ($P \leq 0.01$) depending on the differences in the genotypes and growing environments as well as the interaction between these factors (Table 5). The mean values of WG ranged from 32.39 to 46.94%, 28.63 to 46.53%, and 35.5 to 44.26% for the wheat genotypes grown at Wad Medani, Hudeiba, and Dongola, respectively. Regardless of the growing environment, the WG contents of all wheat genotypes in the current study are more than 28% and are, therefore, at a high to the very high range. Recently, in a multienvironment trial for Turkish wheat genotypes the wet gluten content was varied from 28 to 37% depending

on the variation in the environment, genotype, and their interaction (Kaya and Akcura 2014). The highest mean value for WG was obtained for G12 (46.94%) and G20 (46.53%) grown at Wad Medani and Hudeiba, respectively, while the lowest value was recorded for G14 (28.63%) cultivated at Hudeiba. Throughout the growing environment, both Dongola and Hudeiba are suitable conditions for WG content compared to Wad Medani. This result indicates that the growing environment influence WG content of these genotypes and hence the gluten quality. The variation in WG could be attributed to the differences in the genotypes, agronomical practices, and environmental conditions such as temperature and soil fertility. Similarly, significant variation in WG content due to the difference in wheat genotypes and growing environment has been reported (Kaya and Akcura 2014).

The results showed significant differences ($P \leq 0.01$) in dry gluten (DG) among genotypes, growing environments, and the genotype-environment interaction (Table 5). The DG values for the genotypes in the three environments ranged from 10.71 to 15.66% at Wad Medani, 8.96 to 16.76% at Hudeiba, 11.60 to 15.3% at Dongola. Concerning growing area, DG content was higher at Hudeiba followed by Dongola and then Wad Medani. Regardless of the growing environment and throughout all genotypes, the highest mean for DG was obtained for G12 while the lowest value was obtained for G14. Regarding the interaction, the highest value was obtained for G20 (16.76%) and G3 (16.16%) at Hudeiba and G12 (15.66%) at Wad Medani, while the lowest value (8.96%) was obtained for G14 in Hudeiba. The yield of DG was closely associated with the total protein of these wheat lines. These results agreed with those reported previously for other of wheat genotypes (Makawi et al. 2013).

The gluten index (GI) is a predictive method of gluten strength and thus it is a good indicator for gluten quality and quantity (Vida et al. 2014). Wide variations ($P \leq 0.01$) in the GI of 20 wheat genotypes grown in three different environments were explicitly noted (Table 5) and associated with genotypes, growing environments, and the interaction between these factors. The range values of GI for the genotypes were 57.86–92.17%, 36.36–84.39%, and 49.22–92.77% of the growing environments Wad Medani, Hudeiba, and Dongola, respectively. Strikingly, the gluten index of all wheat genotypes fall within the optimal range (55–100) for breadmaking (Har Gil et al. 2011; Makawi et al. 2013) when they grow in Wad Medani. Throughout the three growing environments, the highest mean for GI obtained were 92.77% and 92.17% at Dongola and Wad Medani, respectively, while the lowest value (36.36%) was observed at Hudeiba. Among genotypes and regardless of the environment, the results showed that G14 has the highest (86.66%) mean value of GI while G5 has the

Table 5. The values (%) of wet and dry gluten, gluten index, and water holding capacity of twenty local wheat genotypes grown in three environments.

Genotypes	Wet gluten (WG)				Dry gluten (DG)				Gluten index (GI)				Water holing capacity (WHC)			
	Wad Medani	Hudeiba	Dongola	Mean	Wad Medani	Hudeiba	Dongola	Mean	Wad Medani	Hudeiba	Dongola	Mean	Wad Medani	Hudeiba	Dongola	Mean
G1	37.32^{klmn}	38.78^{hijk}	38.43^{hijk}	38.18^{fg}	12.67^{efghi}	13.58^{cdefg}	12.60^{fghi}	12.95^{fgh}	64.04^{lmno}	50.42^{qr}	69.22^{ijkl}	61.23^{i}	194.43^{defgh}	185.66^{efghi}	205.16^{abcd}	195.08^{ef}
G2	34.47^{pq}	43.55^{bcd}	35.50^{op}	37.84^{ghi}	11.09^{kl}	14.10^{bcdef}	11.73^{ijkl}	12.31^{hij}	85.12^{bcd}	56.47^{opq}	92.77^{a}	78.12^{bcd}	210.68^{abcd}	209.33^{abcd}	202.60^{abcde}	207.54^{bc}
G3	36.20^{nop}	45.00^{b}	36.13^{nop}	39.11^{de}	11.54^{jkl}	16.16^{a}	11.66^{jkl}	13.12^{efg}	73.86^{ghij}	63.44^{lmno}	69.61^{ijkl}	68.97^{bh}	213.79^{abcd}	185.37^{efghi}	209.70^{abcd}	202.95^{bcde}
G4	36.50^{mon}	44.06^{bc}	38.38^{hijk}	39.65^{de}	11.90^{ijkl}	14.96^{bc}	12.46^{ghij}	13.11^{efgh}	71.04^{hijkl}	53.04^{pqr}	64.42^{klmno}	62.83^{ij}	206.60^{abcd}	194.96^{defgh}	207.86^{abcd}	203.14^{bcde}
G5	40.23^{fgh}	40.63^{fgh}	41.60^{def}	40.82^{c}	14.06^{bcdef}	13.70^{cdefg}	13.53^{cdefg}	13.76^{cde}	64.03^{lmno}	36.36^{s}	49.22^{r}	49.87^{l}	186.33^{efghi}	196.87^{cdefg}	207.33^{abcd}	196.84^{def}
G6	36.29^{no}	34.61^{pq}	37.43^{klmn}	36.11^{k}	11.50^{jkl}	12.09^{hijk}	11.70^{ijkl}	11.76^{jk}	76.75^{efgh}	67.10^{jklm}	67.94^{jklm}	70.60^{fgh}	216.03^{abc}	186.63^{efghi}	218.16^{a}	206.94^{bcde}
G7	33.18^{qr}	39.86^{ghij}	39.30^{ghij}	37.45^{ef}	11.13^{jkl}	14.30^{bcde}	12.86^{efghi}	12.76^{ghi}	92.17^{a}	61.00^{mmnop}	90.90^{ab}	81.36^{b}	197.96^{cdefg}	179.45^{ghi}	205.45^{abcd}	194.29^{efg}
G8	38.54^{hijk}	36.85^{lmno}	41.66^{def}	39.02^{ef}	13.69^{cdefg}	12.25^{ghij}	14.10^{bcdef}	13.34^{defg}	71.59^{hij}	82.87^{cdef}	64.36^{klmno}	72.94^{efg}	181.50^{ghi}	200.85^{bcdef}	195.83^{defgh}	192.73^{fgh}
G9	43.10^{cd}	42.75^{cd}	42.53^{de}	42.79^{cd}	13.66^{cdefg}	14.33^{bcd}	14.10^{bcdef}	14.03^{bcd}	64.38^{klmno}	66.52^{klm}	74.72^{ghij}	68.54^{hi}	215.60^{abc}	198.40^{bcdefg}	201.99^{abcde}	205.33^{bcd}
G10	37.27^{klmn}	38.86^{hijk}	38.36^{hijk}	38.16^{fgh}	12.47^{ghij}	13.43^{defgh}	12.26^{ghij}	12.72^{ghi}	73.47^{ghij}	68.07^{ijklm}	78.81^{defg}	73.45^{ef}	198.77^{bcdefg}	189.62^{efgh}	212.80^{abcd}	200.39^{cdef}
G11	33.20^{qr}	34.24^{pq}	36.33^{no}	34.59^{l}	11.03^{kl}	11.78^{jkl}	11.70^{ijkl}	11.50^{kl}	83.11^{cde}	75.25^{efgh}	81.37^{cdef}	79.91^{bc}	201.03^{abcde}	190.66^{defgh}	210.52^{abcd}	200.74^{bcdef}
G12	46.94^{a}	42.00^{def}	43.26^{cd}	44.07^{a}	15.66^{a}	15.00^{b}	15.30^{b}	15.32^{a}	57.86^{nopq}	64.53^{klmno}	66.09^{klmn}	62.83^{ij}	170.92^{j}	180.03^{hi}	182.80^{fghi}	177.92^{i}
G13	40.96^{ef}	40.00^{fgh}	39.20^{ghij}	40.05^{cd}	14.44^{bcd}	14.36^{bcd}	14.60^{bcd}	14.46^{bc}	59.28^{nop}	48.05^{r}	55.96^{opq}	54.43^{k}	168.42^{j}	178.66^{hi}	184.30^{fghi}	177.12^{i}
G14	34.37^{pq}	28.63^{s}	37.53^{klmn}	33.51^{m}	11.39^{jkl}	8.96^{m}	12.40^{ghij}	10.92^{l}	83.43^{bcde}	84.39^{bcd}	92.17^{a}	86.66^{a}	201.43^{abcde}	219.32^{a}	202.79^{abcde}	207.84^{b}
G15	39.34^{ghij}	38.66^{hijk}	39.20^{ghij}	39.07^{ef}	13.45^{defgh}	14.02^{bcdef}	13.56^{cdefg}	13.67^{cdef}	68.18^{ijkl}	64.42^{klmno}	67.26^{ijklm}	66.62^{hi}	194.06^{defgh}	175.92^{i}	188.92^{efgh}	186.30^{ghi}
G16	32.39^{r}	35.93^{nop}	35.53^{no}	34.61^{l}	10.71^{l}	12.16^{hijk}	11.96^{ijkl}	11.61^{jkl}	85.99^{abc}	79.02^{defgh}	80.58^{cdefg}	81.86^{b}	202.18^{abcde}	195.43^{defgh}	196.38^{cdefg}	197.99^{cdef}
G17	38.32^{jiklm}	37.53^{klmn}	40.83^{efg}	38.89^{ef}	13.31^{defgh}	13.56^{cdefg}	14.16^{bcdef}	13.68^{def}	73.54^{ghij}	77.02^{efgh}	75.02^{fghi}	75.19^{de}	188.23^{efghi}	177.25^{hi}	188.35^{efghi}	184.61^{hi}
G18	38.06^{klm}	36.36^{no}	36.80^{lmno}	37.07^{ij}	12.48^{ghij}	11.83^{jkl}	11.60^{jkl}	11.97^{ijk}	80.33^{cdefg}	63.72^{lmno}	85.05^{bcd}	76.37^{cde}	205.10^{abcd}	207.32^{abcd}	217.20^{ab}	209.87^{a}
G19	36.74^{lmno}	35.03^{op}	37.50^{klmn}	36.42^{jk}	11.93^{ijkl}	11.90^{ijkl}	12.30^{ghij}	12.04^{ijk}	86.87^{abc}	72.51^{ghij}	75.88^{efgh}	78.42^{bcd}	208.43^{abcd}	195.35^{defgh}	205.07^{abcd}	202.95^{bcde}
G20	41.60^{def}	46.53^{a}	44.26^{bc}	44.13^{a}	13.45^{defgh}	16.76^{a}	14.03^{bcdef}	14.75^{ab}	72.33^{hijk}	75.58^{efgh}	71.67^{hijk}	73.19^{ghi}	209.43^{abcd}	179.05^{ghi}	215.43^{abc}	201.30^{bcde}
Mean	37.75^{b}	38.99^{a}	38.99^{a}	38.58	12.58^{c}	13.46^{a}	12.93^{b}	12.99	74.37^{a}	65.49^{b}	73.65^{a}	71.17	194.43^{a}	191.31^{b}	202.14^{a}	197.59

Means followed by the same letter are not significantly different ($P \leq 0.01$) from each other, according to Duncan's multiple range test.

lowest value. Our findings demonstrated that genotypes, growing environments, and their interaction significantly affected GI, with the highest effect being from the genotypes. In agreement with our findings, Vida et al. (2014) reported that the gluten index had the greatest dependence on the genotype compared to environmental factors and agronomic treatments. Furthermore, the more significant effect of genotype on the gluten index compared to the impact of environment and fertilizer application was recently reported (Bouacha et al. 2014). GI correlated positively ($*P < 0.05$) with WHC ($r = 0.53$) at Wad Medani and with SV ($r = 0.50$) at Dongola, while it showed negative correlations with WG at Wad Medani ($***P < 0.001$, $r = -0.83$), Hudeiba ($*P < 0.05$, $r = -0.47$) and Dongola ($*P < 0.05$, $r = -0.48$) (Table 3). GI also correlated negatively with DG at Wad Medani ($***P < 0.001$, $r = -0.86$) and Dongola ($*P < 0.05$, $r = -0.51$), and with PC at Hudeiba ($**P < 0.01$, $r = -0.61$). These results indicate a contradicting response between GI and the three major wheat quality parameters (PC, WG, and DG) and therefore much concern has to be considered when using GI for wheat quality evaluation (Bonfil and Posner 2012; Kaur et al. 2013).

The results showed significant differences ($P \leq 0.01$) in the water holding capacity (WHC) among environments and genotypes as well as the interaction between genotypes and environments (Table 5). The mean values of WHC were 168.42–216.03%, 175.92–219.32%, and 182.8–218.16% for the genotypes grown at Wad Medani, Hudeiba, and Dongola, respectively. Throughout the three growing environments, WHC showed extremely negative ($***P < 0.001$) correlation with DG ($r = -0.64$ to -0.70), whereas the correlations between WG and DG were highly positive ($***P < 0.001$, $r = 0.87$–0.96) (Table 3). The highest mean percentages for WHC of gluten was obtained for G14 (219.32%) and G6 (218.16%) grown at Hudeiba and Dongola, respectively, while the lowest value was observed for G13 (168.42%) cultivated at Wad Medani. Throughout all the three environments, the results indicated highest WHC for G18 while the lowest value was obtained for G13. Between environments, not all varieties varied in the same manner; however, some had the same general score in all areas, whereas others varied. This variation may be due to the effect of environmental conditions such as heat stress, soil conditions, and agronomical practices.

Falling number

Statistical analysis revealed significant differences ($P \leq 0.01$) in the mean falling number (FN) of 20 wheat genotypes grown in three different environments (Table 6), indicating that environmental conditions had influenced the flour FN. In addition, the interaction between

Table 6. Falling number and sedimentation value of 20 local wheat genotypes grown in three environments.

Genotypes	Falling number (sec)				Sedimentation value (mL)			
	Wad Medani	Hudeiba	Dongola	Mean	Wad Medani	Hudeiba	Dongola	Mean
G1	609.67[r]	508.00[z]	650.33[n]	589.33[k]	23.00[k]	22.00[l]	23.33[k]	22.78[l]
G2	633.33[p]	636.33[op]	974.67[a]	748.11[a]	25.33[i]	30.00[d]	24.00[j]	26.44[f]
G3	693.00[gh]	556.33[wx]	648.67[n]	632.67[h]	27.00[g]	29.00[e]	24.00[j]	26.67[e]
G4	715.00[f]	656.67[mn]	665.00[kl]	678.89[c]	24.00[j]	27.00[g]	26.00[h]	25.67[g]
G5	563.33[w]	524.00[z]	600.00[rs]	562.44[m]	23.00[k]	24.00[j]	24.00[j]	23.67[k]
G6	672.33[jk]	528.33[z]	808.67[b]	669.78[d]	19.00[n]	22.00[l]	24.00[j]	21.67[m]
G7	673.33[jk]	540.33[y]	730.33[e]	648.00[f]	25.00[i]	24.00[j]	40.33[a]	29.78[b]
G8	685.33[hi]	595.33[st]	795.00[c]	691.89[b]	25.00[i]	32.00[b]	27.00[g]	28.00[c]
G9	589.00[tu]	556.67[wx]	668.67[kl]	604.78[j]	32.00[b]	28.33[f]	33.00[b]	31.11[a]
G10	594.67[st]	508.33[z]	604.33[rs]	569.11[l]	28.00[f]	26.00[h]	29.00[e]	27.67[d]
G11	687.33[hi]	547.00[xy]	733.00[e]	655.78[e]	23.00[k]	24.00[j]	24.33[j]	23.78[k]
G12	604.00[rs]	555.67[wx]	653.33[mn]	604.33[j]	27.00[g]	24.00[j]	24.00[j]	25.00[i]
G13	607.00[r]	623.67[q]	654.67[mn]	628.44[h]	23.00[k]	23.00[k]	23.00[k]	23.00[l]
G14	668.33[kl]	512.00[z]	759.67[d]	646.67[f]	19.00[n]	23.00[k]	28.00[f]	23.33[l]
G15	636.00[op]	576.67[v]	681.67[i]	631.44[h]	25.00[i]	23.00[k]	25.00[i]	24.33[j]
G16	583.67[uv]	577.00[v]	701.33[g]	620.67[i]	20.00[m]	27.00[g]	29.00[e]	25.33[h]
G17	532.67[z]	583.00[uv]	594.33[st]	570.00[l]	28.00[f]	26.00[h]	29.33[e]	27.78[d]
G18	681.33[ji]	580.33[uv]	600.33[rs]	620.67[i]	27.00[g]	27.00[g]	29.00[e]	27.67[d]
G19	638.33[op]	596.67[st]	662.67[lm]	632.56[h]	26.00[h]	23.00[k]	28.67[ef]	25.89[g]
G20	626.67[pq]	646.00[no]	648.00[n]	640.22[g]	31.00[c]	28.00[f]	30.33[cd]	29.78[b]
Mean	634.72[b]	570.42[c]	691.73[a]	632.29	25.02[c]	25.62[b]	27.27[a]	25.97

Means followed by the same letter are not significantly different ($P \leq 0.01$) from each other, according to Duncan's multiple range test.

Effect of environment and genotypes on the physicochemical quality of the grains...

39

genotypes and growing environments was also significantly affected the α-amylase activity. The FN values were ranged from 532.67 to 715.0 sec, 508.00 to 656.67 sec, and 594.33 to 974.67 sec for the genotypes cultivated at Wad Medani, Hudeiba, and Dongola, respectively. These results were in good agreement with the data reported by Kaur et al. (2013) who found that the falling number of Indian wheat cultivars was high and ranged from 485 to 967 sec. Through the three environments, the highest FN was obtained at Dongola while the lowest value was obtained at Hudeiba. FN correlated positively with SV (*$P < 0.05$, $r = 0.47$), WG (*$P < 0.05$, $r = 0.48$), and FC (***$P < 0.001$, $r = 0.67$) at Hudeiba, while it showed negative correlation with DG (*$P < 0.05$, $r = -0.45$) at Wad Medani (Table 3). Among all genotypes, the highest falling number recorded for G2 while the lowest value was obtained for G5. The Sudanese wheat genotypes possess very high FN and thus indicate flours with a little α-amylase activity. This could be attributed to the dry weather during grain filling and harvesting time, which consequently affect the activity of α-amylase (Erekul and Kohn 2006; Hamad et al. 2013). Thus, the seasonality and the environment, storage conditions of wheat grain (moisture and temperature), had a significant impact on the α-amylase activity. Previous reports indicate that the FN is diverse among different genotypes that cultivated in various environment (Hamad et al. 2013) with the environmental impact on the FN being higher than the genotype and the genotype-environment interaction (Erekul and Kohn 2006).

Sedimentation values

The sedimentation value (SV) assessment provides information on the protein quantity and the quality of wheat flour (Makawi et al. 2013). It is thus used as a screening tool in wheat breeding programs as well as in milling and breadmaking processes. Our results revealed significant differences ($P \leq 0.01$) in SV among environments and genotypes (Table 6). The SV of the genotypes in the three environments was in the range of 19.00–32.00 mL at Wad Medani, 22.00–32.00 mL at Hudeiba and 23.00–40.33 mL at Dongola. Similarly, Makawi et al. (2013) stated that the sedimentation value of the three Sudanese cultivars (Debaira, WadiElneel, and Elneelain) ranged between 19.6 and 37.4 mL. Additionally, Kaya and Akcura (2014) found the sedimentation value of 24–33 mL for Turkish wheat genotypes grown in different environments. The highest mean SV (40.33 mL) was obtained from G7 grown at Dongola while the lowest value (19.00 mL) was obtained from G6 and G14 grown at Wad Medani. Throughout the three environments, genotypes grown at Dongola showed the highest SV followed by Hudeiba and then Wad Medani. SV on the other hand revealed

positive correlation with WG at Wad Medani (*$P < 0.05$, $r = 0.55$) and Hudeiba (*$P < 0.05$, $r = 0.45$), and with PC (***$P < 0.001$, $r = 0.69$) at Wad Medani, and FC (**$P < 0.01$, $r = 0.59$) at Hudeiba (Table 3). The positive association of SV with PC and WG is consistent with the fact that this value depends mainly on the wheat protein composition and gluten quality and is frequently correlate with these quality attributes (Ozturk and Aydin 2004; Tahir et al. 2006; Kaya and Akcura 2014). Unrelatedly with the growing area, G9 expressed the highest mean SV. As for other traits, the SV of the current study also depend mainly on the genotype, the growing environment (temperature and soil fertility) and their interaction. Similar observation on the effect of environment (temperature, rainfall, and soil quality) and agronomical treatments on the sedimentation value of many wheat genotypes has been previously reported (Erekul and Kohn 2006; Tahir et al. 2006; Kaya and Akcura 2014).

Biplot analysis

To profoundly determine the multivariate relationships between the grain end-use quality traits and the growing environments of 20 wheat genotypes, biplot analysis was carried out by comparing the eigenvalues of PC1 and PC2 of principal component analysis (PCA) for both the genotypes and the quality traits (Fig. 1A–C). Regarding the interrelation between the traits and genotypes, the results of the first two PC axes (PC1, 39.89% and PC2, 23.37%) accounted for about 63.26% of the total variability reflecting the complexity of the variation between the plotted components (Fig. 1A). In the biplot, vectors of traits (variables) showing acute angle are positively correlated, whereas those formed obtuse or straight angles are negatively correlated, and those with right angle have no correlation. The distance between the raw (genotypes) is interpreted in terms of similarity. Regarding the traits, PC1 had the breadmaking quality parameters (DG, WG, PC, GI, and WHC) as the principal components, and FN and MC to a lesser extent while, PC2 had the SV, FC, and TW as the primary elements. The cosine of the angles between vectors indicated a high positive correlation between WHC, FN, and GI in the positive direction. These three traits were also positively correlated with FC in the positive direction and AC in the negative direction. High positive correlation was also observed between PC, WG, and DG and between SV and FC, and similarly between TW, TKW, and MC. In contrast, WHC, FN, and GI were negatively correlated with other breadmaking quality parameters mainly PC, WG, and DG and with grain physical characteristics such as TW, TKW, and MC. The SV was also negatively correlated with AC, TW, and TKW. Overall, the biplot analysis exhibits three groups of the traits based

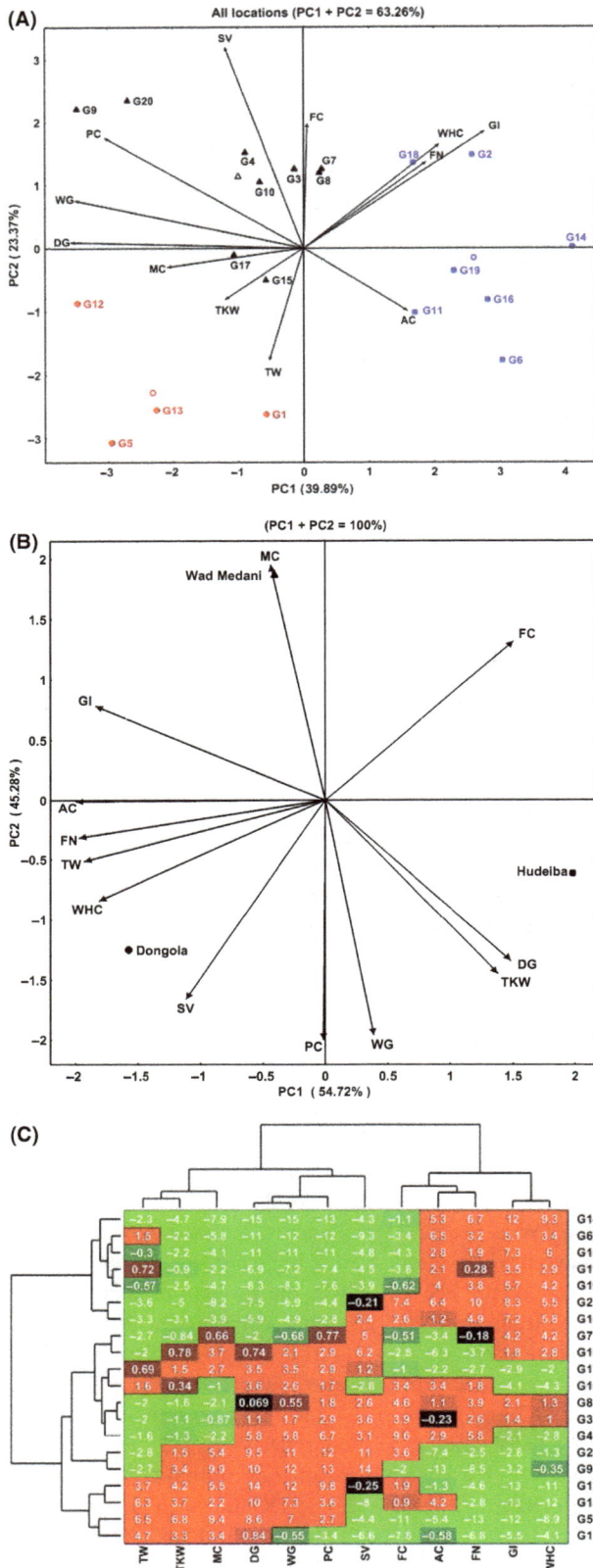

Figure 1. Biplot based on principal component analysis for grain quality traits in 20 wheat genotypes (G1–G20) grown in three different environments (Wad Medani, Hudeiba, and Dongola). The biplots showed the interrelations between the quality traits (A) and the environments (B). Bidimensional clustering analysis is presenting the relationships between the genotypes (C). TKW, Thousand kernel weight (g); TW, Test weight (kg/hL); AC, Ash content; FC, Fat content; PC, Protein content; FN, Falling number; WHC, Water holding capacity; GI, Gluten index; SV, Sedimentation value; WG, Wet gluten; DG, Dry gluten.

on their phenotypic associations, those include; gluten, starch, and milling quality characteristics (GI, WHC, FN, and AC) group, breadmaking quality attributes (SV, PC, WG, and DG) group, and grain physical and marketing characteristics (MC, TKW, and TW). These results shows some differences from that of the correlation analysis among pairs of characters as the biplot describes the interrelationships among all characters concurrently based on the overall contribution of the data (Yan and Fregeau-Reid 2008).

Additionally, the biplot could indicate the interrelation among the genotypes as well as their stability and contribution toward an individual trait (Morris et al. 2004). In this regards, hierarchical clustering clearly distinguished three groups of genotypes according to their quality characteristics in all growing environments. The first group (right half of the graph, square symbol) is formed by the genotypes (G2, G8, G11, G14, G16, G18, and G19) with the highest values of WHC, GI, FN, and AC compared to other genotypes. Within this group, G14 had the highest values for these traits, followed by G6, G16, G12, G11, and G18. The second group (upper left, triangle symbol) consists of G3, G4, G7, G8, G9, G10, G15, G17, and G20. This group characterized by its high breadmaking quality parameters such as PC, WG, DG, and SV with G9 and G20 outscore all others genotypes for these traits. However, G9 and G20 are less stable for these quality traits compared to the other genotypes in this group as well as they are not contributed to the other quality traits such as FN, GI, and WHC. By contrast, G3, G4, G7, G8, and G10 are more stable and well-associated with all end-use quality attributes. The last group (lower left, circle symbol) contain G1, G5, G12, and G13, those characterized mainly by their high values of grain milling and marketing characteristics especially TW, TKW, and MC. Like that of traits, the results of biplot analysis display three distinguished groups of the genotypes based on their performance for one or more quality traits. Saint Pierre et al. (2008) stated that the grouping of genotypes in the biplot indicated that the genotypes of the quality groups show similar performance to numbers of the quality traits.

The biplot is an appropriate method for the analysis of the interaction between the traits and environments. Thus, it can identify the effects of the environment on one or more characters through a range of genotypes. In this study, a biplot was formed by using the average means of each trait for all genotypes growing in three environments to find the better wheat-growing environment for the end-use quality attributes (Fig. 1B). The results showed extremely high variability (100%), arising from the first two principal component PC1 (54.92%) and PC2 (45.28%), which indicating an excellent contribution of these two axes to the data presentation. Interestingly, the association between the traits shows some variations especially in MC, FC, TW, and FN compared to those presented in Figure 1A, suggesting the effect of the growing environment on grain end-use quality traits. The high MC and GI characterize the genotypes grown at Wad Medani compared to the same genotypes when cultivated in the other two environments. Higher MC at Wad Medani could be attributed to the soil type in this environment which it could retain more water than that of the two other environments. The genotypes grown at Hudeiba had higher DG, TKW, and FC compared to the same genotypes grown at Wad Medani and Dongola. Interestingly, most of the end-use quality parameters are great in the genotypes when cultivated at Dongola compared to the other two growing environments. Based on the end-use quality traits, the environment in Dongola is most suitable followed by that of Hudeiba, whereas, the environment in Wad Medani is not suitable for wheat cultivation as the end-use quality attributes were significantly reduced in this environment. The chief difference between the three environments is the temperature. Therefore, the inferior quality of the genotypes at Wad Medani could be attributed to the high temperature during the growing season.

To select the best genotype based on its quality performance throughout the growing environments, we generated a bidimensional cluster from the mean of the quality attributes of each genotype across all environments (Fig. 1C). The horizontal axis groups the genotypes based on phenotypic similarity concerning their quality traits. The differences in the color intensity indicated the values of each feature with the red color being the highest and green is the lowest. The two major branches of the horizontal cluster (traits) discrete the genotypes in the upper clusters, in which most of the green color (small values) appears for the attributes TW, TKW, MC, DG, WG, PC, and SV, from those in the lower clusters as they showed red color (high values) for the same attributes. With view exceptions, this results suggests that the upper branch includes genotypes (G14, G6, G16, G11, and G19) with poor grain filling and end-use quality, whereas, the lower branch contains G9, G20, G10, G15, and G17 with real

grain weight and end-use quality. Despite their poor grain filling and moisture content, G3 and G4 have a good end-use quality attributes with G3 outscore all other genotypes in this regard. Strikingly, this genotype (G3) also shows good stability for these end-use quality traits (Fig. 1A). These results indicate the potentiality of G3 as an excellent and stable genotype for end-use quality attributes under hot environments.

Conclusion

In conclusion, the results of this study demonstrate that the genotypes, the environment, and their interaction have a high impact on the end-use quality attributes of Sudanese wheat genotypes grown in three different environments. Throughout the three growing environments, Dongola is most appropriate for producing wheat grains with adequate end-use quality characteristics while Wad Medani is the least in this regard due to its high temperature. Among wheat genotypes, G3 and G4 exhibit good performance and reasonable stability for most of the tested quality traits. These genotypes are potentially excellent candidates for cultivation in the hot environments of the Sudan for producing wheat grains with good breadmaking quality. In addition, the crossing of these genotypes with high yield and milling quality genotypes will improve the adaptability, productivity, quality, and marketability of Sudanese wheat grains.

Conflict of Interest

The authors declare to have no conflict of interest.

References

AACC. 2000. Approved methods of the american association of cereal chemists, 10th edn. American Association of Cereal Chemists, St. Paul, MN, USA.

Ali, A. M., H. M. Mustafa, I. S. A. Tahir, A. B. Elahmadi, M. S. Mohamed, M. A. Ali, et al. 2006. Two doubled haploid bread wheat cultivars for irrigated heat-stressed environments. Sudan J. Agric. Res. 6:35–42.

Bonfil, D. J., and E. S. Posner. 2012. Can bread wheat quality be determined by gluten index? J. Cereal Sci. 56:115–118.

Bouacha, O. D., S. Nouaigui, and S. Rezgui. 2014. Effects of N and K fertilizers on durum wheat quality in different environments. J. Cereal Sci. 59:9–14.

El Siddig, M. A., S. Baenziger, I. Dweikat, and A. A. El Hussein. 2013. Preliminary screening for water stress tolerance and genetic diversity in wheat (Triticum aestivum L.) cultivars from Sudan. J. Gen. Engin. Biotechnol. 11:87–94.

Elahmadi, A. B. 1996. Review of wheat breeding in the Sudan. Pp. 33–53 in O. A. Ageeb, A. B. Elahmadi,

M. B. Salh, M. C. Saxena, eds. Wheat production and improvement in the Sudan. Proceedings of the national research review workshop, 27–30 August 1995, Wad Medani, Sudan. ICARDA/Agricultural Research Corporation, ICARDA, Aleppo, Syria.

Elmobarak, A., A. Elnaem, A. Adam, and C. Richter. 2004. Effect of different nitrogen sources to wheat on two soils in Sudan. Deutscher Tropentag. October 5-7, Berlin.

Elsheikh, O. E., A. A. Elbushra, and A. A. A. Salih. 2015. Economic impacts of changes in wheat's import tariff on the Sudanese economy. J. Saudi Soc. Agric. Sci. 14:68–75.

Erekul, O., and W. Kohn. 2006. Effect of weather and soil conditions on yield components and bread-making quality of winter wheat (*Triticum aestivum* L.) and winter Triticale (*Triticosecale* Wittm.) varieties in North-East Germany. J. Agron. Crop Sci. 192:452–464.

FAO. 2012. FAO statistical yearbook 2012. World Food and Agriculture. Food and agriculture organization of the united nations, Rome, Italy.

FAO. 2014. Global information and early warning system on food and agriculture, GIEWS country brief, Sudan. Food and Agriculture Organization of the United Nations, Rome, Italy.

Gomez, K. A., and A. A. Gomez. 1984. Statistical procedures for agricultural research, Pp. 13–175, 2nd edn. John Wiley and Sons, Inc., London, UK.

Hamad, S. A. A., A. H. R. Ahmed, and I. A. Mohamed Ahmed. 2013. Effect of production location and addition of guar gum on the quality of a Sudanese wheat cultivar for bread making. Int. J. Innov. Appl. Stud. 7:317–328.

Har Gil, D., D. J. Bonfil, and T. Svoray. 2011. Multi-scale analysis of the factors influencing wheat quality as determined by Gluten Index. Field Crops Res. 123:1–9.

Huseynova, I. M., and S. M. Rustamova. 2010. Screening for drought stress tolerance in wheat genotypes using molecular markers. Proc. ANAS (Biol. Sci.) 65:132–139.

Kaur, A., N. Singh, A. K. Ahlawat, S. Kaur, A. M. Singh, H. Chauhan, et al. 2013. Diversity in grain, flour, dough and gluten properties amongst Indian wheat cultivars varying in high molecular weight subunits (HMW-GS). Food Res. Inter. 53:63–72.

Kaya, Y., and M. Akcura. 2014. Effects of genotype and environment on grain yield and quality traits in bread wheat (*Triticum aestivum* L.). Food Sci. Technol. Campinas. 34:386–393.

Li, Y., Y. Wu, N. Hernandez-Espinoza, and R. J. Pena. 2013. The influence of drought and heat stress on the expression of end-use quality parameters of common wheat. J. Cereal Sci. 57:73–78.

Lopes, M. S., M. P. Reynolds, M. R. Jalal-Kamali, M. Moussa, Y. Feltaous, I. S. A. Tahir, et al. 2012. The yield correlations of selectable physiological traits in a population of advanced spring wheat lines grown in warm and drought environments. Field Crops Res. 128:129–136.

Makawi, A. B., A. I. Mustafa, and I. A. Mohamed Ahmed. 2013. Characterization and improvement of flours of three Sudanese wheat cultivars for loaf breadmaking. Innov. Roman Food Biotechnol. 13:30–44.

Mohammadi, M. 2012. Effects of kernel weight and source-limitation on wheat grain yield under heat stress. Afri. J. Biotechnol. 11:2931–2937.

Morris, C. F., K. G. Campbell, and G. E. King. 2004. Characterization of the end-use quality of soft wheat cultivars from the eastern and western US germplasm 'pools'. Plant Genet. Res. 2:59–69.

Ozturk, A., and F. Aydin. 2004. Effect of water stress at various growth stages on some quality characteristics of winter wheat. J. Agron. Crop Sci. 190:93–99.

Saint Pierre, C., C. J. Peterson, A. S. Ross, J. B. Ohm, M. C. Verhoeven, M. Larson, et al. 2008. Winter wheat genotypes under different levels of nitrogen and water stress: changes in grain protein composition. J. Cereal Sci. 47:407–416.

Shiferaw, B., M. Smale, H.-J. Braun, E. Duveiller, M. Reynolds, and G. Muricho. 2013. Crops that feed the world 10. Past successes and future challenges to the role played by wheat in global food security. Food Security 5:291–317.

Tahir, I. S. A., N. Nakata, A. M. Ali, H. M. Mustafa, A. S. I. Saad, K. Takata, et al. 2006. Genotypic and temperature effects on wheat grain yield and quality in a hot irrigated environment. Plant Breeding 125:323–330.

Tolbert, E. S. 2004. Influence of genotype, environment, and nitrogen management on spring wheat quality. Crop Sci. 44:425–432.

Vicente-Villardón, J. L.. 2010. MULTBIPLOT: a package for multivariate analysis using biplots. Department of Statistics, University of Salamanca, Spain.

Vida, G., L. Szunics, O. Veisz, Z. Bedo, L. Lang, T. Arendas, et al. 2014. Effect of genotypic, meteorological and agronomic factors on the gluten index of winter durum wheat. Euphytica 197:61–71.

Yan, W., and J. A. Fregeau-Reid. 2008. Breeding line selection based on multiple traits. Crop Sci. 48:417–423.

6

The effects of protein isolates and hydrocolloids complexes on dough rheology, physicochemical properties and qualities of gluten-free crackers

Natthakarn Nammakuna[1], Sheryl A. Barringer[2] & Puntarika Ratanatriwong[1]

[1]Department of Agro-Industry, Faculty of Agriculture Natural Resources and Environment, Naresuan University, Phitsanulok 65000, Thailand
[2]Department of Food Science and Technology, The Ohio State University110 Parker Hall2015 Fyffe Road, Columbus, Ohio 43210

Keywords
Gluten-free rice crackers, hydroxypropylmethylcellulose, pea protein isolate, soy protein isolate, whey protein isolate, xanthan gum

Correspondence
Puntarika Ratanatriwong, Department of Agro-Industry, Faculty of Agriculture Natural Resources and Environment, Naresuan University, Phitsanulok 65000, Thailand.

E-mail: puntarikar@nu.ac.th, rikaja@yahoo.com

Funding Information
This project is supported by the Naresuan University Research Funding, and the Thailand Research Fund (TRF) for grant number MRG5380217.

Abstract

To understand the suitability of protein-hydrocolloid complexes as replacement for wheat protein in rice crackers, and the effect of protein source, carboxylmethylcellulose (CMC) and hydroxylpropylmethylcellulose (HPMC) at 1.0%, 1.5%, and 2.0% w/w, and 0.25%, 0.50%, and 0.75% w/w of xanthan gum (XN) were added to flour-blendedrice crackers (FF). A variety of protein isolates was added to 2.5%, 5.0%, and 10% w/w combinations of protein isolates and hydrocolloids were investigated. The controls were FF, 100% rice crackers (RF), and wheat crackers (WF). About 1.5% CMC samples had the closest hardness to WF, followed by 0.5%XN and 1.5%HPMC, and 0.5%XN crackers had the highest moisture content and water activities followed by 0.75%XN, 1.5%CMC, and 1.5%HPMC. Increasing % of hydrocolloids also increased puffiness. Protein isolate crackers had higher moisture content and water activity. Protein isolates improved puffiness. Whey protein improved elasticity, while hydrocolloids added to leguminous protein increased loss tangent.

Introduction

Celiac disease affects about 0.2–1.0% of the world population and the number is steadily increasing worldwide (Abdel-Aal 2009; Toft-Hansen et al. 2014). Patients who have celiac disease are unable to consume products made from wheat flour (WF). The replacement of gluten in a cereal-based food system poses a major technological challenge due to gluten's structure-forming capacity. Gluten substitutes must be able to form cohesive elastic dough that can be baked into a food product with pleasant taste and acceptable texture (Abdel-Aal 2009). The removal of wheat proteins from gluten-free cracker products causes significant changes in the volume, brittleness, and rheological properties.

Rice flour (RF) has been used as the basic ingredient in gluten-free bread because it lacks gluten and contains low levels of sodium and high amounts of easily digested carbohydrates (Gallagher et al. 2002). However, rice proteins have relatively poor functional properties for food processing. Due to their hydrophobic nature, rice proteins are insoluble and unable to form the viscoelastic dough necessary to hold the carbon dioxide produced during proofing of yeast-leavened bread-like products (He and

Hoseney 1991). Flour blends (FFs) consisting of pregelatinized starch have been traditionally used as a replacement for WF in gluten-free crackers, however, the crackers produced with these blends have been deemed to be of lower eating and overall organoleptic quality than wheat crackers. Proteins are added to gluten-free applications to increase elastic modulus by cross linking, to improve perceived quality, to improve structure with gelation, and to aid in foaming (Crockett et al. 2011). The high quality of dairy ingredients-containing breads was attributed to the ability of the dairy ingredient to form a network similar to gluten. Hydrocolloids are used in a wide range of food applications to impart texture and appearance as well as to improve product stability. In the baking industry, hydrocolloids are of increasing importance as bread-making improvers, and help increase dough properties such as water absorption, gas retention, and improve product properties such as texture and to retard starch retrogradation (Bárcenas and Rosell 2005; Lazaridou et al. 2007). Hydroxyl propylmethyl cellulose (HPMC) were reported to provide dough stability during proofing with increasing gas retention by adding strength to gas cells, resulting in better gas retention and higher product volume (Bell 1990). In bread crumbs, addition of xanthan help increase the hardness of the bread crumb by acting as a thickener (Rosell et al. 2001). Carboxylmethylcellulose (CMC) and HPMC help increase the crispness of partial wheat-substitute crackers by reducing the hardness and increase the fracture of crackers (Nammakuna et al. 2009).

Despite much research into gluten-free baked products, information on gluten-free crackers made from RF is rarely found. No research on optimization of hydrocolloids or their combination on the qualities of gluten-free crackers from RF has been reported. Available information is still limited, in particular research on suitable types of protein isolates and the optimum amounts to add. An understanding of the effects of combinations of protein isolates and hydrocolloids on cracker properties is still needed. Thus, this research aims to understand the relationship between the type, concentration, and interaction of hydrocolloids and protein isolates on the overall quality and rheology of gluten-free cracker products made from rice FF.

Material and Method

Gluten-free cracker preparation from rice FF

The gluten-free cracker formulation was created by using the mixed FF with RF as major raw material to completely substitute WF in the cracker formulation. The mixed FF consisted of three flour types including low-amylose content RF (Chaijalearn Co., Ltd., Bangkok, Thailand), waxy RF (Jewhoksang Co., Ltd., Lampang, Thailand) and tapioca

pregelatinized starch (Bangkok Starch Co., Ltd., Bangkok, Thailand). This mixed-FF without addition of either hydrocolloids or protein isolates was used as control for treatment samples whereas 100%RF and 100%WF were also used as negative control and benchmark, respectively.

The basic ingredients for making crackers for all formulations were salt, sugar, milk powder, palm oil, glucose syrup, baking powder, margarine, dry yeast, ammonia powder, lecithin, corn flour, and tapioca flour. These ingredients were used as basic ingredients. Firstly, flour, sugar, yeast, and water were blended in a mixer (Kitchen-Aid model 5SS, St. Joseph, MI, USA) for 5 min, and then the rest of the ingredients were added. The mixture was blended for 7 min before proofing the dough at 25–30°C for 60 min. After proofing, the dough was kneaded, sheeted, layered, cut into a cracker size of 2.5 × 5 cm, and subsequently baked in the oven at 180–200°C for 15 min. Baked crackers were then cooled, packed in sealed polyproplyene bags, and stored at room temperature before further experiment.

Experimental design

Three types of hydrocolloids (Bronson & Jacobs International Co., Ltd., Bangkok, Thailand) including CMC and HPMC and xanthan gum (XN) were added to FF. These hydrocolloids were added at different concentrations. HPMC and CMC were added at 1.0%, 1.5% and 2.0%, respectively; XN was added at 0.25%, 0.50%, and 0.75% (total RF basis) to mixed FF. Three types of protein isolates including, soy protein isolate (F.A. Unity Co., Ltd, Bangkok, Thailand), pea protein isolate (Roquette Singapore PTE, LTD, Bangkok, Thailand), and whey protein isolate (Grande Custom Ingredients Group, Lomira, WI 53048, United States) were added at 2.5%, 5.0%, and 10.0% (total RF basis) on mixed FF. The qualities of all treatment samples were determined according to 2.3 and compared to controls. Moreover, the impact of combinations of protein isolates and hydrocolloids, added at optimum level to FF, on gluten-free rice crackers were also investigated. The qualities of all treatment samples were determined according to Section Quality determination of rice crackers and compared with the three controls.

Quality determination of rice crackers

Moisture content and water activity (a$_w$) determination

The moisture content and the water activity (a_w) of samples were determined in three replicates using a moisture meter model Sartorius MA40 (Sartorius, Inc., Goettingen, Germany) and a water activity meter model Novasina RS 200 (Novasina, Axair Ltd., Pfaffikon, Switzerland), respectively.

Texture properties

Texture was determined in ten replicates using a Texture Analyzer model QTS25 (Brookfield Engineering Labs., Inc. Middleboro, MA 02346 U.S.A.) equipped 25 kg load cell. The three-point bend fixture test method was used. Texture profile analysis was performed. The textural characteristics were expressed in terms of hardness (the height of the force peak on the first compression cycle [first bite]), cohesiveness, chewiness, and springiness (the spring back after it has been deformed during first compression) of gluten-free rice crackers. Each sample was analyzed in 10 replicates.

Color measurement

Cracker samples were measured for color in CIE system (L^*, a^*, b^*, hue angle, and chroma) by a color reader model CR-10 (Konica Minolta sensing Inc., Osaka, Japan). The analysis was performed four replicates for each sample.

Rheological properties of gluten-free rice cracker

The dough rheological measurement was studied by using the dynamic oscillatory test. The test was performed by a controlled stress–strain rheometer (Physica MCR 301, Physica/Anton Paar, Germany), using a parallel-plate geometry (PP25/TG 6866) with plate diameter and plate gap of 25 and 2 mm, respectively. A frequency sweep test provided the information of dough rheological changes including the structure, molecular structure, and viscoelastic behavior (Angioloni and Rosa, 2007). The measurements were conducted at 25°C by using a frequency sweep from 0.1 to 10 Hz at 0.1% strain.

Puffiness (%)

Thicknesses of the cracker before and after baking were measured by vernier calipers in five replicates then the sample puffiness (%) was calculated from the difference of cracker thicknesses as shown in Eq. 1.

$$\% \text{ puffiness} = \left(\frac{\text{thickness of baked cracker} - \text{thickness of cracker dough}}{\text{thickness of cracker dough}} \right) \star 100 \quad (1)$$

Scanning electron microscope

The dough microstructure was studied using a scanning electron microscope (SEM) Model 1455VP (Leo Electric Systems, Cambridge, UK). Prior to the SEM study, cracker dough samples were cut to size 10 × 10 mm, freeze dried,

and kept in a desiccator until further use. Dough samples were mounted on a slide and separately placed on a sample holder using double-sided scotch tape. The internal structure was upward facing and sputter-coated with gold (2 min, 2 mbar) before being transferred on to a microscope where it was observed in vacuum at an accelerating voltage of 5 kV.

Statistical analysis

The qualities of samples were analyzed in two replicates. The experimental design used in this research was a Completely Randomized Design. All treatments were done in duplicate. Data were statistically analyzed by Ducan's Multiple Range test at 95% confidence level.

Results and Discussion

Effect of hydrocolloids on physicochemical properties of gluten-free dough and crackers

All the treatment samples had significant differences in moisture content, water activity (a_w), and puffiness ($P \leq 0.05$) as observed in Table 1.

Gluten-free crackers were made with 0.5%XN and had the highest moisture content and water activity ($P \leq 0.05$). Gluten-free crackers containing 1.5%CMC had moisture content closest to that of wheat crackers compared to other treatment samples. The addition of 1.0–2.0%HPMC lead to water activity values close to those of the wheat control ($P \leq 0.05$). Increasing the concentration of HPMC had no corresponding, statistically significant effect on the water activity value (a_w) of the dough. The addition of 0.5%, 0.75%XN, and 1.5%, 2.0%HPMC increased both the water activity and moisture content of the finished baked product when compared to the hydrocolloids-free control.

These results indicated the benefits of hydrocolloids as a dough improver as the hydrocolloids help increase the water-holding capacity of samples due to the chemical structure of hydrocolloids and their interaction with the food ingredients (Rosell et al. 2001; Lazaridou et al. 2007).

In addition, the barrier formed by hydrocolloids is illustrated in Figure 1. This barrier is formed by hydrocolloids near or at the surface during heating, which leads to reduction in water loss, thus increasing the final moisture content of the crackers. (Khalil 1999; Albert and Mittal 2002; Mellema 2003; Akdeniz et al. 2006).

The addition of hydrocolloids to the FF formula caused an increase in puffiness for all hydrocolloids tested with the exception of the 1.0% HPMC sample (Table 1). The treatment sample with CMC-produced puffiness values

Table 1. Effect of hydrocolloids on physicochemical properties of gluten-free rice cracker.

Samples	Moisture content (%)	Water activity (a_w)	Puffiness (%)
WF	5.99 ± 0.16[c]	0.343 ± 0.005[c]	84.91 ± 2.58[a]
RF	5.13 ± 0.32[g,h]	0.302 ± 0.001[e]	20.08 ± 1.84[i]
FF	5.41 ± 0.16[e,f]	0.326 ± 0.007[d]	37.98 ± 4.29[g]
0.25%XN	5.00 ± 0.17[h]	0.276 ± 0.015[f]	42.60 ± 1.62[f]
0.5%XN	7.16 ± 0.21[a]	0.430 ± 0.000[a]	46.06 ± 2.20[e,f]
0.75%XN	6.21 ± 0.26[b]	0.365 ± 0.003[b]	43.99 ± 1.64[f]
1.0%CMC	5.08 ± 0.18[g,h]	0.277 ± 0.006[f]	59.28 ± 1.24[c]
1.5%CMC	6.08 ± 0.17[b,c]	0.363 ± 0.001[b]	73.03 ± 2.11[b]
2.0%CMC	5.34 ± 0.09[f]	0.295 ± 0.002[e]	52.49 ± 0.98[d]
1%HPMC	5.23 ± 0.28[f,g]	0.342 ± 0.001[c]	34.03 ± 2.59[h]
1.5%HPMC	5.70 ± 0.46[d]	0.344 ± 0.007[c]	44.17 ± 2.52[f]
2.0%HPMC	5.58 ± 0.17[d,e]	0.339 ± 0.001[c]	49.04 ± 2.59[e]

Different letters in each column indicate statistical differences ($P \leq 0.05$). Rice flour (RF), wheat flour (WF), and flour blend (FF) were controls and were made of 100%RF, 100%WF, and formulated FF, respectively.

was closest to those of wheat crackers. The extent of the puffiness increase was directly proportional to the usage level for some hydrocolloids such as HPMC, but was not significantly affected by the usage level for XN. Bell (1990) explained that HPMC gives some stability to the interface of a dough system during proofing and confers additional strength to the gas cells through baking, which increases gas retention and thus leads to higher volume. This was the same case as shown in Table 1, the dough of gluten-free cracker with HPMC was stronger with better air retention thus an increase in puffiness of HPMC-added crackers were observed. However, the optimum amount of HPMC was also crucial for puffiness because 1.0% HPMC did not have enough effect on gluten-free crackers. In addition, Marco and Rosell (2008) found that the volume of rice bread increases with the addition of hydrocolloids except XN, and reported that high values

of crumb porosity were obtained when 1.0%CMC and β-glucans were added.

Samples treated with hydrocolloids had hardness and fracture force values lower than the WF control but higher than the RF and FF controls (Table 2). The increase in fracture force was directly proportional to the increase in usage level of the hydrocolloids for samples made with XN and CMC, Addition of hydrocolloids caused an increase in hardness values for all samples tested, so treated samples were less crumbly than RF which had the lowest hardness. The WF control had the highest fracture force which indicated that the texture of WF crackers were light crisp and brittle. This was caused by gluten known as the protein-forming structure in wheat dough (Fig. 1).

The film-like veil witnessed in the SEM scans of the treatment samples in Figure 1 and believed to be

Table 2. Effect of hydrocolloids on texture characteristics of gluten-free rice cracker.

Samples	Attributes			
	Hardness (g)	Cohesiveness	Chewiness (gmm)	Springiness (mm)
WF	325.25 ± 5.66[h]	0.64 ± 0.02	1.04 ± 0.01	0.06 ± 0.04
RF	156.20 ± 9.52[i]	0.12 ± 0.05	11.6 ± 1.86	0.51 ± 0.03
FF	396.67 ± 14.2[f]	0.02 ± 0.01	0.14 ± 0.02	0.01 ± 0.00
0.25%XN	481.69 ± 17.9[d]	0.21 ± 0.07	−102.01 ± 0.39	−2.81 ± 0.34
0.5%XN	367.73 ± 8.95[g]	0.35 ± 0.02	−126.02 ± 0.65	−2.11 ± 0.72
0.75%XN	653.18 ± 22.6[a]	6.61 ± 1.14	−124.19 ± 0.51	−2.45 ± 0.55
1.0%CMC	462.11 ± 9.99[d]	0.12 ± 3.09	−151 ± 5.22	−2.22 ± 0.06
1.5%CMC	386.90 ± 6.36[f,g]	1.15 ± 4.05	−138.22 ± 3.05	−2.08 ± 0.10
2.0%CMC	618.13 ± 12.4[b]	18.87 ± 7.1	−145.09 ± 6.01	−2.05 ± 0.81
1%HPMC	418.40 ± 14.1[e]	2.47 ± 1.07	−182 ± 4.39	−2.61 ± 0.26
1.5%HPMC	516.36 ± 8.52[c]	1.65 ± 0.03	−26.22 ± 0.43	−2.41 ± 0.99
2.0%HPMC	460.75 ± 8.98[d]	1.96 ± 0.10	−184.29 ± 3.40	−2.55 ± 0.85

Different letters in each column indicate statistical differences ($P \leq 0.05$). Rice flour (RF), wheat flour (WF), and flour blend (FF) were controls and were made of 100%RF, 100%WF, and formulated FF, respectively.

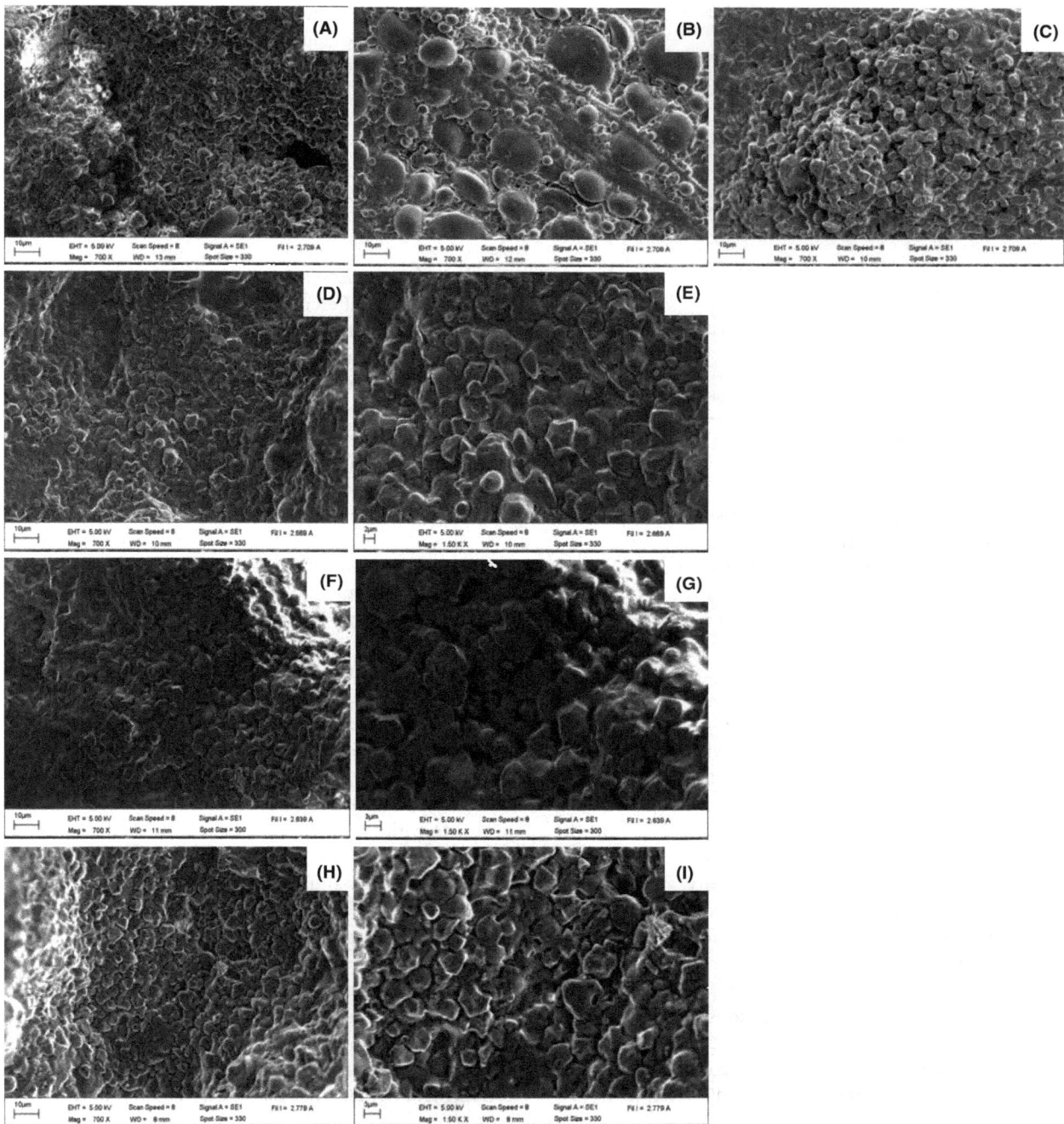

Figure 1. Effect of hydrocolloids on the dough microstructure of gluten-free rice crackers. Rice flour (RF), wheat flour (WF), and flour blend (FF) were controls and were made of 100%RF (A), 100%WF (B), formulated FF (C), FF + 1.5%HPMC (×700) (D), (×1500) (E), FF + 0.5%XN (×700) (F), (×1500) (G), FF + 1.5%CMC (×700) (H) and (×1500) (I), respectively.

formed by hydrocolloids, can partially explain the increase in fracture force and hardness. As these films dehydrate during the baking process, they create a type of tough skin at the surface of the cracker which increases its resistance to breakage and thus increases the fracture force and hardness values. The hydrocolloids also helped retain gas in samples during baking so the crispness of gluten-free crackers with hydrocolloids did not rely only on starch gelatinization as in the case of RF.

Table 3. Effect of hydrocolloids on color values of gluten-free rice crackers.

Sample	Color		
	L^*	Hue angle	Chroma
WF	45.8 ± 2.7^e	78.6 ± 2.3^f	34.8 ± 2.1^a
RF	$51.6 \pm 1.5^{a,b,c}$	85.1 ± 1.2^a	32.6 ± 1.2^d
FF	50.0 ± 2.0^d	85.3 ± 1.1^a	31.7 ± 1.1^e
0.25%XN	$50.7 \pm 1.8^{c,d}$	81.6 ± 2.0^e	$34.1 \pm 1.5^{a,b}$
0.5%XN	49.7 ± 1.5^d	$83.9 \pm 1.7^{b,c}$	$34.2 \pm 1.6^{a,b}$
0.75%XN	$51.6 \pm 1.3^{a,b,c}$	$83.6 \pm 2.0^{c,d}$	$34.3 \pm 1.4^{a,b}$
1.0%CMC	$51.5 \pm 2.3^{a,b,c}$	85.0 ± 2.3^a	$33.1 \pm 1.8^{c,d}$
1.5%CMC	$51.1 \pm 2.0^{b,c}$	$84.7 \pm 2.1^{a,b}$	31.8 ± 1.2^e
2.0%CMC	$51.8 \pm 2.5^{a,b}$	$84.6 \pm 2.3^{a,b}$	32.7 ± 1.4^d
1%HPMC	52.5 ± 2.3^a	85.4 ± 1.5^a	$34.0 \pm 1.5^{a,b}$
1.5%HPMC	$50.5 \pm 2.3^{c,d}$	$84.0 \pm 1.9^{b,c}$	$33.6 \pm 1.6^{b,c}$
2.0%HPMC	$51.3 \pm 2.2^{b,c}$	82.9 ± 1.8^d	$34.3 \pm 2.1^{a,b}$

Different letters in the same column indicate statistical differences found on each parameter of color ($P \leq 0.05$).

The addition of the hydrocolloids results in an increase in the rigidity as a consequence of the decrease in the swelling of starch granules and amylose lixiviation (Biliaderis et al. 1997). In addition as Rosell et al. (2001) found the addition of xanthan increased the hardness of the bread crumbs which could be a consequence of the thickening effect of hydrocolloids on the crumb walls surrounding the air spaces.

The microstructures of gluten-free cracker dough were obtained by performing SEM analysis with 700× magnification and 1500× magnification (Fig. 1). Compared to the 100%RF control, samples made with hydrocolloids displayed a more irregular starch matrix structure, with the starch granules appearing somewhat deformed. However, not all starch granules lost their identity and they did not disintegrate completely.

Dough made from the two negative controls RF & FF tended to be porous and to have gaps of varying sizes and frequency within their matrix, whereas the WF positive control had a continuous matrix with no visible gaps. The dough microstructure of samples containing hydrocolloids had a more continuous matrix than the negative controls RF & FF. Hydrocolloids-added dough seemed to hold the constituent starch granules and matrix covering them within a veil-like film.

These findings agree with the result that Bárcenas and Rosell (2005) reported that the gas cell walls of the crumb-containing HPMC showed a smooth structure with a fewer number of cavities than wheat bread without HPMC. They had a continuous structure with the appearance of a veil where the bread components could be observed. In addition, the use of hydrocolloids, such as XN in gluten-free bread produces a web-like structure similar to that of standard wheat bread (Ahlborn et al. 2005). However, the hydrocolloid-containing samples were able to produce fairly stable gas cells and fairly continuous matrixes on their own without the addition of proteins (Fig. 1). These results appear to contradict others (Ahlborn et al. 2005; Rosell and Marco 2008) who had suggested that hydrocolloids alone do not seem to do enough to stabilize gas cells.

Table 3 shows the effect of different hydrocolloids and usage levels on the color components of rice crackers. The lightness values (L^*) of all treatment samples were not significantly different when compared with formulated flour without hydrocolloids added (FF). The L^* values of gluten-free rice crackers containing 0.5%XN, 1.5%HPMC, and 1.5%CMC were the closest to WF; the rest of the treated samples had a lighter color and tended to be closer to FF.

The surface color of 100% wheat crackers (WF) was more reddish-brown than yellow-brown, whereas 100% rice crackers (RF) had a more pale yellow character. Inclusion of XN and HPMC produced treated samples with similar yellowness (b^*) and color intensity (Chroma) to the WF control. Despite the color values obtained by instrumental methods, all crackers evaluated showed slight differences in color and surface appearance from each other as illustrated in Figure 2. The darker color of the WF samples is due to a higher level of protein in the wheat samples which leads to higher amounts of free-amino acids available to participate in maillard browning. Thus, samples made with wheat have a higher concentration of maillard browning reaction products which lead to a more reddish brownish color for these samples when compared to the treatment and negative control samples which have a lower level of proteins.

The addition of hydrocolloid to samples produced lighter colored crackers compared to the FF control. This can be explained by the increase in moisture content; as the moisture content increased, the Maillard browning reaction rate slowed and the reaction products responsible for brown color became further diluted thus producing a lighter finished cracker color (Mezaize et al. 2010).

Effect of protein isolates on physicochemical properties of gluten-free dough and crackers

The moisture content, water activity (a_w), and % puffiness of samples with added protein isolates are shown in Table 4. The addition of different protein isolates increased the moisture content of gluten-free crackers. Significant differences were observed in the moisture content of the

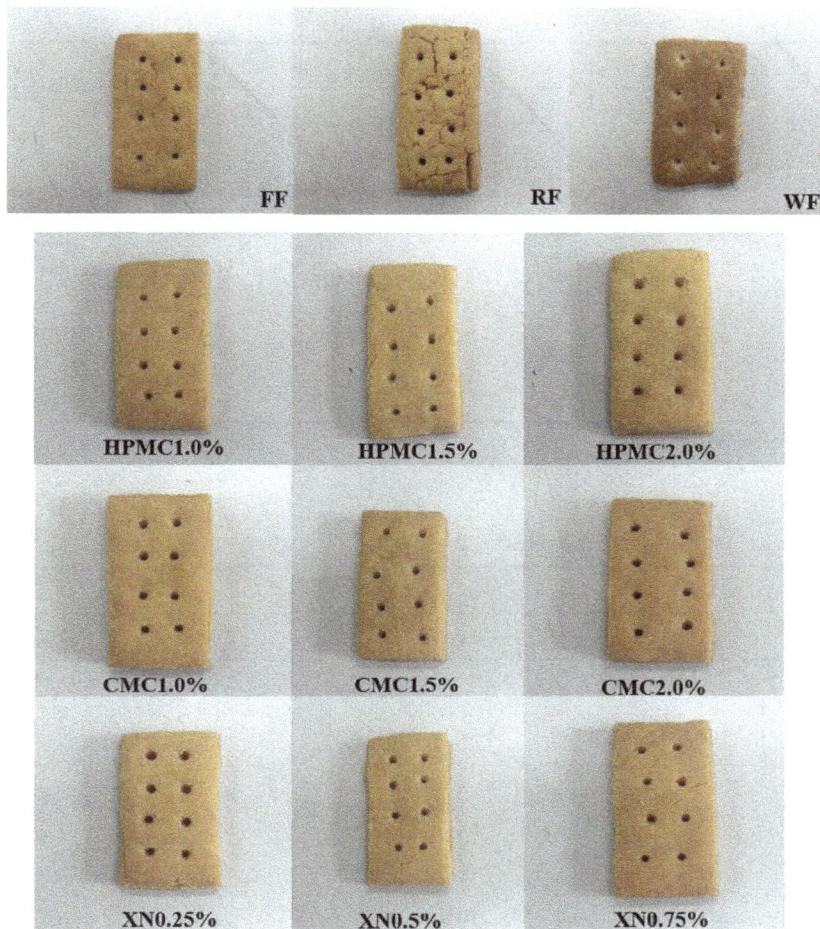

Figure 2. Effect of hydrocolloids on surface appearance of gluten-free rice crackers. Rice flour (RF), wheat flour (WF), and flour blend (FF) were controls and were made of 100%RF, 100%WF, and formulated FF, respectively.

treated samples ($P \leq 0.05$). The treated crackers samples containing 5% and 10% soy protein isolate had the highest moisture content of all samples measured. The moisture content of the treated crackers samples containing 5% and 10% whey protein isolates were closest to those of wheat crackers.

The puffiness of all protein-isolate-treated samples appeared to be higher than controls made from RF were statistically significantly different ($P \leq 0.05$). The addition of protein isolates increased the puffiness of treated samples, but these samples were still less puffy than wheat crackers. The puffiness of wheat crackers was directly related to the protein matrix created in wheat crackers. The gluten matrix contains many inner layers which slows down the rate of gas diffusion and allows its retention before and after baking as well as determines the puffiness and crumb structure of the wheat crackers (Faubion and Hoseney 1990). In contrast, a continuous protein network with starch granule embedded was not found in the negative control (RF). Rice cracker texture was mostly created by starch gelatinization (Sozer 2009) and rice dough cannot retain the gas produced during fermentation which led to a crumbly rice cracker (He and Hoseney 1991). The addition of protein isolates to replace the gluten in the protein matrix produced protein matrixes with qualities between those of the wheat cracker and the negative control.

The texture characteristics of gluten-free rice crackers are shown in Table 5. The addition of protein isolates decreased the fracture forces required to fracture the crackers when compared to the FF cracker so samples had a closer texture to wheat crackers. The hardness of samples containing 2.5 or 5.0% soy protein isolates and 5.0% pea protein isolates were close to those of wheat crackers (WF), and their texture characteristics were also similar to wheat crackers. In contrast, the

Table 4. Effect of protein isolates from different sources physicochemical properties of gluten-free rice crackers.

Samples	Moisture content (%)	Water activity (a_w)	Puffiness (%)
WF	4.43 ± 0.040[c]	0.347 ± 0.005[a]	63.63 ± 2.80[a]
RF	2.80 ± 0.031[e]	0.134 ± 0.001[d]	28.00 ± 2.17[c]
FF	3.35 ± 0.087[d]	0.303 ± 0.007[b]	36.18 ± 1.29[b,c]
FF + 2.5%SP	5.11 ± 0.23[b]	0.221 ± 0.001[c]	41.58 ± 1.12[b]
FF + 5.0%SP	5.76 ± 0.37[a]	0.240 ± 0.008[c]	43.36 ± 2.45[b]
FF + 10.0%SP	5.84 ± 0.13[a]	0.318 ± 0.001[b]	44.42 ± 1.67[b]
FF + 2.5%PP	5.20 ± 0.16[b]	0.266 ± 0.001[c]	40.63 ± 4.22[b]
FF + 5.0%PP	5.29 ± 0.05[b]	0.248 ± 0.001[c]	41.77 ± 2.21[b]
FF + 10.0%PP	5.23 ± 0.07[b]	0.285 ± 0.001[b,c]	40.89 ± 1.34[b]
FF + 2.5%WP	5.29 ± 0.18[b]	0.319 ± 0.008[b]	43.78 ± 0.98[b]
FF + 5.0%WP	4.82 ± 0.122[b,c]	0.315 ± 0.002[b]	41.87 ± 1.67[b]
FF + 10.0%WP	4.62 ± 0.144[b,c]	0.251 ± 0.002[c]	42.08 ± 2.14[b]

Different letters in the same column indicate statistical differences ($P \leq 0.05$). Rice flour (RF), wheat flour (WF), and flour blend (FF) were controls and were made of 100%RF, 100%WF, and formulated FF, respectively.

Table 5. Effect of protein isolates from different sources on texture characteristics of gluten-free rice crackers.

Samples	Attributes			
	Hardness (g)	Cohesiveness	Chewiness (gmm)	Springiness (mm)
WF	309.6 ± 7.9[f,g]	0.64 ± 0.02	1.04 ± 0.01	0.06 ± 0.04
RF	165.8 ± 5.7[k]	0.12 ± 0.05	11.6 ± 1.86	0.51 ± 0.03
FF	459.0 ± 4.97[c]	0.02 ± 0.01	0.14 ± 0.02	0.01 ± 0.00
FF+2.5%SP	297.8 ± 5.3[h]	0.16 ± 0.02	2.03 ± 0.03	0.32 ± 0.01
FF+5.0%SP	319.0 ± 5.0[f]	0.03 ± 0.04	6.53 ± 0.00	0.31 ± 0.02
FF+10.0%SP	226.5 ± 4.6[j]	0.03 ± 0.05	8.52 ± 0.00	0.19 ± 0.03
FF+2.5%PP	252.6 ± 6.0[i]	0.69 ± 0.01	2.03 ± 0.04	0.82 ± 0.004
FF+5.0%PP	320.2 ± 7.4[f]	0.12 ± 0.01	2.82 ± 0.01	0.71 ± 0.004
FF+10.0%PP	555.0 ± 28.6[b]	0.07 ± 0.02	1.88 ± 0.01	0.57 ± 0.003
FF+2.5%WP	402.2 ± 12.3[e]	0.17 ± 0.02	0.01 ± 0.001	0.02 ± 0.001
FF+5.0%WP	433.6 ± 6.32[d]	0.11 ± 0.03	0.05 ± 0.001	0.16 ± 0.001
FF+10.0%WP	794.6 ± 56.15[a]	0.15 ± 0.08	0.01 ± 0.002	0.23 ± 0.001

Different letters in the same column indicate statistical differences ($P \leq 0.05$). Rice flour (RF), wheat flour (WF), and flour blend (FF) were controls and were made of 100%RF, 100%WF, and formulated FF, respectively.

hardness of the control rice cracker (RF) was the lowest, and samples were the most crumbly and dense, illustrated as lack of layers. This was due to the differences between rice protein and wheat gluten. Unlike wheat gluten, rice-flour dough is not cohesive and lacks good viscoelastic properties and is not strong enough to entrap gas produced in the food system (Sozer 2009). Rice cracker structure is mostly created from starch gelatinization that cannot act as a strong backbone to support crackers' structure.

The formulated flour-blended control (FF) contained pregelatinized-tapioca starch in the FF that facilitated dough formation from more starch gelatinization (Sozer 2009), consequently resulting in better texture than rice control (Nammakuna et al. 2009). However, the texture of nonprotein-added control still had fewer layers and was less puffy, which made its texture tougher and less brittle than wheat crackers, despite its lower values of hardness. Besides wheat crackers, the protein-added samples were perceived as relatively more brittle and crisper than other controls because there were more layers and puffiness. These results agreed with (Sozer 2009) that addition of proteins helps starch granules to adhere to one another, and water is more distributed through the system because of the polymeric structure of proteins (Sivaramakrishnan et al. 2004).

The viscoelastic behavior of dough was illustrated in terms of storage modulus (G′), loss modulus (G″), and tanδ value. They were used as the indicators of dough rheology and characteristics. The loss tangent or tanδ is the tangent of the phase angle which is the ratio of viscous modulus (G″) to elastic modulus (G′) that shows the presence of fluid's elasticity. The loss tangent values less than unity indicate an elastic-dominant behavior, whereas values

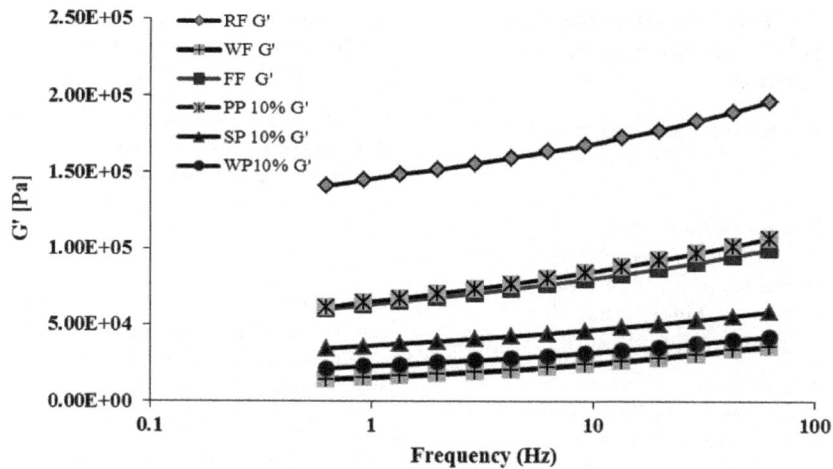

Figure 3. Effect of protein isolates from different source on rheological properties of gluten-free rice cracker based on the frequency sweep test. Rice flour (RF), wheat flour (WF) and flour blend (FF) were controls and were made of 100%RF, 100%WF and formulated FF, respectively.

greater than unity indicate viscous-dominant behavior. Simply put, these values could be used to describe the balance between elastic properties such as film formation and gas retention in dough, and viscous properties such as protein absorption to the liquid lamella and flexibility for gas expansion in dough (Lazaridou et al. 2007).

The viscoelastic behavior of dough samples with addition of protein isolates are illustrated in Figure 3 and Table 6. The control rice dough had the highest storage modulus value which made its structure more rigid and less elastic. This also made its structure harder to poor stretch during kneading and sheeting process. On the other hand, wheat dough with gluten network formation had the lowest storage modulus value which caused a unique viscoelastic property in dough, being more elastic, easier to handle in kneading, and in sheet- making (Rosell and Marco 2008).

In doughs made with protein isolates, the storage modulus values were lower than the control rice dough; in particular dough with a 10% whey protein added had a storage modulus closest to that of the wheat cracker dough (Fig. 3), showing an improvement in dough elasticity. This was probably due to the polymeric structure of protein that facilitated better water-holding capacity and water distribution in dough when compared with rice control. Sozer (2009) also noted that a decrease in storage modulus value was observed in pasta dough with more gelatinized RF because of an increase in water absorption and starch granule swelling. The addition of protein isolates to replace the gluten in the protein matrix produced protein matrixes with qualities between those of the wheat cracker and the negative control.

Table 6 illustrates the loss tangent (tanδ) of dough samples with addition of protein isolates and hydrocolloids. All dough samples had tan delta values less than

unity which means their behaviors were more elastic than viscous. Among all samples, wheat crackers had the highest loss tangent value (0.44) which indicated more elasticity than others. As compared, among the other protein-isolate dough samples, the loss tangent values of dough with whey protein isolates were closer to wheat dough. Those values were also higher than dough of RF or FF, indicating samples were more elastic and stretchy. This behavior could be explained that the addition of protein isolates increased the amount of polymers in the dough system, resulting in an improvement in elastic properties of dough samples (Sozer 2009).

Table 6. The effect of hydrocolloids and protein isolates from different source on storage modulus (G′), loss modulus (G″), and tanδ on gluten-free rice cracker dough.

Samples	Viscoelastic parameter (overall mean)		
	G′ (Pa)	G″ (Pa)	tanδ
WF	23538.46	10500.77	0.446
RF	95500.00	22884.62	0.239
FF	114700.00	28384.61	0.244
FF + 10%SP	45400.00	10766.15	0.237
FF + 10%SP + 0.5%XN	60276.92	16646.15	0.276
FF + 10%SP + 1.5%HPMC	110915.21	32318.87	0.264
FF + 10%PP	82030.77	19615.38	0.239
FF + 10%PP + 0.5%XN	52515.38	13040.77	0.248
FF + 10%PP + 1.5%HPMC	95812.93	23211.62	0.242
FF + 10% WP	30807.69	10586.92	0.344
FF + 10%WP + 0.5%XN	94684.61	28284.61	0.298
FF + 10%WP + 1.5%HPMC	99628.12	27587.95	0.277

Rice flour (RF), wheat flour (WF), and flour blend (FF) were controls and were made of 100%RF, 100%WF, and formulated FF, respectively.

Effect of protein isolate and hydrocolloids on physicochemical properties of gluten-free dough and crackers

The effects of hydrocolloids and protein isolate combinations on the physical and chemical properties of gluten-free crackers are shown in Table 7. Significant differences in moisture content and water activity (a_w) in all treatments were observed ($P \leq 0.05$). The addition of hydrocolloids and protein isolates produced moisture contents and water activity (a_w) values higher than that of the formulated flour- blended control (FF), with the exception of the 1.5%HPMC and 10% soy protein sample which was not significantly different from the formulated FF control.

The treated samples with 10% pea protein isolate and 0.5%XN had the highest moisture content and water activity (a_w). The addition of 1.5%HPMC in the samples containing 10% pea protein isolate and 10% whey protein isolate caused the moisture content to become closer to that of wheat crackers and higher than the FF control.

This resulted in water retention ability of this dough due to its hydrophilic nature (Christianson et al. 1981; Twillman and White 1988; Bell 1990; Dziezak 1991; Armero and Collar 1998; Gurkin 2002; Guarda et al. 2004). The hydrocolloids in the dough held on to a fraction of the water. These findings are consistent with prior research on the effects of hydrocolloids on the properties of baked bread loaves, for example, as reported by Guarda et al. (2004), the HPMC network formed during baking could act as a barrier to gas diffusion, decreasing the water vapor losses, and increasing the final moisture content of the loaves. However, the addition of 1.5%HPMC into dough containing 10% soy protein isolate and 10% pea protein isolate produced a decrease in the moisture content of the samples compared to dough containing only 10% pea protein and 10% soy protein. This effect can be attributed to the unique interaction between HPMC and

soy protein according to Rosell and Marco (2008), the reduction in the moisture content induced by the HPMC was partially masked, when part of the RF was replaced by soybean protein, where rice starch molecules were replaced by protein molecules. No significant difference in puffiness was found among treatments ($P > 0.05$). The hydrocolloids-added samples showed higher puffiness than control formulated flour (FF) and 100%RF.

The texture characteristics of the combinations of protein isolates and hydrocolloids are shown in Table 8. The hardness of all treatment samples was higher compared to RF and FF control ($P \leq 0.05$), thus the samples were less crumbly and crispier. These results agreed with previous studies which determined that hydrocolloids such as HPMC increased water binding in the rice cassava dough, and the amphiphilic nature of the hydrocolloids acted as a surfactant stabilizing the gas–liquid interface around the gas bubble and resulting in increased loaf volume, improved crumb structure, and reduced crumb firmness (Crockett et al. 2011). The viscoelastic behavior of protein and hydrocolloids-added dough samples was investigated by the oscillation frequency sweep test using a frequency sweep from 0.1 to 10 Hz. The results are shown in Figure 4.

All samples showed an increase of the storage modulus (G′) with increasing frequency. The storage modulus of rice cracker dough was the highest followed by dough samples with hydrocolloids. In dough containing both soy protein isolate and HPMC, the protein altered the HPMC functionality due to competition for water. This weakened the HPMC interactions with the starch matrix and reduced form stability (Crockett et al. 2011). Addition of 0.5%XN resulted in dough with viscoelastic properties close to 100% wheat dough, especially cracker dough containing 10% pea protein and 0.5%XN, thus dough had G′ values closer to 100% wheat dough than all other treatments.

Table 7. Physicochemical properties of gluten-free rice crackers with protein isolates from different sources and hydrocolloids addition.

Samples	Moisture content (%)	Water activity (a_w)	Puffiness (%)
WF	4.43 ± 0.040[c]	0.347 ± 0.005[a]	63.63 ± 4.80[a]
RF	2.80 ± 0.031[e]	0.134 ± 0.001[g]	28.00 ± 2.17[c]
FF	3.35 ± 0.087[d]	0.303 ± 0.007[d]	38.18 ± 3.29[b]
FF + 10%SP + 0.5%XN	4.96 ± 0.049[b]	0.324 ± 0.002[c]	42.54 ± 3.04[b]
FF + 10%SP + 1.5%HPMC	3.76 ± 0.123[d]	0.287 ± 0.002[f]	41.35 ± 2.47[b]
FF + 10%PP + 0.5%XN	5.28 ± 0.040[a]	0.337 ± 0.002[b]	41.81 ± 1.14[b]
FF + 10%PP + 1.5%HPMC	4.42 ± 0.100[c]	0.298 ± 0.006[e]	41.31 ± 1.79[b]
FF + 10%WP + 0.5%XN	4.88 ± 0.016[b]	0.331 ± 0.003[b]	40.45 ± 2.01[b]
FF + 10%WP + 1.5%HPMC	4.67 ± 0.040[b,c]	0.318 ± 0.005[c]	41.76 ± 1.73[b]

Different letters in the same column indicate statistical differences ($P \leq 0.05$). Rice flour (RF), wheat flour (WF), and flour blend (FF) were controls and were made of 100%RF, 100%WF, and formulated FF, respectively.

Table 8. The texture characteristic of gluten-free rice crackers with protein isolates from different sources and hydrocolloids addition.

Samples	Attributes			
	Hardness (g)	Cohesive	Chewiness (gmm)	Springiness (mm)
WF	336 ± 6.24[d]	0.066 ± 0.014	4.14 ± 0.012	1.09 ± 0.008
RF	68.75 ± 2.68[g]	0.122 ± 0.05	11.59 ± 15.86	0.81 ± 0.010
FF	401.75 ± 7.60[b]	0.069 ± 0.07	21.89 ± 0.018	2.51 ± 0.012
FF + 10%SP + 0.5%XN	244.50 ± 5.42[f]	0.53 ± 0.013	5.03 ± 0.002	4.18 ± 0.034
FF + 10%SP + 1.5%HPMC	278.0 ± 5.55[e]	0.66 ± 0.023	6.13 ± 0.002	6.09 ± 0.018
FF + 10%PP + 0.5%XN	332.0 ± 2.56[d]	0.78 ± 0.044	4.93 ± 0.002	3.22 ± 0.001
FF + 10%PP + 1.5%HPMC	274.5 ± 5.89[e]	0.18 ± 0.050	3.02 ± 0.002	2.39 ± 0.001
FF + 10%WP + 0.5%XN	358.4 ± 8.20[c]	0.18 ± 0.034	6.22 ± 0.002	3.32 ± 0.002
FF + 10%WP + 1.5%HPMC	454.8 ± 5.02[a]	0.12 ± 0.008	7.23 ± 0.002	3.11 ± 0.002

Different letters in the same column indicate statistical differences ($P \leq 0.05$). Rice flour (RF), wheat flour (WF), and flour blend (FF) were controls and were made of 100%RF, 100%WF, and formulated FF, respectively.

The dough-containing combinations of hydrocolloids and protein isolates had tanδ values higher than 100% rice cracker dough and formulated flour dough controls (Table 6). The dough with 10% whey protein isolate had the highest tanδ values, followed by a combination of 0.5%XN and 10% whey protein isolated sample, respectively. The G′ values of all treatment samples were slightly increased, and the dough became more rubbery when hydrocolloids were added which facilitated better dough kneading and sheeting when compared to the RF dough. According to (Sánchez et al. 2004), this may have been caused by a specific interaction between proteins and hydrocolloids leading to the formation of a more viscoelastic dough. The addition of hydrocolloids into protein isolates-treated doughs caused the G′ values to increase for all treatments, especially when HPMC was added to soy protein isolates and pea protein isolates dough, which had G′ values close to formulated FF dough. The addition of hydrocolloids to leguminous protein isolate-treated

dough induced an increase in the loss tangent (tanδ). Crockett et al. (2011) also found that the addition of hydrocolloids and soy protein isolate increased G′ in gluten-free bread from rice and cassava which may have been caused by the action of the two main globulins in soy protein isolate, including β-conglycinin and glycinin (Lampart-Szczapa 2001; Crockett et al. 2011). Glycinin forms a thermoplastic gel at 80°C, but above 100°C, the further unfolding of protein exposes more hydrophobic regions, further stabilizing the gel and preventing denaturing of the protein (Lampart-Szczapa 2001).

Conclusion

In crackers, wheat gluten plays a very unique role compared to its functionality in other baked products as it acts to make the cracker weaker and stronger at the same time. While the strong protein structure created by gluten–protein interaction creates rigid, crispy, and continuous

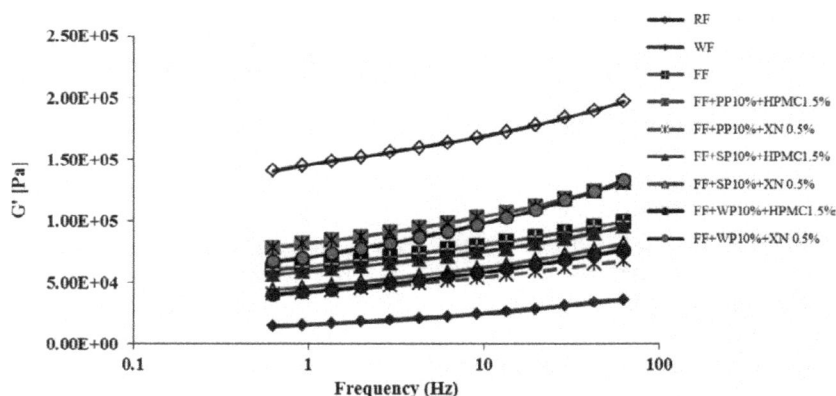

Figure 4. Effect of protein isolates from different sources and hydrocolloids addition on rheological properties of gluten-free rice crackers based on the frequency sweep test. Rice flour (RF), wheat flour (WF) and flour blend (FF) were controls, and were made of 100%RF, 100%WF, and formulated FF, respectively.

layers in the cracker, at the same time it aides in the formation of a relatively small number of relatively large gas bubbles, which separates these layers thus reducing the density of the overall cracker and increasing its overall surface area, thus increasing the moisture loss rate, and therefore resulting in a low moisture, crispy finished texture.

Addition of protein isolates helps improve the performance of gluten-free rice crackers, with the amount of improvement being dependent on the nature and the functional properties of the added protein. The gluten in wheat-based crackers is able to form elastic structures that entrap air and expand to relatively large sizes without bursting during the baking process. As a result, crackers made with WF have a comparatively small number of large gas cells, whereas dough made with rice FF has a high number of relatively small gas cells as the dough cannot produce the same type of elastic structures found in wheat-based crackers, due to significantly lower levels of protein in RF and thus cannot produce large gas cells. The microstructure of dough-containing hydrocolloids had a more continuous matrix with fewer and larger gas cells when compared to FF and 100% rice crackers, the dough seemed to hold the constituent, starch granules, and matrix covering resembling a veil-like film.

Cracker doughs made with hydrocolloids and protein isolates have structures containing gas cell numbers and sizes in between those seen in 100% wheat crackers and those seen in the rice-blended crackers. This more compact structure combined with the higher water-holding ability of hydrocolloids lead to higher moisture contents in the finished cracker. This higher finished moisture content and the lower protein content, compared to 100%WF crackers, resulted in lighter colored crackers. The added hydrocolloids and protein isolates interacted with the native proteins of the rice formula to create more elastic structures that could partially mimic those created by the gluten-forming protein in wheat-based crackers. The combination of hydrocolloids and protein isolates create cracker dough which have higher G′ values and are more rubbery than dough made from the FF blend, this facilitates dough kneading and sheeting as it forms a more viscoelastic gel. However, soy protein isolates and pea protein isolates lead to undesirable flavors in the finished cracker.

Thus, this research shows that combinations of hydrocolloids and protein isolates can be used to partially or fully replace wheat gluten in crackers with acceptable organoleptic results, however, no combination produced gluten-free crackers with organoleptic properties identical to those of wheat gluten-containing crackers. The cracker containing 10.0% whey protein isolate had the best texture characteristics and rheological properties closest to the 100% wheat control.

Acknowledgments

This project is supported by the Naresuan University Research Funding, and the Thailand Research Fund (TRF) for grant number MRG5380217. Sample and equipments supports by F.A. Foods Co., Ltd, Roquette Singapore PTE, Ltd, Grande Custom Ingredients Group, Faculty of Agriculture Natural Resources and Environment, and Metrohm Siam, Ltd are gratefully acknowledged.

Conflict of Interest

None declared.

References

Abdel-Aal, E.-S. M. (2009). 11 Functionality of Starches and Hydrocolloids in Gluten-Free Foods. Pp. 200 in E. Gallagher (Ed), Gluten-free food science and technology. Wiley-Blackwell, Iowa.

Ahlborn, G. J., O. A. Pike, S. B. Hendrix, W. M. Hess, and C. S. Huber. 2005. Sensory, mechanical, and microscopic evaluation of staling in low-protein and gluten-free breads. Cereal Chem. 82:328–335.

Akdeniz, N., S. Sahin, and G. Sumnu. 2006. Functionality of batters containing different gums for deep-fat frying of carrot slices. J. Food Eng. 75:522–526.

Albert, S., and G. S. Mittal. 2002. Comparative evaluation of edible coatings to reduce fat uptake in a deep-fried cereal product. Food Res. Int. 35:445–458.

Armero, E., and C. Collar. 1998. Crumb firming kinetics of wheat breads with anti-staling additives. J. Cereal Sci. 28:165–174.

Bárcenas, M. E., and C. M. Rosell. 2005. Effect of HPMC addition on the microstructure, quality and aging of wheat bread. Food Hydrocolloid. 19:1037–1043.

Bell, D. 1990. Methylcellulose as a structure enhancer in bread baking. Cereal Foods World 35:1001–1006.

Biliaderis, C., I. Arvanitoyannis, M. Izydorczyk, and D. Prokopowich. 1997. Effect of hydrocolloids on gelatinization and structure formation in concentrated waxy maize and wheat starch gels. Starch-Stärke 49:278–283.

Christianson, D., J. Hodge, D. Osborne, and R. W. Detroy. 1981. Gelatinization of wheat starch as modified by xanthan gum, guar gum, and cellulose gum. Cereal Chem. 58:513–517.

Crockett, R., P. Ie, and Y. Vodovotz. 2011. Effects of soy protein isolate and egg white solids on the physicochemical properties of gluten-free bread. Food Chem. 129:84–91.

Dziezak, J. D. 1991. A focus on gums. Food Technol. 45:116–132.

Faubion, J., and R. C. Hoseney (1990). 2-The viscoelastic properties of wheat flour doughs. Pp. 29–66 in H. Faridi and J. M. Faubion (Eds), Dough rheology and baked product texture. Van Nostrand Reinhold, New York.

Gallagher, E., O. Polenghi, and T. Gormley. 2002. Improving the quality of gluten-free breads. Farm Food 12:8–13.

Guarda, A., C. Rosell, C. Benedito, and M. Galotto. 2004. Different hydrocolloids as bread improvers and antistaling agents. Food Hydrocoll. 18:241–247.

Gurkin, S. 2002. Hydrocolloids: ingredients that add flexibility to tortilla processing. Cereal Foods World 47:41–43.

He, H., and R. Hoseney. 1991. Differences in gas retention, protein solubility, and rheological properties between flours of different baking quality. Cereal Chem. 68:526–530.

Khalil, A. H. 1999. Quality of french fried potatoes as influenced by coating with hydrocolloids. Food Chem. 66:201–208.

Lampart-Szczapa, E. (2001). 14-Legume and oilseed proteins. Pp. 407–436 in Z. E. Sikorski (Ed), Chem. Funct. Proper. Food Proteins. CRC Press, Florida.

Lazaridou, A., D. Duta, M. Papageorgiou, N. Belc, and C. G. Biliaderis. 2007. Effects of hydrocolloids on dough rheology and bread quality parameters in gluten-free formulations. J. Food Eng. 79:1033–1047.

Marco, C., and C. M. Rosell. 2008. Functional and rheological properties of protein enriched gluten free composite flours. J. Food Eng. 88:94–103.

Mellema, M. 2003. Mechanism and reduction of fat uptake in deep-fat fried foods. Trends Food Sci. Technol. 14:364–373.

Mezaize, S., S. Chevallier, A. Le-Bail, and M. de Lamballerie. 2010. Gluten-free frozen dough: influence of freezing on dough rheological properties and bread quality. Food Res. Int. 43:2186–2192.

Nammakuna, N., S. Suwansri, P. Thanasukan, and P. Ratanatriwong. 2009. Effects of hydrocolloids on quality of rice crackers made with mixed-flour blend. Asian J. Food Agro. Ind. 2:780–787.

Rosell, C. M., and C. Marco (2008). 4—Rice. Pp. 81–III in K. A. Elke, B. Fabio Dal (Eds.), Gluten-Free Cereal Products and Beverages. Academic Press, San Diego.

Rosell, C., J. Rojas, and C. Benedito De Barber. 2001. Influence of hydrocolloids on dough rheology and bread quality. Food Hydrocoll. 15:75–81.

Sánchez, C. C., M. R. Rodríguez Niño, S. E. Molina Ortiz, M. C. Añon, and J. M. Rodríguez Patino. 2004. Soy globulin spread films at the air–water interface. Food Hydrocoll. 18:335–347.

Sivaramakrishnan, H. P., B. Senge, and P. Chattopadhyay. 2004. Rheological properties of rice dough for making rice bread. J. Food Eng. 62:37–45.

Sozer, N. 2009. Rheological properties of rice pasta dough supplemented with proteins and gums. Food Hydrocoll. 23:849–855.

Toft-Hansen, H., K. S. Rasmussen, A. Staal, E. L. Roggen, L. M. Sollid, S. T. Lillevang, et al. 2014. Treatment of both native and deamidated gluten peptides with an endo-peptidase from Aspergillus niger prevents stimulation of gut-derived gluten-reactive T cells from either children or adults with celiac disease. Clin. Immunol. 153:323–331.

Twillman, T., and P. White. 1988. Influence of monoglycerides on the textural shelf life and dough rheology of corn tortillas. Cereal Chem. 65:253–257.

7

Use of the wetting method on cassava flour in three konzo villages in Mozambique reduces cyanide intake and may prevent konzo in future droughts

Dulce Nhassico[1], James Howard Bradbury[2], Julie Cliff[3], Rita Majonda[4], Constantino Cuambe[4], Ian C. Denton[2], Matthew P. Foster[2], Arlinda Martins[5], Adelaide Cumbane[1], Luis Sitoe[1], Joao Pedro[4] & Humberto Muquingue[1]

[1]Department of Biochemistry, Faculdade de Medicina, Universidade Eduardo Mondlane, Maputo, Mozambique
[2]EEG, Research School of Biology, Australian National University, Canberra, ACT 2601, Australia
[3]Department of Community Health, Faculdade de Medicina, Unversidade Eduardo Mondlane, Maputo, Mozambique
[4]Instituto de Investigacao Agraria de Mocambique (IIAM), Nampula, Mozambique
[5]Direccao Provincial de Saude, Nampula, Mozambique

Keywords
Cassava flour, cyanide, konzo, urinary thiocyanate, wetting method

Correspondence
J. Howard Bradbury, EEG, Research School of Biology, Australian National University, Canberra, ACT 2601, Australia.

E-mail: Howard.Bradbury@anu.edu.au

Funding Information
This study was funded by Australian Agency for International Development (AusAID), Australian National University.

Abstract

Konzo is an irreversible paralysis of the legs that occurs mainly in children and young women associated with large cyanide intake from bitter cassava coupled with malnutrition. In East Africa outbreaks occur during drought, when cassava plants produce much more cyanogens than normal. A wetting method that removes cyanogens from cassava flour was taught to the women of three konzo villages in Mozambique, to prevent sporadic konzo and konzo outbreaks in the next drought. The intervention was in three villages with 72 konzo cases and mean konzo prevalence of 1.2%. The percentage of children with high (>350 µmol/L) urinary thiocyanate content and at risk of contracting konzo in Cava, Acordos de Lusaka, and Mujocojo reduced from 52, 10, and 6 at baseline to 17, 0, and 4 at conclusion of the intervention. Cassava flour showed large reductions in total cyanide over the intervention. The percentage of households using the wetting method was 30–40% in Acordos de Lusaka and Mujocojo and less in Cava. If the wetting method is used extensively by households during drought it should prevent konzo outbreaks and chronic cyanide intoxication. We recommend that the wetting method be taught in all konzo areas in East Africa.

Introduction

The wetting method is a simple method of removing residual cyanogens (mainly linamarin and acetone cyanohydrin) from cassava flour. The method involves mixing cassava flour with water and then leaving the wet mixture in a thin layer for 2 h in the sun or 5 h in the shade for hydrogen cyanide gas produced by the enzymatic and nonenzymatic breakdown of cyanogens to be evolved (Bradbury 2006; Cumbana et al. 2007; Bradbury and Denton 2010). The wetting method was field-tested in Mozambique in 2005 (Muquingue et al. 2005; Nhassico et al. 2008) and was found to be easy to use and popular

with rural women. The thick porridge (nchima) produced by boiling the cassava flour in the traditional way had lost the bitter flavor, due to residual bitter linamarin (King and Bradbury 1995) remaining in nchima made from flour not treated by the wetting method. The wetting method was also taught to village women in southern Tanzania where there had previously been an outbreak of konzo (Mlingi et al. 2011).

Konzo is a spastic paraparesis that causes irreversible paralysis of the legs, mainly among children and young women. It is associated with a high intake of cyanide from a restricted diet of bitter (high cyanide) cassava combined with malnutrition (Nzwalo and Cliff 2011; Cliff et al. 1985; Howlett et al. 1990; Banea et al. 2015a). Konzo epidemics occur at times of agricultural crisis, such as during drought or war in the poorest rural cassava-staple areas of Africa. Konzo occurs in the Democratic Republic of Congo (DRC), Mozambique, Tanzania, Cameroon, Central African Republic, and Angola. Konzo is a persistent public health problem in the DRC, occurring in at least four provinces and with rapidly increasing incidence in Bandundu Province (Banea et al. 2015b). A recent study in a konzo area found that children with and without konzo had impaired neurocognition compared with controls from a non-konzo area (Boivin et al. 2013). In Mozambique, children in konzo areas have high urinary thiocyanate levels, a measure of their cyanide intake over previous days, at the time of the cassava harvest (August–October). Apparently healthy school children in these areas, have also shown signs of subclinical neurological damage (Ernesto et al. 2002).

In 2010 the wetting method was taught to the women of Kay Kalenge village in Bandundu Province of DRC where there were 34 konzo cases, and they used it daily to remove cyanogens from their cassava flour. No new cases of konzo occurred during the 18 month intervention, the cyanide content of the treated cassava flour reduced to below 10 ppm and the urinary thiocyanate content of the school children, reduced to safe levels below 350 µmol/L (Banea et al. 2012) Konzo had been prevented in Kay Kalenge by the regular use by the village women of the wetting method, which greatly reduced their cyanide intake. Fourteen months after the intervention ceased in Kay Kalenge we found that the women were still using the wetting method, urinary thiocyanate levels in school children were still below 350 µmol/L, there were no new cases of konzo and the wetting method had spread by word of mouth to three adjacent villages (Banea et al. 2014a) The wetting method is still being used there 5 years later. Subsequently, there have been three more interventions to control konzo in DRC and konzo has now been prevented in 13 villages with a total population of nearly 10,000 people (Banea et al. 2013, 2014b, 2015a). The wetting method has recently been recognized by the World Bank, FAO and WHO as a "sensitive intervention" to remove cyanogens from cassava flour.

In Mozambique and Tanzania epidemics of konzo have occurred due to drought, when the cassava plant is stressed and makes 2–4 times the normal amount of linamarin, the major cyanogen of cassava roots and leaves (Bokanga et al. 1994; Cardoso et al. 2005). During drought the cyanogen content of cassava flour is greatly increased, (Cardoso et al. 2005) village people get sick from cyanide poisoning (acute cyanide intoxication) and many change their processing method from sun drying to heap fermentation, which reduces the cyanide content of flour

by about 50% (Ernesto et al. 2002). However, this reduction is insufficient to prevent the occurrence of konzo and ongoing chronic cyanide intoxication during a drought (Cardoso et al. 2005) and the wetting method is therefore needed as an additional processing method to greatly reduce cyanide intake.

In this study, we describe the introduction of the wetting method in three villages in Nampula Province of Mozambique in which there are many konzo cases from previous droughts and/or war, but few sporadic cases in recent years, and we compare these results with the situation in konzo villages in the DRC, where konzo is occurring in increasing numbers every year.

Experimental

Study area and sample collection

In September 2012 representatives of the research teams met with health, administrative authorities and community leaders to identify three villages in Nampula Province which were most affected by konzo. The three villages were Cava, and Acordos de Lusaka (also called Miaja) in Memba District and Mujocojo in Mogincual District, see Figure 1. In Memba, konzo epidemics have been associated with drought, beginning with a severe drought in 1981. In Mogincual, a large epidemic was associated with war in 1992–3. Both Cava and Mujocojo have continued to record sporadic cases, with Cava suffering an epidemic due to drought in 2005. Acordos de Lusaka is now close to a booming economic area and only two new cases have been reported in the past two decades. One health worker in each village was involved in the project. All suspected konzo cases in each village were identified by community leaders and examined by Dr Nhassico to assess their gait, reflexes and the presence of ankle clonus. History of onset was also recorded. They were confirmed as konzo cases if they met the WHO criteria (World Health Organisation, 1996). All konzo cases were encouraged to do basic rehabilitation exercises.

Fifty women in each village were surveyed with regard to the various types of food they had eaten the previous day. Fifty urine samples were obtained from school children in each village with the oral consent of community leaders and school teachers and were analyzed for thiocyanate. Cassava flour samples (42–50 from each village) were obtained from families and were analyzed for total cyanide. The team trained 30 senior women from each village to use the wetting method and they in turn trained 5–8 other women in the village. This resulted in the training of 150–240 women in each village, with a lower number being trained in Cava. To facilitate use of the wetting method, laminated posters in Portuguese describing the

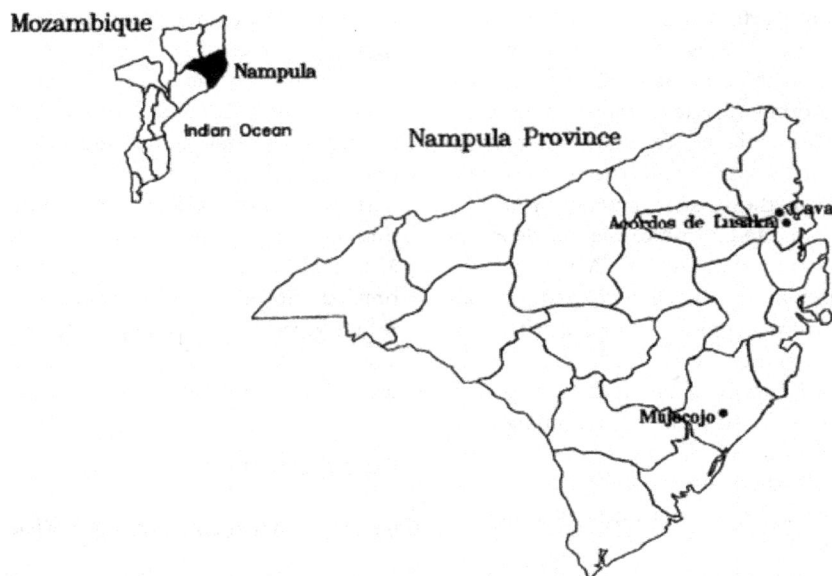

Figure 1. Map of Nampula Province showing district boundaries and the location of the three villages. The inset map shows the location of Nampula Province within Mozambique.

wetting method were distributed to 190–270 women in each village (Bradbury et al. 2011) and an equal number of bowls, knives and mats.

Subsequently over the period September 2012–March 2013 Ms Majonda's team from Nampula made three monitoring visits to the three villages to encourage the women to keep using the wetting method and in each village visited 25–30 households. In March 2013 the second visit of the full team was made with a further check on the use of the wetting method by the women, urinary thiocyanate analyses made on 50 samples of urine from school children in each village and cyanide analyses made on cassava flour samples just before they were used to make the daily porridge. This was followed by three monitoring visits from Ms Majonda's team. In September 2013 the third visit of the full team was made to the three villages and urinary thiocyanate and cassava flour analyses made. This was followed by two monitoring visits to encourage women to keep using the wetting method. The fourth and final visit of the full team to collect flour and urine samples as before was in August 2014.

Urinary thiocyanate analysis

About 50 urine samples were collected from school age children in each village and these samples were analyzed using the simple picrate thiocyanate kit D1 (Haque and Bradbury 1999) http://biology.anu.edu.au/hosted_sites/CCDN/, which contains a color chart with 10 shades of color from yellow to brown that correspond to 0 to 1720 µmol thiocyanate/L.

Flour cyanide analysis

Samples of cassava flour (40–50) were collected from households in each village before teaching the wetting method and at subsequent visits just before they were used to make nchima. Total cyanide analyses were made using kit B2 (Egan et al. 1998; Bradbury et al. 1999). http://biology.anu.edu.au/hosted_sites/CCDN/.

A color chart was used with 10 shades of color from yellow to brown corresponding to 0–800 mg HCN equivalents/kg cassava flour (ppm).

Results and Discussion

In Table 1 is given the number of konzo cases in each village and the percentage konzo prevalences calculated from the population data. There are 77 konzo cases with

Table 1. Population, number of konzo cases, and % konzo prevalence in Nampula Province villages.

Village	Population	Number of konzo cases	% Konzo prevalence
Cava	1654	48	2.9
Acordos de Lusaka	1618	12	0.74
Mujocojo	2918	17	0.58
Total	6190	77	1.2

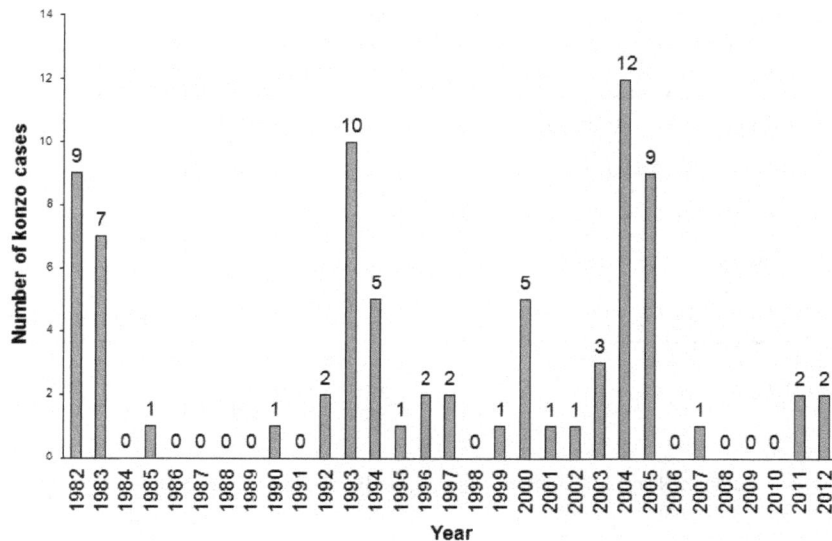

Figure 2. Annual distribution of onset of konzo cases in the three Nampula villages.

a konzo prevalence ranging from 0.58% in Mujocojo to 2.9% in Cava. The annual distribution of konzo cases given in Figure 2 shows that the konzo incidence since the last drought in 2005 is low, with one case in 2007 from Cava and two cases in 2011 (one from Cava and one from Mujocojo) and two cases from Mujocojo in 2012. These are cases of "persistent" or "endemic" konzo (Ernesto et al. 2002) that occur in years of adequate rainfall, but in drought years these areas are at risk of konzo epidemics. In Figure 2, taking into account the passage of time and some unreliability in remembering the year of onset, the peaks in 1982–3 are due to severe drought in Acordos de Lusaka and Cava, (Ministry of Health Mozambique, 1984) in 1993–4 due to war and its aftermath in Mujocojo, in 2000 probably due to a low rainfall period in 1998–9, (Cardoso et al. 2005) and in 2004–5 due to a drought in Cava. Because there has been a period of adequate rainfall since 2005 there are very few new cases of konzo in these villages and hence much less incentive for village women to remove konzo-producing cyanogens, than for village women in DRC where many new konzo cases are occurring every year (Banea et al. 2012, 2013, 2014b). These cases are the result of high cyanide intake from inadequately processed cassava flour made from cassava roots soaked for only 1–2 days instead of the 3–4 days that are needed to remove cyanogens.

Table 2 shows the food consumption patterns in the three villages at the beginning of the intervention in September 2012, at the peak of the cassava harvest. As expected there is a very great dependence on cassava flour eaten as a thick porridge (nchima) with 63–92% of families having consumed nchima on the previous day and 14–40%

Table 2. The percentage of families that ate a particular food on the previous day.[1]

Food	Percentage of families that ate a particular food on the previous day from		
	Cava	Acordos de Lusaka	Mujocojo
"Nchima" from			
Cassava flour	62	72	92
Fish	38	32	36
"Matapa"[2]	14	40	20
Beans	20	20	2
Maize meal	20	16	0
Chicken	12	2	16
Rice	10	10	2
"Minane"[3]	0	0	14
Goat meat	10	0	0

[1]Data collected in September 2012 and arranged in descending order.
[2]"Matapa" consists of pounded and then boiled cassava leaves, cooked with peanuts, and prawns.
[3]Minane" is a local wild tuber (Cardoso et al. 2004).

cassava leaves pounded and boiled to remove cyanogens (Bradbnury and Denton 2014). The combined fish, chicken and goat consumption is lower at 34–60% as expected and other components of the diet are much smaller in amount. The diet is very restricted and heavily focused on cassava flour and cassava leaves. In the monitoring visits it was found that only 30–40% of households were using the wetting method in Acordos de Lusaka and Muocojo and a lower percentage in Cava, possibly due to difficulties in implementation.

The results of the thiocyanate analyses of the urine of school children are shown in Tables 3 and 4. The lower

Table 3. Mean thiocyanate content (μmol/L)[1] of urine of school children in Nampula Province villages before introducing the wetting method in September 2012 and during the intervention.

	Mean urinary thiocyanate content (μmol/L) in			
Village	September 2012[2]	March 2013	September 2013	August 2014
Cava	530 (460)	100 (130)	280 (290)	180 (200)
Acordos de Lusaka	200 (140)	130 (180)	140 (180)	60 (50)
Mujocojo	160 (140)	70 (60)	60 (50)	150 (90)

[1]Standard deviation in brackets.
[2]Before introduction of the wetting method.

Table 4. Percentage of school children in Nampula Province villages with urinary thiocyanate contents of >350 μmol/L.

	Percentage of children with urinary thiocyanate content of >350 μmol/L			
Village	September 2012[1]	March 2013	September 2013	August 2014
Cava	52	2	20	17
Acordos de Lusaka	10	4	3	0
Mujocojo	6	0	0	4

[1]Before introduction of the wetting method.

results obtained in March 2013 are seasonal, due to the fact that the peak cassava harvesting season is August-October and less cassava is being consumed in March (Casadei et al. 1990; Cliff et al. 2011). There is a downward trend from 2012 to 2014 in the mean values for the August-September results, with the exception of Mujocojo which was lowest in September 2013. A similar trend is found in Table 4, which shows the percentage of children with high urinary thiocyanate levels (>350 μmol/L), who are considered to be at risk of contracting konzo (Banea et al. 2012, 2015a). There is also a downward trend in the mean total cyanide content of cassava flour shown in Table 5 for all villages. At the completion of the intervention there are 0% and 4% of children in Acordos de Lusaka and Mujocojo, respectively, with high urinary thiocyanate levels (Table 4), which is consistent with the results obtained at the conclusion of four previous interventions in DRC (Banea et al. 2012, 2013, 2014b, 2015a) all of which prevented new cases of konzo from occurring. Combined with the reduction in the mean total cyanide content in flour in Acordos de Lusaka and Mujocojo to 17 ppm and 9 ppm, respectively, this is a satisfactory result. The FAO/WHO maximum safe level for cassava flour is 10 ppm (FAO/WHO, 1991).

Table 5. Mean total cyanide content (ppm) of cassava flour in the Nampula villages.[1]

	Mean total cyanide content (ppm) of cassava flour samples in			
Village	September 2012[2]	March 2013[3]	September 2013[3]	August 2014[3]
Cava	64 (52)	66 (42)	56 (56)	25 (19)
Acordos de Lusaka	27 (30)	26 (18)	6 (3)	17 (10)
Mujocojo	17 (15)	24 (14)	20 (12)	9 (6)

[1]Standard deviation in brackets.
[2]Samples taken before the wetting method was introduced.
[3]Samples taken just before flour was used to make nchima.

In contrast, the results for Cava are more challenging with 17% of children with high urinary thiocyanate levels of >350 μmol/L at completion of the intervention and a mean total cyanide content in cassava flour of 25 ppm. Nevertheless, these levels show a great improvement for Cava compared with the initial values in September 2012 in Tables 3–5. In the six villages in the DRC it was found that the percentage of families using the wetting method was inversely proportional to the mean urinary thiocyante content of the children and the percentage of households using the method was 68–94% (Banea et al. 2015a). In Acordos de Lusaka and Mujocojo only 30–40% used the wetting method with a lower percentage in Cava, which explains why the results in Cava were not as good as in the other two villages. The much lower percentage use of the wetting method by the women in Mozambique compared with the women in the DRC (Banea et al. 2015a) results from lack of incentive to use the method, because of the relative absence of new cases of konzo in the Mozambique villages compared with the DRC villages.

Conclusion

Konzo outbreaks in East Africa occur under drought conditions, due to large increases by the cassava plant in the cyanide content of cassava roots and hence of cassava flour made from these roots (Cardoso et al. 2005) and also due to war, where postharvest processing may be shortened and processed cassava flour stolen (Cliff et al. 2011). The wetting method has been introduced into three villages in Mozambique with previous konzo outbreaks (see Fig. 2). The uptake of the method by households has been 30–40% in Acordos de Lusaka and Mujocojo and less in Cava, compared with 68–94% in the DRC, (Banea et al. 2015a) because there is little incentive for village women to use the method during a period of normal rainfall, when there are few or no new konzo cases. In contrast, konzo is occurring every year on an increasing scale in Kwango District, Bandundu Province, DRC (Okitundu et al. 2014) and hence village women have more readily accepted the wetting method to reduce cyanide intake and prevent konzo (Banea et al. 2012, 2013, 2015a). We believe that the wetting method should be taught in all those areas in Mozambique and Tanzania where konzo has occurred, so that the wetting method will be fully understood and accepted by village women and be used by them to prevent cyanide intoxication and outbreaks of konzo during future droughts.

Acknowledgments

We thank the Australian Agency for International Development (AusAID) for financial support without which this project would not have been possible. We also thank IIAM, Nampula for working with us in the administration of the grant and Direccao Provincial de Saude, Nampula for their cooperation and help.

Conflict of Interest

None declared.

References

Banea, J. P., D. Nahimana, C. Mandombi, J. H. Bradbury, I. C. Denton, and N. Kuwa. 2012. Control of konzo in DRC using the wetting method on cassava flour. Food Chem. Toxicol. 50:1517–1523.

Banea, J. P., J. H. Bradbury, C. Mandombi, D. Nahimana, I. C. Denton, N. Kuwa, et al. 2013. Control of konzo by detoxification of cassava flour in three villages in the Democratic Republic of Congo. Food Chem. Toxicol. 60:506–513.

Banea, J. P., J. H. Bradbury, C. Mandombi, D. Nahimana, I. C. Denton, N. Kuwa, et al. 2014a. Effectiveness of wetting method for control of konzo and reduction of cyanide poisoning by removal of cyanogens from cassava flour. Food Nutr. Bull. 35:28–32.

Banea, J. P., J. H. Bradbury, C. Mandombi, D. Nahimana, I. C. Denton, M. P. Foster, et al. 2014b. Prevention of konzo in the DRC using the wetting method and correlation between konzo incidence and percentage of children with high urinary thiocyanate level. Afr. J. Food Sci. 8:297–304.

Banea, J. P., J. H. Bradbury, C. Mandombi, D. Nahimana, I. C. Denton, M. P. Foster, et al. 2015a. Konzo prevention in six villages in the DRC and the dependence of konzo prevalence on cyanide intake and malnutrition. Toxicol. Rep. 2:609–615.

Banea, J. P., J. H. Bradbury, D. Nahimana, I. C. Denton, N. Mashukano, and N. Kuwa. 2015b. Survey of the konzo prevalence of village people and their nutrition in Kwilu District, Bandundu Province, DRC. Afr. J. Food Sci. 9:45–50.

Boivin, M. J., D. Okitundu, G. M. Bumoka, M. T. Sombo, D. Mumba, T. Tylleskar, et al. 2013. Neuropsychological effects of konzo: a neuromotor disease associated with poorly processed cassava. Pediatrics 131:e1231.

Bokanga, M., I. J. Ekanayake, A. G. O. Dixon, and M. C. M. Porto. 1994. Genotype-environment interactions for cyanogenic potential in cassava. Acta Hort. 375:131–139.

Bradbnury, J. H., and I. D. Denton. 2014. Mild method for removal of cyanogens from cassava leaves with retention of vitamins and protein. Food Chem. 158:417–420.

Bradbury, J. H. 2006. Simple wetting method to reduce cyanogen content of cassava flour. J. Food Comp. Anal. 19:388–393.

Bradbury, J. H., and I. C. Denton. 2010. Rapid wetting method to reduce cyanogen content of cassava flour. Food Chem. 121:591–594.

Bradbury, M. G., S. V. Egan, and J. H. Bradbury. 1999. Determination of all forms of cyanogens in cassava roots and cassava products using picrate paper kits. J. Sci. Food Agric. 79:593–601.

Bradbury, J. H., J. Cliff, and I. C. Denton. 2011. Uptake of wetting method in Africa to reduce cyanide poisoning and konzo from cassava. Food Chem. Toxicol. 49:539–542.

Cardoso, A. P., M. Ernesto, D. Nicala, E. Mirione, L. Chavane, H. Nzwalo, et al. 2004. Combination of cassava flour cyanide and urinary thiocyanate measurements of school children in Mozambique. Int. J. Food Sci. Nutr. 55:183–190.

Cardoso, A. P., E. Mirione, M. Ernesto, F. Massaza, J. Cliff, M. R. Haque, et al. 2005. Processing of cassava roots to remove cyanogens. J. Food Comp. Anal. 18:451–460.

Casadei, E., J. Cliff, and J. Neves. 1990. Surveillance of urinary thiocyanate concentration after epidemic spastic paraparesis in Mozambique. J. Trop. Med. Hyg. 93:257–261.

Cliff, J., J. Martensson, P. Lundquist, H. Rosling, and B. Sorbo. 1985. Association of high cyanide and low sulphur intake in cassava induced spastic paraparesis. Lancet 11:1211–1213.

Cliff, J., H. Muquingue, D. Nhassico, H. Nzwalo, and J. H. Bradbury. 2011. Konzo and continuing cyanide intoxication from cassava in Mozambique. J. Chem. Toxicol. 49:631–635.

Cumbana, A., E. Mirione, J. Cliff, and J. H. Bradbury. 2007. Reduction of cyanide content of cassava flour in Mozambique by the wetting method. Food Chem. 101:894–897.

Egan, S. V., H. H. Yeoh, and J. H. Bradbury. 1998. Simple picrate paper kit for determination of the cyanogenic potential of cassava flour. J. Sci. Food Agric. 76:39–48.

Ernesto, M., A. P. Cardoso, D. Nicala, E. Mirione, F. Massaza, J. Cliff, et al. 2002. Persistent konzo and cyanide toxicity from cassava in northern Mozambique. Acta Trop. 82:357–362.

FAO/WHO. 1991. Joint FAO/WHO Food Standards Programme, Codex Alimentarius Commission XII, Supplement 4, FAO, Rome, Italy.

Haque, M. R., and J. H. Bradbury. 1999. Simple method for determination of thiocyanate in urine. Clin. Chem. 45:1459–1464.

Howlett, W. P., G. R. Brubaker, N. Mlingi, and H. Rosling. 1990. Konzo, an epidemic upper motor neuron disease studied in Tanzania. Brain 113:223–235.

King, N. L. R., and J. H. Bradbury. 1995. Bitterness of cassava: identification of a new apiosyl glucoside and other compounds that affect its bitter taste. J. Sci. Food Agric. 68:223–230.

Ministry of Health Mozambique. 1984. Mantakassa: an epidemic of spastic paraparesis associated with chronic cyanide intoxication in a cassava staple area in Mozambique. 1. Epidemiology and clinical laboratory findings in patients. Bull. WHO 62:477–484.

Mlingi, N. L. V., S. Nkya, S. R. Tatala, S. Rashid, and J. H. Bradbury. 2011. Recurrence of konzo in southern Tanzania: rehabilitation and prevention using the wetting method. Food Chem. Toxicol. 49:673–677.

Muquingue, H., D. Nhassico, J. Cliff, L. Sitoe, A. Tonela, and J. H. Bradbury. 2005. Field trial in Mozambique of a new method for detoxifying cyanide in cassava products. CCDN News 6:3–4.

Nhassico, D., H. Muquingue, J. Cliff, A. Cumbana, and J. H. Bradbury. 2008. Rising African cassava production, diseases due to high cyanide intake and control measures. J. Sci. Food Agric. 88:2043–2049.

Nzwalo, H., and J. Cliff. 2011. Konzo: from poverty, cassava and cyanogen intake to toxico-nutritional neurological disease. PLoS Negl. Trop Dis. 5:e1051.

Okitundu, D., I. E. Andjafono, G. B. Makila- Mabe, M. T. S. Ayanna, J. K. Kikandau, N. Mashukano, et al. 2014. Persistance des epidemies de konzo a Kahemba, Republique Democratique du Congo: aspects phenomenologiques et socio-economiques. Pan Afr. Med. J. 18:213–221.

World Health Organisation. 1996. Konzo: a distinct type of upper motor neuron disease. Wkly Epidemiol. Rec. 71:225–232.

8

Determination of vitamins, minerals, and microbial loads of fortified nonalcoholic beverage (*kunun zaki*) produced from millet

Olusegun A. Olaoye, Stella C. Ubbor & Ebere A. Uduma

Department of Food Science and Technology, Michael Okpara University of Agriculture, Umudike, Abia State, Nigeria

Keywords
Kunun zaki, microbial loads, nutritional deficiency, sensory evaluation, tigernut milk extract

Correspondence
Olusegun A. Olaoye, Department of Food Science and Technology, Michael Okpara University of Agriculture, Umudike, Abia State, Nigeria.

E-mail: olaayosegun@yahoo.com

Funding Information
No funding information provided.

Abstract

The objective of this study was to evaluate the possibility of fortifying *kunun zaki* with tigernut milk extract due to nutritional deficiency of the former. *Kunun zaki* and tigernut milk extract (TME) were produced using traditional methods, with little modification. They were mixed in respective percentages of 90:10 (KN10), 80:20 (KN20), and 70:30 (KN30) while whole *kunun zaki* without addition of tigernut milk extract (KN00) served as control. The resulting *kunun zaki* samples were analyzed for proximate composition, vitamins, minerals, microbial loads, and sensory evaluation. Results showed improvement in thiamine and riboflavin contents of the fortified samples over the unfortified counterparts, with the KN30 sample having highest values of 1.05 and 0.56 mg/kg thiamine and riboflavin, respectively. Minerals were higher in the samples containing TME than their KN00 counterparts; the KN30 sample had highest values of 23.5, 8.8, 148.9, 63.7, 6.7, and 18.6 mg/100 mL for respective Na, Ca, K, Mg, P, and Fe while lowest values were recorded for the KN00 sample. Microbial analysis indicated that total viable bacteria and yeast and molds were in the range 2.2–2.6 and 2.1–2.7 log CFU/g, respectively, while there was no detection of coliforms and Staphylococcus in the samples. The sensory evaluation of the *kunun zaki* samples indicated that higher mean scores were recorded for samples containing TME than those without it in most of the attributes tested. The KN30 sample was most preferred, having highest mean scores of 7.2, 7.8, 6.9, and 7.4 in the attributes of appearance, flavor, taste, and acceptability, respectively. The study concluded that inclusion of tigernut extract in *kunun zaki* resulted in improved nutritional and sensory qualities.

Introduction

Kunun zaki is an energy-dense beverage normally prepared from germinated cereals and is very popular in Nigeria, especially in the Northern part. The beverage is produced from either one or more of fermented millet, sorghum, guinea-corn, and maize. *Kunun zaki* has thirst quenching properties and is therefore extensively consumed during the dry season, though consumption may be observed throughout the year (Adelekan et al. 2013).

Millet, sorghum, and maize grains are the three principal cereals from which *kunun zaki* can be produced (Adeleke

et al. 2004). It is usually flavored with such spices as ginger, black pepper, and tamarind for improvement in its taste and aroma, and also to serve as purgative and cure for flatulent conditions. It is a considerably cheap beverage drink because of the ingredients used for preparation, and this makes the product readily available (Makinde and Oyeleke 2012). The process of production involves wet milling of the cereal, wet sieving, partial gelatinization of the slurry, mild fermentation, sugar addition, and bottling.

The fermentation process may last for 12–72 h (Gaffa and Ayo 2002), after which it is kept for acidification to develop. Brief fermentation, involving mainly lactic acid bacteria and yeast, usually occurs during steeping of the grains in water over 8–48 h. Wide varieties exist in the methods of preparation depending on taste, cultural norms, and habits (Abulude et al. 2006). Additives may be used in fortification of the beverage to compensate for losses during processing, or add nutrients that are either present at low level or not present at all; thus resulting in improved nutritional quality.

Tigernuts (*Cyperus esculentus*) are cultivated throughout the world including Nigeria, especially in the northern part, and other West Africa Countries like Guinea, Cote d'ivore, Cameroon, Senegal, America and other parts of the World (Belewu and Abodunrin 2006). The nuts are valued for their highly nutritious starch content, dietary fiber and carbohydrate (mono, di and polysaccharides). The nut has also been reported to be rich in sucrose (17.4–20.0%), fat (25.50%), protein (8%), and minerals such as sodium, calcium, potassium, and magnesium (Umerie and Enebeli 1997).

Kunun zaki is often used as a weaning food beverage for infants in Nigeria. However, since the beverage is produced majorly from cereals, it could be deficient in nutritional quality, especially protein, vitamins, and minerals, and hence supplementation with richer sources of nutrients may be required (Adelekan et al. 2013).

As a result of the low nutritional value of this beverage, research efforts are required to ensure its improvement in nutritional quality. A good approach could be the exploitation of tigernuts in complementing nutritional deficiency that is associated with *kunun zaki*. This study was therefore undertaken to incorporate tigernut milk extract into the beverage for possible nutritional improvement.

Materials and Methods

Source of materials

The millet grains and sweet potatoes used in this study were obtained from National Stored Products Research Institute (NSPRI), Port Harcourt, Rivers State, and National Root Crop Research Institute (NRCRI), Umudike, Abia State, respectively. Tigernuts (*Cyperus esculentus*), spices (ginger and pepper), and sugars were purchased from a local market in Umuahia Township, Abia State, Nigeria.

Production of *kunun zaki*

The modified method of Ayo et al. (2013) was adopted in the production of *kunun zaki* (Fig. 1). One kilogram (1 kg) of cleaned millet grains was washed and steeped in clean water for 48 h to soften the seed. The grains were washed to remove stones and wet milled along with added spices (65 g ginger, 10 g red pepper and 15 g

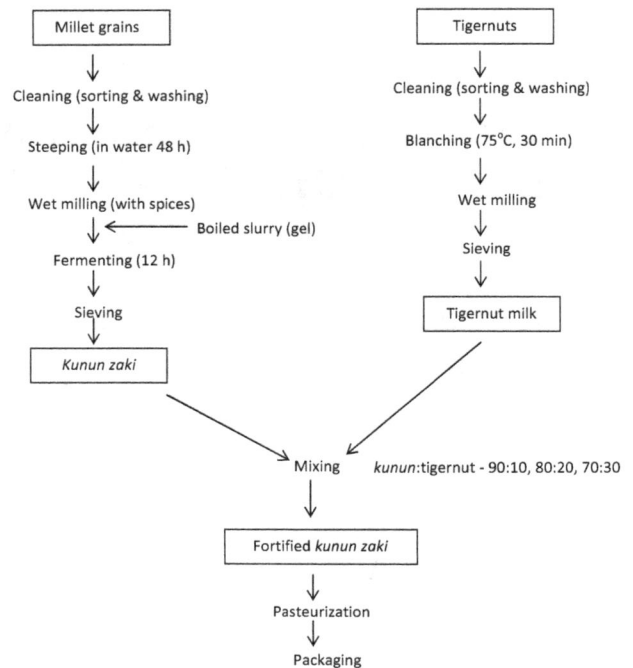

Figure 1. Flowchart for production of fortified *kunun zaki*.

sweet potatoes) into slurry. Two-third of the slurry was mixed with 2500 mL of boiling water and stirred to form a gel; this was allowed to cool for 3 h. The remaining one-third of the slurry was added to the gel, mixed with cold boiled water (1000 mL) and left open to ferment for 12 h. It was then sieved with a muslin cloth and the filtrate was sweetened with sucrose (250 g).

Production of tigernut milk

Tigernut milk extract was produced according to the method of Odunfa and Adeyeye (1985). Dried tigernuts (1 kg) were washed and soaked in tap water (1 L) at room temperature (30°C) for three days; changing of the soaking water was observed 24 hourly. The tigenuts were washed and then boiled in 0.2% (w/v) solution of sodium bicarbonate for 30 min to reduce objectionable flavor. The nuts were drained, mixed with water (ratio 1:4) and milled (Philips kenwood, UK). The homogenous slurry was filtered using a muslin cloth and the resultant filtrate was tigernut milk (Fig. 1).

Formulation of enriched *kunun zaki*

Kunun zaki and tigernut milk were mixed in three different proportions of 90:10, 80:10, and 70:30 (v/v) of the respective products, and coded as KN10, KN20, and KN30, respectively. A sample of *kunun zaki* with no addition of tigernut milk served as control (KN00). The various samples were pasteurized at 70°C for 30 min.

Proximate analysis

Proximate analysis of moisture, ash, fat, and protein contents of *kunun zaki* samples was carried out using the methods of Association of Official Analytical Chemists (AOAC 2005). Carbohydrate was determined by difference.

Physical properties and vitamin contents

Total solid was determined by evaporating 25 mL of *kunun zaki* on a boiling water bath which was followed by drying to constant weight in an oven at 130°C for 2–3 h.

$$\% \text{ Total solid} = \frac{\text{Dry weight} \times 100}{\text{Weight of sample}}$$

Specific gravities of the samples were determined as described by Pearson (1976). Vitamin C, thiamine, riboflavin, and niacin were determined according to the methods of AOAC, Association of Official Analytical Chemists (2005).

Mineral analysis

Analysis of potassium content of the samples was carried out using flame photometry, while phosphorus was determined by the phosphovanado-molybdate method (AOAC, Association of Official Analytical Chemists 2005). The other elemental contents (Na, Ca, Mg, and Fe) were determined, after wet digestion of sample ash with an Atomic Absorption Spectrophotometer (AAS, Hitachi Z6100, Tokyo, Japan). All determinations were carried out in triplicates.

Microbial analysis

Ten milliliter of *kunun zaki* samples was thoroughly mixed in 90 mL sterile distilled water to obtain 10^{-1} dilution, from which further dilutions were made. One milliliter of appropriate dilutions was mixed with molten medium (45°C)

using potato dextrose agar (PDA) for molds; malt extract agar (MEA) supplemented with streptomycin for yeast; MacConkey agar for coliforms; Nutrient agar (NA) for total viable bacteria; and Mannitol salt agar for *Staphylococcus*. Incubation period was 48 h at 37°C except for yeast and molds (25°C, 72 h). Determinations were carried out in triplicates and counts were expressed in logarithmic of colony-forming unit per mL of sample (log CFU/mL)

Sensory evaluation

The *kunun zaki* samples were subjected to sensory evaluation for the attributes of appearance, viscosity, aroma, taste, and acceptability. A semitrained twenty member panel was used and scores were allocated to the attributes based on a 9-point hedonic scale ranging from 1 (dislike extremely) to 9 (like extremely). The data collected were subjected to statistical analysis to determine possible differences among samples.

Statistical analysis

Data which depended on different percentages of *kunun zaki* and tigernut milk extract were analyzed according to a completely randomized design with three replicates. Data were subjected to variance analyses and differences between means were evaluated by Duncan's multiple range test using SPSS statistic programme, version 10.01 (SPSS 1999). Significant differences were expressed at $P < 0.05$.

Results and Discussion

Table 1 shows the result of proximate analysis carried out on the *kunun zaki* samples. pH values of the products were between 5.7 and 6.0. No significant differences ($P < 0.05$) were recorded in the pH values among samples. A number of the pH values recorded in this study were slightly higher than those reported in previous studies

Table 1. Proximate analysis of the *kunun zaki* samples.

Kunun	Proximate parameters					
	pH	MC (%)	Protein (%)	Ash (%)	Fiber (%)	Carbohydrate (%)
KN00	$5.7^a \pm 0.0$	$87.27^b \pm 0.14$	$4.96^c \pm 0.10$	$2.09^b \pm 0.02$	$0.84^a \pm 0.02$	$3.8^a \pm 0.23$
KN10	$5.9^a \pm 0.0$	$87.71^b \pm 0.13$	$5.31^b \pm 0.20$	$2.93^{ab} \pm 0.02$	$0.81^a \pm 0.04$	$2.99^b \pm 0.38$
KN20	$6.0^a \pm 0.0$	$88.41^a \pm 0.12$	$5.72^b \pm 0.10$	$3.31^a \pm 0.03$	$0.78^a \pm 0.11$	$1.72^c \pm 0.12$
KN30	$6.0^a \pm 0.0$	$89.37^a \pm 0.08$	$6.14^a \pm 0.15$	$3.44^a \pm 0.02$	$0.75^a \pm 0.05$	$0.14^d \pm 0.00$

MC, moisture content; KN00, unfortified *kunun zaki*; KN10, *kunun zaki* fortified with 10% tigernut milk extract; KN20 *kunun zaki* fortified with 20% tigernut milk extract; KN30, *kunun zaki* fortified with 30% tigernut milk extract.
Values are mean scores of three replicated samples. Values in columns with different superscript letters are significantly different ($P < 0.05$).

(Agarry et al. 2010; Ayo et al. 2013); probably due to slight difference in fermentation period and cereal types as well as processing conditions used in the studies. Agarry et al. (2010) recorded pH value of 5.44 in *kunun zaki* produced from millet and this is similar to those recorded in the present study. Lower values were obtained by the authors when cereal marsh was treated with cultures of lactic acid bacteria prior to fermentation. Moreover, Belewu and Abodunrin (2006) recorded a pH of about 6 which is also similar to those recorded in the present study. The pH of about 6.0 or slightly below is usually associated with the beverage as a result of slight fermentation that is associated during production. *Kunun zaki* enriched with tigernut milk extracts (TME) recorded higher protein contents than the control containing no tigernut milk extract (KN00). The highest protein content of 6.14% was observed in sample with 30%v/v TME while lowest value of 4.96 was obtained for control sample. Increase in protein content was noticed as percentage tigernut milk extract increased, suggesting that the tigernut milk extract contributed to the protein contents of *kunun zaki* samples. This is as a result of significant differences ($P < 0.05$) recorded in the protein contents of the samples containing TME in comparison to the control. The protein content of KN00 was higher than the value reported by Ayo et al. (2013) and this may be attributed to the different cereals adopted during the production of the beverage. The improved protein contents of the *kunun zaki* samples containing TME recorded in this study may be beneficial to consumers of the product as majority of people in Nigeria could not afford protein sources of foods such as meat and egg for economic reasons.

The result of the ash contents of the beverage samples was similar to those of the protein contents; ash contents increased correspondingly with increase in incorporation of TME. The highest ash content of 3.44 was observed in sample containing 30% tigernut extract while control sample had the lowest value (3.09). It is interesting to note that tigernut contributed significantly to ash contents

of the beverage as a results of higher values recorded in samples containing TME than the control sample; significant differences were also noted between the samples. In a related study, Makinde and Oyeleke (2012) reported increase in the ash contents of *kunun zaki* enriched with extract of sesame seeds over the control sample without the extract. Values of the ash contents in the *kunun zaki* recorded by the research workers are similar to those obtained in the present study. However, Ogbonna et al. (2013) and Adelekan et al. (2013) obtained ash contents of higher values in their findings on *kunun zaki* than those recorded in this study. The difference could be attributed to the different types of cereals used in the production of the beverage in the different studies. Different cereal types have abilities to contribute to the ash contents of *kunun zaki* as a result of the differences in their ash compositions.

The crude fibers of the samples remain statistically insignificant ($P > 0.05$). Although values seemed to decrease with increase in incorporation of TME, however the addition of the extract did not have any significant effect. The control sample had the highest value of 0.84 compared to the lowest (0.75) obtained for sample containing 30% TME. This could be as a result of higher content of fiber contained in millet than tigernut (Belewu and Abodunrin 2006).

The physicochemical properties and vitamin contents (Table 2) of the *kunun zaki* samples indicate that total solids ranged between 10.66 and 12.31, with the control having the highest value while sample containing 30% TME had the lowest. In a study carried out by Ayo et al. (2013) on the production of *kunun zaki*, higher values of total solids were recorded than those obtained in the present study. The difference may be as a result of the difference in the recipes used during production; lower quantity of water was used by Ayo et al. (2013) than that used in the present study. A contrary observation to the trend of total solids was made for the specific gravity which ranged from 0.73 and 0.85; specific gravity

Table 2. Physical properties and vitamin contents of the *kunun zaki* samples.

Kunun	Physical properties and vitamins					
	TS (%)	SG (g/cm³)	Vitamin C(mg/100 g)	Thiamine (mg/kg)	Riboflavin (mg/kg)	Niacin (mg/kg)
KN00	12.31[a] ± 0.66	0.73[a] ± 0.03	18.77[a] ± 1.02	0.71[bc] ± 0.13	0.35[c] ± 0.06	1.17[a] ± 0.12
KN10	12.28[a] ± 0.13	0.77[a] ± 0.00	13.49[b] ± 1.02	0.69[c] ± 0.04	0.36[c] ± 0.05	0.45[c] ± 0.20
KN20	11.64[b] ± 0.14	0.81[a] ± 0.00	12.91[c] ± 2.02	0.78[b] ± 0.06	0.43[b] ± 0.06	0.61[b] ± 0.02
KN30	10.66[c] ± 0.08	0.85[b] ± 0.00	12.91[c] ± 1.02	1.05[a] ± 0.06	0.56[a] ± 0.06	0.59[b] ± 0.04

TS, total solids; SG, specific gravity; KN00, unfortified *kunun zaki*; KN10, *kunun zaki* fortified with 10% tigernut milk extract; KN20 *kunun zaki* fortified with 20% tigernut milk extract; KN30, *kunun zaki* fortified with 30% tigernut milk extract.
Values are mean scores of three replicated samples. Values in columns with different superscript letters are significantly different ($P < 0.05$).

Table 3. Mineral contents of the *kunun zaki* samples.

Kunun	Minerals (mg/100 mL)					
	Na	Ca	K	Mg	P	Fe
KN00	$21.5^b \pm 1.2$	$5.6^c \pm 1.2$	$140.2^b \pm 12.9$	$34.9^d \pm 3.2$	$2.1^d \pm 0.0$	$9.3^b \pm 5.4$
KN10	$22.0^b \pm 2.1$	$5.8^c \pm 0.9$	$143.8^{ab} \pm 3.1$	$44.5^c \pm 0.1$	$3.5^c \pm 0.2$	$13.8^b \pm 0.8$
KN20	$22.1^b \pm 0.8$	$6.5^b \pm 1.7$	$144.5^{ab} \pm 7.2$	$49.1^b \pm 1.7$	$4.9^b \pm 1.1$	$15.2^a \pm 3.2$
KN30	$23.5^a \pm 4.1$	$8.8^a \pm 1.0$	$148.9^a \pm 9.4$	$63.7^a \pm 3.8$	$6.7^a \pm 0.6$	$18.6^a \pm 2.7$

KN00, unfortified *kunun zaki*; KN10, *kunun zaki* fortified with 10% tigernut milk extract; KN20 *kunun zaki* fortified with 20% tigernut milk extract; KN30, *kunun zaki* fortified with 30% tigernut milk extract.
Values are mean scores of three replicated samples. Values in columns with different superscript letters are significantly different ($P < 0.05$).

increased with increase in addition of TME. Adelekan et al. (2013) recorded a similar value of about 0.75 for specific gravity in *kunun zaki* samples.

Moreover, increase in the contents of thiamine and riboflavin was recorded in the *kunun zaki* containing TME (Table 2), and this may contribute to nutritional intake of consumers. The control sample had lowest values of 0.71 and 0.35 of thiamine and riboflavin while the sample with 30% TME had highest values of 1.05 and 0.56 for the respective vitamins. The reverse was, however, recorded in the values of vitamin C and niacin. The increase recorded in some of the vitamins, especially thiamine and riboflavin, in the samples could be due to incorporation TME (Bernat et al. 2014).

Analysis of mineral contents of the *kunun zaki* samples indicates that sample without TME recorded the lowest value (21.5) for sodium while the highest (23.5) was obtained for sample containing 30% TME (Table 3). Similar trends were observed for calcium, potassium, magnesium, phosphorus, and iron in the *kunun zaki* samples; the sample with 30% TME recorded highest values while the lowest was obtained for the control sample. The values recorded for the beverage samples were slightly different from those reported by Makinde and Oyeleke (2012) in *kunun zaki*; this could be due to the different cereals adopted for production of the products. Calcium and iron contents were, however, similar to those reported by Adelekan et al. (2013). The report of Ogbonna et al. (2013) also corroborates the values of magnesium and iron recorded in this study. Higher values of the minerals recorded in this study than those reported in other studies could obviously be attributed to the incorporation of TME into the *kunun zaki* beverage. The increase in the minerals contents in the samples containing TME could therefore justify the need to enrich the beverage with sources that are rich in other nutrients lacking in cereals normally adopted in its production (Bernat et al. 2014). Minerals are of great importance in diet as they play important roles in body metabolism. For example calcium helps in the regulation of muscle contractions and transmission of nerve impulses as well as bone and teeth development (Cataldo et al. 1999). Phosphorus has also been reported to be required for bone growth, kidney function, cell growth, and maintaining the body's pH balance (Fallon and Enig 2001). Furthermore, potassium is essential for its important role is the synthesis of amino acids and proteins. Moreover, magnesium helps in relaxation of the muscle and in the formation of strong bones and teeth. It also plays fundamental roles in most reactions involving phosphate transfer, believed to be essential in the structural stability of nucleic acid and intestinal absorption while its deficiency can cause severe diarrhea, hypertension, and stroke (Appel 1999). The increase in the contents of the minerals recorded in the *kunun zaki* samples containing TME could therefore be of nutritional advantage to consumers of the products.

The microbial loads of the *kunun zaki* samples are presented in Table 4. The total viable bacteria were between 2.2 and 2.6 while yeast and molds ranged from 2.4 to 2.7. No coliforms and *Staphylococcus* were detected in the beverage samples. The low values of microbial counts recorded in the *kunun zaki* samples could be due to heat treatment (pasteurization) given to the products during production. The total viable counts recorded in this study

Table 4. Microbial loads (log CFU/g) of the *kunun zaki* samples.

Kunun	Microbial loads			
	TVB	Y & M	Coliforms	*Staphylococcus*
KN00	2.2 ± 0.1^b	2.7^a	ND	ND
KN10	2.4 ± 0.9^{ab}	2.1^b	ND	ND
KN20	2.1 ± 0.6^b	ND	ND	ND
KN30	2.6 ± 0.1^a	2.4^b	ND	ND

TVB, total viable bacteria; Y & M, yeast and molds; KN00, unfortified *kunun zaki*; KN10, *kunun zaki* fortified with 10% tigernut milk extract; KN20 *kunun zaki* fortified with 20% tigernut milk extract; KN30, *kunun zaki* fortified with 30% tigernut milk extract; ND, nondetectable
Values are mean scores of three replicated samples. Values in columns with different superscript letters are significantly different ($P < 0.05$).

Table 5. Sensory attributes of the *kunun zaki* samples.

Kunun	Attributes				
	Appearance	Viscosity	Flavour	Taste	Acceptability
KN00	$6.9^a \pm 0.83$	$7.1^b \pm 0.85$	$4.6^c \pm 0.11$	$5.1^c \pm 0.65$	$6.1^c \pm 0.55$
KN10	$6.8^a \pm 0.80$	$6.7^b \pm 0.85$	$5.9^b \pm 0.90$	$6.0^b \pm 0.69$	$6.7^b \pm 0.63$
KN20	$6.9^a \pm 0.83$	$5.9^b \pm 0.90$	$6.8^{ab} \pm 0.81$	$6.1^b \pm 0.55$	$6.7^b \pm 0.63$
KN30	$7.2^a \pm 0.52$	$5.7^a \pm 0.03$	$7.8^a \pm 0.75$	$6.9^a \pm 0.85$	$7.4^a \pm 0.40$

KN00, unfortified *kunun zaki*; KN10, *kunun zaki* fortified with 10% tigernut milk extract; KN20 *kunun zaki* fortified with 20% tigernut milk extract; KN30, *kunun zaki* fortified with 30% tigernut milk extract.
Values are mean scores of three replicated samples. Values in columns with different superscript letters are significantly different ($P < 0.05$).

were similar to those reported in findings of Ayo et al. (2013) and Adelekan et al. (2013); the authors reported average microbial loads of 3.0 and 2.5, respectively, in *kunun zaki*. The latter also reported an average fungal count of about 2.2 in the beverage and this corroborates those recorded in the present study. The nondetection of coliforms and *Staphylococcus* in the samples could be as a result of good manufacturing and hygiene practices observed during production. Coliforms are majorly of fecal origin and their presence in foods indicates contamination from fecal sources which is highly undesirable; this is because some coliforms such as *Escherichia coli* can cause diseases such as gastroenteritis, diarrhea, and urinary tract infections (Pelczar et al. 1993).

The means scores obtained for sensory evaluation of the *kunun zaki* samples indicate that the sample containing 30% TME recorded highest scores of 7.1, 7.8, 6.9, and 7.4 in the respective attributes of appearance, flavor, taste, and acceptability (Table 5). Significant differences ($P < 0.05$) were recorded between the samples without TME and those with the extract.

From the results of this study, it could be concluded that incorporation of tigernut milk extract into *kunun zaki* gave some degree of fortification of the product as a result of enhanced quantities of protein, ash, and vitamins. There was also improvement in the sensory quality. Inclusion of tigernut milk extract should therefore be encouraged among producers of the beverage drink as a result of the derivable nutritional benefits that consumers can gain. However, good manufacturing and good hygiene practices should be given utmost importance during production to avoid microbial contamination that may cause foodborne illness.

Acknowledgments

Authors appreciate National Stored Products Research Institute (NSPRI), Port Harcourt, Rivers State, and National Root Crop Research Institute (NRCRI), Umudike, Abia State for providing the respective millet grains and sweet potatoes used in this study.

Conflict of Interest

None declared.

References

Abulude, F. O., M. O. Ogunkoya, and V. A. Oni. 2006. Mineral composition, shelf life, and sensory attributes of fortified 'Kunuzaki' beverage. Acta Sci. Pol. Technol. Aliment. 5:155–162.

Adelekan, A. O., A. E. Alamu, N. U. Arisa, Y. O. Adebayo, and A. S. Dosa. 2013. Nutritional, microbiological and sensory characteristics of malted soy-kunu zaki: an improved traditional beverage. Adv. Microbiol. 3:389–397.

Adeleke, O. E., J. O. Olaitan, and O. Olubile. 2004. Microbial isolates from Kunnu-zaki and their antibiotic sensitivities. Adv. Food Sci. 26:168–170.

Agarry, O. O., I. Nkama, and O. Akoma. 2010. Production of *Kunun-zaki* (A Nigerian fermented cereal beverage) using starter culture. Int. Research J. Microbiol. 1:018–025.

AOAC, Association of Official Analytical Chemists. 2005. Official methods of analysis of the Association of Analytical Chemists International, 18th ed. AOAC, Gaithersburg, MD.

Appel, L. J. 1999. Nonpharmacologic therapies that reduce blood pressure: a fresh perspective. Clin. Card. 22:1–5.

Ayo, J. A., V. A. Ayo, B. Yelmi, G. Onuoha, and M. O. Ikani. 2013. Effect of preservatives on microbiological qualities of *kunu zaki*. Int. J. Agric. Sci. Research 2:124–130.

Belewu, M. A., and O. A. Abodunrin. 2006. Preparation of Kunnu from unexploited rich food source: tiger nut (*Cyperus esculentus*). World J. Dairy Food Sci. 1:19–21.

Bernat, N., N. Chafer, A. Chiralt, and C. Gonzalez-Martinez. 2014. Vegetable milks and their fermented derivative products. Int. J. Food Stud. 3:93–124.

Cataldo, C. B., L. K. DeBruyne, and E. N. Whitney. 1999. Nutrition and diet therapy principles and practise 5th edition. Wadsworth Publishing Company and International Thompson Publishing Company, Belmont, Clifornia. pp: 35–204.

Fallon, S., and M. G. Enig. 2001. Nourishing traditions. The cookbook that challenges politically correct nutrition and the diet dictocrats. Revised 2nd Edn. New Trends Publishing, Inc., Washington DC. pp: 40–45.

Gaffa, T., and J. A. Ayo. 2002. Innovations in the traditional *kunun zaki* production process. Pak. J. Nutr. 1:202–205.

Makinde, F., and O. Oyeleke. 2012. Effect of sesame seed addition on the chemical and sensory qualities of sorghum based *kunun- zaki* drink. Afr. J. Food Sci. Technol. 3:204–212.

Odunfa, S. A., and S. Adeyeye. 1985. Microbiological changes during the traditional production of ogibaba, a West African fermented sorghum gruel. J. Cereal Sci. 3:173–180.

Ogbonna, A. C., C. I. Abuajah, M. F. Akpan, and U. S. Udofia. 2013. A comparative study of the nutritional values of palmwine and kunu-zaki. Annals Food Sci. Technol. 14:39–43.

Pearson, D. 1976. The chemical analysis of food. 7th ed. Churchill Livingstone, London.

Pelczar, M. J., C. S. Chane, and R. K. Noel. 1993. Microbiology. 5th ed. McGraw-Hill Publishing Company, New Delhi. p. 272.

SPSS, Statistical Package for Social Sciences. 1999. SPSS 1001 for windows. SPSS Inc., Chicago, Illinois.

Umerie, S. C., and J. N. Enebeli. 1997. Malt caramel from the nuts of *Cyperus esculentus*. J. Bio. Resource Technol. 8:215–216.

9

Quality attributes of sweet potato flour as influenced by variety, pretreatment and drying method

Ganiyat O. Olatunde[1], Folake O. Henshaw[1], Michael A. Idowu[1] & Keith Tomlins[2]

[1]Department of Food Science and Technology, Federal University of Agriculture, Abeokuta, Ogun State, Nigeria
[2]Foods and Markets Department, Natural Resources Institute, University of Greenwich, Medway, Kent, United Kingdom

Keywords
Drying, food quality, pretreatment, sweet potato flour, variety

Correspondence
Ganiyat O. Olatunde, Department of Food Science and Technology, Federal University of Agriculture, Abeokuta, Ogun State, Nigeria.

E-mail: olatundego@funaab.edu.ng

Funding Information
Funding for this study was provided by the Education Trust Fund of the Federal Republic of Nigeria.

Abstract

The effect of pretreatment methods (soaking in water, potassium metabisulphite solution, and blanching) and drying methods (sun and oven) on some quality attributes of flour from ten varieties of sweet potato roots were investigated. The quality attributes determined were chemical composition and functional properties. Data obtained were subjected to descriptive statistics, multivariate analysis of variance, and Pearson's correlation. The range of values for properties of sweet potato flour were: moisture (8.06–12.86 ± 1.13%), starch (55.76–83.65 ± 6.82%), amylose (10.06–21.26 ± 3.92%), total sugar (22.39–125.46 ± 24.68 μg/mg), water absorption capacity (140–280 ± 26), water solubility (6.89–26.18 ± 3.80), swelling power (1.66–5.00 ± 0.50), peak viscosity (24.50–260.92 ± 52.61 RVU), trough (7.08–145.83 ± 34.48 RVU), breakdown viscosity (11.00–125.33 RVU), final viscosity (10.21–225.50 ± 60.55 RVU), setback viscosity (3.04–92.21 RVU), peak time (6.07–9.06 min) and pasting temperature (69.8–81.3°C). Variety had a significant ($P < 0.001$) effect on all the attributes of sweet potato flour. Pretreatment did not significantly ($P > 0.05$) affect moisture, fat and lightness (L*). Drying method did not significantly ($P > 0.05$) affect fiber and L*. The interactive effect of variety, pretreatment and drying method had a significant ($P < 0.001$) effect on all the attributes except fat and fiber. Total sugar correlated significantly ($P < 0.01$) with water solubility ($r = 0.88$) of the flour samples. Variety was a dominant factor influencing attributes of sweet potato flour and so should be targeted at specific end uses.

Introduction

Sweet potato [*Ipomoea batatas* L. (Lam.)] is among the world's most important, versatile and underexploited food crops. Nigeria is the leading producer of sweet potato (SP) in Africa with an estimated average production (1993–2013) of 3.45 million metric tonnes (FAOSTAT, 2013). A large number of SP varieties exist and they differ from one another in the color of flesh, and root skin amongst other attributes (Woolfe 1992; Aina et al. 2009). In Nigeria, the two common local varieties are the purple skin–white fleshed and the yellow skin-yellow fleshed. However, improved varieties including orange-fleshed varieties, with varying genetic and agronomic characteristics are been developed in Nigerian research institutions and released to farmers (Afuape 2009, 2013; Egeonu and Akoroda 2009).

SP has been recognised as having an important role to play in improving household and national food security, health, and livelihoods of poor families in sub-Saharan Africa (CIP, 2013). This may be due to its wide range of agronomic and nutritional advantages such as high yield even in marginal soil conditions, wide ecological adaptability, low input requirements, and shorter growing period than other root crops (Horton et al. 1989). SP produces the highest amount of edible energy per hectare per day (Horton et al. 1989). Despite its high carbohydrate content, it has a low glycemic index, indicating low digestibility of the starch (ILSI, 2008). It is the only starchy staple, which contains appreciable amounts of β-carotene (especially the orange-fleshed varieties), ascorbic acid and amino acid lysine that is deficient in cereal-based diets like rice (Bradbury and Singh, 1986; Bradbury et al., 1985).

Processing of SP roots into stable forms such as chips, flour, or starch, have been recommended as an alternative to the difficulties associated with storage and transportation of the raw roots in developing countries (Peters and Wheatley, 1997; Owori and Agona 2003). Flour produced from SP has the potential for making a variety of food products such as baked goods (bread, cakes, cookies, biscuits); doughnuts, breakfast foods (instant porridge, crisp, flake-type products); noodles or pasta-type products; sauces (soy sauce, ketchup); and brewing adjuncts (van Hal 2000; Mais and Brennan 2008).

Functional quality of flour is important to determine its usefulness in food applications. Processing conditions have been shown to influence functional properties of flour; for example, heat processing have been found to affect the functional properties of taro flour (Tagodoe and Nip 1994), while drying temperature, milling procedure and particle size influenced gelatinization profiles of cassava flour (Fernandez et al. 1996). Factors that have been reported to influence the quality of SP flour are variety (Osundahunsi et al. 2003; Aina et al. 2009), processing steps (van Hal 2000), as well as processing methods such as parboiling (Osundahunsi et al. 2003), blanching (Jangchud et al. 2003), drying techniques (Yadav et al. 2006) and peeling, pretreatments and drying temperatures (Maruf et al. 2010a,b).

Most technical research on SP flour has focused on the development of new food products using SP flour rather than on efficient methods to produce the flour (van Hal 2000). Meanwhile, researchers have reported different characteristics of SP flour processed from different varieties and under different conditions (van Hal 2000; Jangchud et al. 2003; Osundahunsi et al. 2003; Yadav et al. 2006; Maruf et al., 2010 a, b). There is need however to harmonize and apply some common processing conditions of pretreatment and drying that has been studied, and report the quality of flour in relation to several varieties. This is important in order to understand the effect of interactions among these independent variables on quality attributes of SP flour. This study was conducted to evaluate the effect of variety, pretreatment, and drying methods on chemical composition and functional properties of SP flour. The functional characterization of flour from varieties of SP roots used in this study was also conducted.

Methodology

Sweet potato roots

Ten varieties of mature sweet potato roots (Table 1) were used for this study. Sweet potato vines were transplanted into the main field in June/July and mature roots were harvested in 4 months after planting in October/November.

Table 1. Characteristics of sweet potato roots

Variety	Status	Color of flesh
LWF[1]	Local	White
LYF[2]	Local	Yellow
Arrowtip	Improved	White
TIS 87/0087	Improved	White
TIS 2531 OP-1-13	Improved	White
TIS 8250	Improved	White
Shaba	Improved	Yellow
CIP Tanzania	Improved	Yellow
199034.1	Improved	Orange
Ex-Oyunga	Improved	Orange

[1]LWF (local white-fleshed) referred to as "Anama funfun" in local Yoruba dialect.
[2]LYF (local yellow-fleshed) referred to as "Anama funfun" in local Yoruba dialect.

Sweet potato flour

Sweet potato roots were washed with tap water to remove soil particles. The cleaned roots were air-dried, peeled manually with stainless steel kitchen knife and sliced into 2 mm thickness (Jangchud et al. 2003), using a domestic plantain slicer. Each variety was subjected to four different pretreatment conditions (1) soaked in water for 90 min (Owori and Hagenimana 2000), (2) soaked in 0.5% potassium metabisulphite solution for 30 min followed by soaking in water for 45 min (Oduro et al. 2003), (3) blanched at 70°C for 5 min (modified from Jangchud et al. 2003), and (4) no pretreatment (control). Each pretreated sample was subjected to each of two drying methods; (i) sun for 3–4 days and (ii) oven at 50°C for 5 h (Woolfe 1992; van Hal 2000). The dried SP chips were stored in heat-sealed polyethylene bags at room temperature (30 ± 2°C) and transported to the laboratory of the Natural Resources Institute, University of Greenwich, United Kingdom. The chips were initially reduced into smaller particles in a blender (VORWERK Thermomix 31-1, France), followed by milling into flour using a laboratory mill (Perten 3600) and sieving through a 250 μ aperture screen (Endecotts, London, UK.). The flour samples were packed in resealable polyethylene bags and stored at −10 ± 3°C prior to analysis.

Flour analyses

Proximate composition

Standard methods were used to determine moisture, protein, fat, fiber (AOAC, 2000), and ash (AOAC, 1984). Total carbohydrate was determined by difference (James 1995) as: Total carbohydrate (%) = 100 − (% moisture + % protein + % fat + % fiber + % ash).

Chemical properties of flour

Starch content was determined using the Phenol-sulphuric acid method (Dubois et al. 1956).

A quantity of 50 g of flour was extracted with hot 80% ethanol to separate the sugar. 1.0 mL of the sugar extract was pipetted into a test tube and diluted to 2.0 mL with distilled water. 1.0 mL of 5% phenol was added and mixed thoroughly. 5.0 mL of concentrated sulphuric acid was added and the tube was allowed to stand for 10 min. The mixture was vortexed and allowed to stay for another 20 min. Absorbance was read at 490 nm. A standard curve was plotted using 0–100 μg glucose. A standard solution of glucose was prepared by dissolving 10 mg of glucose in 100 mL distilled water. 0.20, 0.40, 0.60, 0.80, and 1.00 mL of the standard glucose solution was pipetted into a test tube and treated following the procedure for sugar extract. The amount of sugar in the flour sample was determined by reference to the standard curve, while taking the dilution factor and weight of sample into consideration.

Starch was calculated using this formula:

$$\text{Starch}(\%) = \frac{0.05 \times A \times 1/M}{\text{Weight of sample}} \times 0.9$$

where A = Absorbance, M = Slope of standard curve.

Amylose content was estimated by the rapid colorimetric method (Williams et al.1970). Here 20 mg of flour was weighed into a 50 mL beaker and 10 mL of 0.5 N KOH solution was added. The mixture was stirred with a stirring rod until the flour was fully dispersed in the solution. The dispersed mixture was transferred into a 100 mL volumetric flask and diluted to the mark with distilled water. 10 mL of the diluted test solution was pipetted into a 50 mL volumetric flask, 5 mL of 0.1 N HCl was added, followed by 0.5 mL of iodine reagent. The volume was diluted to 50 mL and the absorbance was measured at 625 nm after 5 min. Amylose was calculated as follows: Amylose (%) = (85.24 × A)–13.19 where A = Absorbance.

Sugar content was determined by a reversed-phase high performance liquid chromatography as described by Picha (1985) by extraction of sugars from the samples of flour; values were expressed as μg/mg of fructose, glucose, sucrose and total sugar. Sugar was extracted from 0.2 g of flour sample using 100 mL of 80% ethanol in 2 mL eppendorf tubes. Each tube was vortexed and placed in a water bath at 70°C with agitation for 2 h. The tubes were vortexed again after 30, 60 and 90 min. The tubes were then centrifuged for 4 min at 10,000 rpm (7,826 g) in a microcentrifuge. Thereafter, the extract was filtered into glass High Performance Liquid Chromatography (HPLC) vials using special syringe filters. The sugars were separated through Agilent Zorbax Carbohydrate column (150 mm ×

4.6 mm × 5 μm), Agilent Zorbax NH$_2$ guard column (12.5 mm × 4.6 mm × 5 μm), with a flow rate of 2 mL/min, at a column temperature of 30°C and an injection volume of 5 μL. The solvent used was a mixture of 75% acetonitrile and 25% water. The detector was a refractive index type while the data system was Agilent EZChrom Elite version 3.3. (Santa Clara, California, United States).

The pH and total titratable acidity (TTA) were measured according to Pearson's (1976). A quantity of 10 g of flour sample was suspended in 50 mL of deionized water for 5 min and pH measured, using a digital pH meter. The flour sample was treated as above and 10 Ml aliquot was titrated with 0.1 mol\L NaOH. The titre value was recorded. The TTA value was then calculated as the citric acid equivalent as follows: 1 mL NaOH = 0.009 mg citric acid.

Color properties were measured using a Minolta Chromameter (Collado et al. 1997). Thereafter , 10 g of flour was placed in a 1 cm high cylindrical Petri dish and measurement of the color was taken on the flattened surface of the flour. The chromometer was calibrated before the measurements, using the white calibration plate provided. Hunter L* values range from 100 (white) to 0 (black), a* values range from +a (green) to –a (red), and b* values range from +b (yellow) to –b (blue).

Minerals (Na, K, Mg, Ca, Fe, Zn, Cu, and Mn) were determined using an Atomic Absorption Spectrophotometer (James 1995). Phosphorus was determined colorimetrically using ammonium vanadate reaction (James 1995).

Functional properties of flour

Water absorption capacity was determined by the method of Sosulki et al. (1976) as described by Akubor (1997). Then 2 g of flour sample was mixed with 20 mL distilled water and allowed to stand at room temperature (30 ± 2°C) for 30 min, then centrifuged for 30 min at 2000 rpm (537 g). The volume of decanted supernatant fluid was measured and volume of water retained/bound per g of sample calculated. WAC was expressed as g of water bound/100 g of flour.

Swelling power and water solubility of the flour were estimated as described by Aina et al. (2009). A flour–water slurry (0.35 flour in 12.5 mL of distilled water) was heated in a water bath at 60°C for 30 min, with constant stirring. The slurry was centrifuged at 3000 rpm (1207 g) for 15 min, the supernatant was decanted into a weighed evaporating dish and dried at 100°C to constant weight. The difference in weight of the evaporating dish was used to calculate the water solubility. Swelling power was obtained by weighing the residue after centrifugation and dividing by original weight of the flour on dry weight basis.

Pasting profile was determined using a Rapid Visco Analyzer (RVA-4; Newport Scientific Pty. Ltd., Australia) as described by Shittu et al. (2007) and also by following the instructions of the manufacturer. The RVA was interfaced with a personal computer equipped with the Thermocline for Windows software provided by the same manufacturer. 3 g of the sample (14% moisture basis) and 25 mL of distilled water was used. The equivalent sample mass (S) and water mass (W) corrected for 14% moisture basis was calculated using the formulae: $S = 86 \times A/100 - M$, $W = 25 + (A - S)$, where S = corrected sample mass, A = sample weight at 14% moisture basis, M = actual moisture content of the sample (% as is), W = corrected water mass. A programed heating and cooling cycle was used at constant shear rate, where the slurry was held at 50 °C for 1 min, heated to 95°C within 7.5 min, and then held at 95°C for 5 min. It was subsequently cooled to 50°C within 8.5 min and held at 50°C for 2 min, while maintaining a rotation speed of 160 rpm. Total cycle time was 23 min. Duplicate tests were performed in each case. The viscosity is expressed as rapid visco units (RVU). The parameters measured automatically by the RVA were: peak viscosity (the highest viscosity of the paste during the heating phase), trough (lowest viscosity of the paste during the heating phase), breakdown viscosity (the difference between the peak viscosity and the trough), setback viscosity (the difference between the final viscosity and the trough), final viscosity (the viscosity at the end of the cycle), pasting temperature (°C) (the temperature at which there is a sharp increase in viscosity of flour suspension after the commencement of heating), and peak time (min) (time taken for the paste to reach the peak viscosity).

Data analyses

Descriptive analysis was performed to explore the general trend of the data. Multivariate analysis of variance was performed to compare the means of the samples for the several response variables. Significant difference was established at $P < 0.05$. Duncan's Multiple Range test was performed to separate the means where a significant difference exists. A multivariate General Linear Model (GLM) analysis was performed to determine the individual and interactive effects of the treatments (variety, pretreatment and drying methods) on the attributes measured. Significant effects were established at $P < 0.05$, 0.01 and 0.001 levels. Pearson's correlation coefficient among the quality attributes was calculated. Statistical packages used were Microsoft Excel and SPSS Version 17.0 (SPSS Inc., Chicago, IL).

Results and Discussion

Chemical properties of sweet potato flour

All the sweet potato (SP) flour differ significantly ($P \leq 0.05$) in all the chemical properties investigated. Table 2 shows the range of values for proximate composition of flour as affected by variety, pretreatment, and drying methods, with moisture content (8.06–12.86%), protein (0.55–5.87%), fat (0.04–1.45%), fiber (0.08–5.54%), ash (0.15–2.09%) and carbohydrate (74.55–90.92%). Moisture content of the flour is within the range of 2.50–13.2% reported for sweet potato flour (van Hal 2000; Osundahunsi et al. 2003; Aina et al. 2009). Local white-fleshed (LWF), Local yellow-fleshed (LYF) and CIP Tanzania had the highest moisture contents (10.84–12.86%). A value of 12.5% has been considered as critical moisture content of flour within a locality with an ambient temperature of 27–29°C while a value of 10% has been recommended for long term storage (van Hal 2000). Moisture content of SP flour is

Table 2. Range of values for proximate composition of sweet potato flour (n = 80) as affected by variety, pretreatment and drying method.

	Moisture (%)	Protein (%)	Fat(%)	Fiber (%)	Ash (%)	Carbohydrate (%)
Minimum	8.06	0.55	0.04	0.08	0.15	74.55
Maximum	12.86	5.87	1.45	5.54	2.09	90.92
Mean	10.79	2.42	0.49	1.70	1.51	83.13
SD	1.13	1.28	0.42	1.43	0.61	2.85
CV (%)	10.44	52.66	86.96	84.17	40.51	3.43
Main effects						
Variety (V)	***	***	***	***	***	***
Pretreatment (P)	NS	***	NS	***	***	***
Drying (D)	***	***	*	NS	***	***
Interactive effects						
V × P	***	***	***	***	***	***
V × D	***	***	***	NS	***	***
P × D	***	***	NS	NS	***	***
V × P × D	***	***	*	NS	***	***

CV, Coefficient of variation.

*, **, *** Indicate significant effects at $P < 0.05$, 0.01, 0.001, respectively; NS indicate not significant.

considered a quality characteristic where storage is concerned, since water can accelerate chemical or microbiological deterioration (van Hal 2000). Since moisture content is directly related not only to drying method, but also conditions of temperature and time (Falade and Solademi 2010), the moisture content could be controlled to a desired level bearing in mind other quality requirements other than storage alone. In terms of pretreatment, the order of moisture content range for the flour was: potassium metabisulphite (8.06–12.74 ± 1.16) < water (8.36–12.80 ± 1.29) < blanched (8.97–12.80 ± 1.03) < untreated (9.35–12.86 ± 1.09). Sun–dried flour had lower moisture range (8.36–12.78 ± 1.26) than oven-dried flour (8.54–12.86 ± 1.12).

In SP flour, carbohydrates account for the bulk of the flour and hence serve as a good energy source (Woolfe 1992). Flour from the two local varieties and TIS 8250 had the highest carbohydrate values (84.50–90.92%). Flour from 199034.1 could be targeted for low calorie, high nutrient foods because they had the lowest carbohydrate values (74.55–79.12%) but had the highest protein (3.77–5.87%), crude fiber (3.85–5.54%) and ash (1.68–2.02%) contents. Flour from the local varieties was generally high in carbohydrate but low in protein, fat, fiber, and ash. Protein is essential in the human diet for growth. Although SP is regarded as a high-energy, low-protein food, sweet potato protein in both fresh and flour form has been reported to be of good biological value (van Hal 2000; International Life Sciences Institute (ILSI) 2008). Hence, it could serve as a fairly important protein source among low-income consumers in developing countries whose diets contain protein derived mostly from foods of vegetable origin. Ash content is a reflection of the mineral content

of a food material. Lower values of protein, crude fiber and ash observed with flour from some varieties used in this study may be due to varietal differences (Woolfe 1992; van Hal 2000). It may also be due to the effect of some pretreatments as Jangchud et al. (2003) and Osundahunsi et al. (2003) reported that pretreatments involving leaching such as blanching and parboiling decreased the protein, crude fiber, and ash contents of SP flour.

The range of values for starch, sugar profile, and total titratable acidity (TTA) of SP flour as influenced by variety, pretreatment and drying method are shown in Table 3. The range of values of these components were starch (55.76–83.65%), amylose (10.06–21.26%), fructose (0.46–29.41 μg/mg), glucose (1.19–32.14 μg/mg), sucrose (4.90–113.18 μg/mg), total sugar (22.39–125.46 μg/mg) and TTA (0.02–0.13%). Although flour from the two local varieties of SP had the least starch content, however, they contained the highest amylose and medium range of total sugar. Flour from TIS 87/0087 and 2532 OP-1-13 had the highest starch while TIS 8250 flour had the least total sugar. In terms of pretreatment, the order of starch content of the flour was: water (55.76–73.18 ± 5.55%) < potassium metabisulphite (59.76–78.81 ± 6.02%) < untreated (60.17–79.55 ± 5.95%) < blanched (60.21–83.65 ± 6.68%). Blanched flour had the lowest ranges of fructose (0.46–12.75 ± 3.59%), glucose (1.19–12.47 ± 3.48%), sucrose (12.74–73.54 ± 17.53%), and total sugar (22.39–96.17 ± 20.36%). The range of total sugar for the flour was in the order: blanched (22.39–96.17 ± 20.36 μg/mg) < water (46.29–111.49 ± 18.31 μg/mg) < untreated (48.29–119.82 ± 18.57 μg/mg) < potassium metabisulphite (54.55–125.46 ± 21.34 μg/mg). Sun-dried flour had slightly lower

Table 3. Range of values for chemical composition of sweet potato flour (n = 80) as affected by variety, pretreatment and drying method.

	Starch (%)	Amylose (%)	Fructose (μg/mg)	Glucose (μg/mg)	Fructose (μg/mg)	Total sugar (μg/mg)	TTA (%)
Minimum	55.76	10.06	0.46	1.19	4.90	22.39	0.02
Maximum	83.65	21.26	29.41	32.14	113.18	125.46	0.13
Mean	69.05	15.97	5.90	7.31	57.66	70.87	0.07
SD	6.82	3.92	5.01	5.82	23.01	24.68	0.03
CV(%)	9.87	24.54	84.84	79.64	39.91	34.82	36.32
Main effects							
Variety (V)	***	***	***	***	***	***	***
Pretreatment (P)	***	***	***	***	***	***	***
Drying (D)	***	***	***	***	***	***	*
Interactive effects							
V × P	***	***	***	***	***	***	***
V × D	***	***	***	***	***	***	***
P × D	***	***	***	***	***	**	**
V × P × D	***	***	***	***	***	***	***

CV, Coefficient of variation; NS, indicates not significant.
*, **, *** indicate significant effects at $P < 0.05$, 0.01, 0.001, respectively.

starch content (55.76–78.45 ± 6.69%) than oven-dried flour (56.38–79.55 ± 6.67%). Sun-dried flour also had lower sucrose (4.09–110.67 ± 21.75 μg/mg) and total sugar (46.29–124.47 ± 19.96 μg/mg).

The starch content of the flour (56–84%) was similar to the range of values (57–85%) reported by van Hal (2000). Starch is the predominant fraction of the dry matter of SP tubers (Ravindran et al. 1995). It is a rich source of carbohydrate and hence energy in the diet. It is a dominant factor that determines the physicochemical, rheological, and textural properties of starch-based food products. It is also an important source of industrial raw materials for products such as noodles and glucose syrup, hence TIS 87/0087 and 2532 OP-1-13 flour with the highest starch content could be useful in these respect.

Amylose content of the flour (10.06–21.26%) was lower than the range of values (15.3–31.2%) reported by Aina et al. (2009) for flour from twenty-one Caribbean SP cultivars. Report of amylose content of SP flour in literature is limited; however values obtained in this study are similar to those reported for SP starches; 19.1% (Collado et al., 1999) and 13.9–21.1% (Noda et al., 1995). Osundahunsi et al. (2003) and Garcia and Walter (1998) however reported higher values of 20.48–25.54%, and 32–34%, respectively, for SP starches. Osundahunsi et al. (2003) suggested that the high values reported in their study may be due to the highly sensitive Differential Scanning Calorimetry method used. Flour from LWF and LYF had the highest amylose content. The starch and amylose composition of staple food materials determines the processing and consumption characteristics of food products (Shittu et al. 2007).

The higher variability in fructose and glucose exhibited by the flour compared to the sucrose and total sugar suggests that fructose or glucose could be used to classify the flour into distinct groups. Sucrose is the dominant sugar contributing to the total sugar in SP flour, hence a good indicator to estimate one another. For staple foods in Nigeria, which are usually characterized by a bland taste, TIS 8250 flour with the least total sugar would be preferred. Flour from EX-OYUNGA had the highest total sugar content and hence could be suitable to provide natural sweetness in the manufacture of food products. Generally, among the varieties, except for EX-OYUNGA, blanched flour had the lowest starch and sugar content. This suggests that during blanching which takes place at an elevated temperature, the soluble sugars and starch in the sweet potato slices leached into the water, resulting in a reduction in the starch and sugar content of the flour. The result in this study is in agreement with that of Jangchud et al. (2003) which reported that blanching caused a reduction in starch; however it differs from the same report with respect to an increase in reducing sugar content of SP flour. The order observed for range of total sugar among the flour with respect to flesh color of the fresh roots is: white-fleshed < yellow-fleshed < orange-fleshed. Significant ($P < 0.01$) positive correlations between total sugar and two tristimulus color parameters a* (0.416) and b*(0.588) were obtained.

Table 4 shows the mineral composition of the SP flour. The range values of the minerals were: Na (0.06–0.18%), K (0.76–1.22%), Mg (0.04–0.15%), Ca (0.09–0.29%), P (0.07–0.19%), Fe (20.65–45.35 mg/kg), Zn (18.85–33.75 mg/kg), Cu (3.60–8.50 mg/kg) and Mn (8.80–16.55 mg/kg). Flour from LWF and TIS 8250 were high in Fe, with LWF containing the highest. Arrowtip flour contained the highest Cu and Mn, while TIS 8250 contained the highest Zn. LYF flour was not only low in Fe,

Table 4. Range of values for mineral composition of sweet potato flour (n = 80) as affected by variety, pretreatment and drying method.

	Na (%)	K (%)	Mg (%)	Ca (%)	P (%)	Fe (mg/kg)	Zn (mg/kg)	Cu (mg/kg)	Mn (mg/kg)
Minimum	0.06	0.76	0.04	0.09	0.07	20.65	18.85	3.60	8.80
Maximum	0.18	1.22	0.15	0.29	0.19	45.35	33.75	8.50	16.55
Mean	0.10	0.94	0.09	0.19	0.13	33.13	25.78	5.54	12.73
SD	0.03	0.10	0.03	0.04	0.03	7.23	3.65	1.05	1.68
CV(%)	28.94	10.59	29.67	23.12	22.68	21.83	14.14	18.97	13.19
Main effects									
Variety (V)	***	***	***	***	***	***	***	***	***
Pretreatment (P)	***	***	*	***	NS	***	***	***	***
Drying (D)	**	***	NS	***	NS	**	***	NS	***
Interactive effects									
V × P	***	***	***	***	***	***	***	***	***
V × D	***	***	***	***	***	***	***	**	***
P × D	NS	*	***	***	***	***	***	**	***
V × P × D	***	***	***	***	***	***	***	***	***

CV: Coefficient of variation; NS, indicates not significant.
*, **, *** indicate significant effects at $P < 0.05$, 0.01, 0.001, respectively.

but they also contained the least Zn and Cu. The least Mn content was found among the 199034.1 flour. Generally, for all the pretreatments, the ranges of minerals in the flour were: (0.04–1.22%) for Na, K, Mg, Ca and P, (20.65–45.35 mg/kg) for Fe, (18.85–33.75 mg/kg) for Zn, (3.60–8.50 mg/kg) for Cu, and (8.80–16.55 mg/kg) for Mn. For oven-dried flour the values were: (0.05–1.22%) for Na, K, Mg, Ca and P, (20.65–45.35 mg/kg) for Fe, (18.35–33.75 mg/kg) for Zn, (3.60–8.50 mg/kg) for Cu, and (9.10–16.55 mg/kg) for Mn. Sun-dried flour however contained (0.06–0.25%) of Na, K, Mg, Ca and P, (21.45–42.45 mg/kg) of Fe, (19.70–31.60 mg/kg) of Zn, (4.15–7.85 mg/kg) for Cu, and (8.80–15.40 mg/kg) for Mn.

The individual mineral composition of SP flour is rarely studied; rather the ash content which is an estimate of the total mineral content is usually reported. Potassium was the major mineral present in the SP flour confirming the report by Ravindran et al. (1995). The results in this study showed that SP roots are moderately good sources of minerals particularly essential micronutrients such as Fe, Cu, Zn, and Mn. According to van Hal (2000), SP flour is estimated to contribute 20–40% of the Recommended Daily Allowance of Fe.

Main and interactive effects of variety, pretreatment and drying methods on chemical properties of sweet potato flour

The interactive effects of variety, pretreatment and drying methods on the chemical properties of SP flour are presented in Tables 2–4. The factors that had the most significant ($P < 0.001$) effect on all the chemical properties of the SP flour were variety and the interaction between variety and pretreatment. Protein, ash, and carbohydrate were significantly (0.001) affected by each of the factors and all the interactions (Table 2). Pretreatment had no significant effect ($P > 0.05$) on moisture and fat content of the flour. Starch, amylose and all the sugars were significantly ($P < 0.001$) affected by each of variety, pretreatment, drying method as well as all the interactions among these factors (Table 3). Ca, Zn and Mn were significantly ($P < 0.001$) affected by each of the factors and all the interactions (Table 4). Drying did not have a significant ($P \geq 0.05$) effect on Mg, P and Cu.

Functional properties of sweet potato flour

Table 5 shows the Hunter L* a* b* color values of SP flour as affected by variety, pretreatment and drying methods. The range of values for all the flour were: L* (79.90–101.48 ± 5.44), a* (−0.27–3.54 ± 0.82) and b* (9.89–27.94 ± 4.02). In terms of variety, the order of mean L* values was white-fleshed (87.79 ± 4.76) < orange-fleshed (95.17 ± 2.03) < yellow-fleshed (95.91 ± 2.66), while that of mean a* values was white-fleshed (14.22 ± 2.64) < yellow-fleshed (15.79 ± 1.47) < orange-fleshed (21.92 ± 4.23). The mean L*: a*: b* values showed considerable difference between flour made from the two orange-fleshed varieties; Ex-oyunga (96.15: 2.49: 25.60) and 199034.1 (94.18: 0.39: 18.25). In terms of pretreatment, the order of mean L* values was: potassium metabisulphite (91.48 ± 5.19) < untreated (91.58 ± 5.23) < water (91.71 ± 6.54) < blanched (92.05 ± 5.07), and b* values was: blanched (14.56 ± 3.99) < potassium metabisulphite (16.71 ± 4.19) < water (16.73 ± 3.74) < untreated (16.92 ± 3.95). Sun-dried flour had higher mean values of L* (91.75 ± 5.58) and a* (0.96 ± 0.75) than oven-dried flour (L* 91.66 ± 5.37, a* 0.73 ± 0.88), however,

Table 5. Range of values for functional properties of sweet potato flour (n = 80) as affected by variety, pretreatment and drying method.

	L*	A*	B*	WAC	WS	SWP
Minimum	79.90	−0.27	9.89	140.00	6.89	1.66
Maximum	101.48	3.54	27.94	280.00	26.18	5.00
Mean	91.70	0.84	16.23	194.13	15.30	3.31
SD	5.44	0.82	4.02	25.97	3.80	0.50
CV(%)	5.93	97.21	24.74	13.38	24.80	14.97
Main effects						
Variety (V)	***	***	***	***	***	***
Pretreatment (P)	NS	***	***	***	***	***
Drying (D)	NS	***	***	***	***	*
Interactive effects						
V × P	***	***	***	***	***	***
V × D	***	***	***	***	***	***
P × D	***	***	NS	*	***	***
V × P × D	***	***	***	***	***	***

CV, Coefficient of variation; WAC, Water absorption capacity; WS, Water solubility; SWP, Swelling power; L*,Lightness; a*,Redness; b*, Yellowness.
*, **, *** indicate significant effects at $P < 0.05$, 0.01, 0.001, respectively; NS indicates not significant.

oven-dried flour had higher b* values (16.61 ± 4.33) than sun-dried flour (15.85 ± 3.69).

The tristimulus color parameter L* indicate whiteness of the flour, as reported by Collado et al. (1997). In the present study however, flour from white-fleshed varieties had the least mean L* values (87.79). This confirms the works of other authors (van Hal 2000) which have shown that the whiteness of the flour is not always directly related to the flesh color of the roots. They suggested that this is an indication of the high level of browning that occurs during drying of SP chips and processing the flour. The b* parameter may be a better measure of the color and intensity of flour from colored varieties, this is because flour from the two orange-fleshed varieties, 199034.1 and EX-OYUNGA, had the highest b* values (17.15–27.94) which increased with intensity. Flour from these orange-fleshed varieties could add natural color to food products. Flour from the white varieties, LWF and TIS 87/0087, would be suitable where whiteness is desired because they have the least b* values (10.25–13.66). Flour produced by blanching showed the least mean b* values (14.56 ± 3.99), this is supported by the report of Jangchud et al. (2003). Aina et al. (2009) reported that flour from the orange-fleshed varieties were characterized by higher a* and b* values, lower amylose, but higher total sugar levels. In the present study however, only flour from EX-OYUNGA (an intensely-colored orange-fleshed variety) agrees with this description, flour from 199034.1, a less intensely-colored orange-fleshed variety did not show these characteristics.

The water absorption capacity (WAC), water solubility (WS), and swelling power (SWP) of SP flour as affected by variety, pretreatment, and drying methods is shown in Table 5. The ranges of values for all the flour were: WAC (140–280 ± 25.97), WS (6.89–26.18 ± 3.80) and SWP (1.66–5.00 ± 0.50). In terms of variety, flour from EX-OYUNGA had the highest WAC (215–280) and WS (18.30–26.18) while LYF had the highest SWP (3.05–5.00) but the least WAC (140–170). Flour from TIS 8250 and SHABA had the least WS (7.49–16.83), while 2532 OP-1-13 flour had the least SWP (2.69–2.99). Generally, blanched flour had the least mean values of WAC, WS and SWP while potassium metabisulphite-treated flour had the highest mean values of WAC and SWP. Untreated flour had the highest WS. Oven-dried flour had higher mean values of WAC (197.50 ± 26.94), WS (16.18 ± 3.45) and SWP (3.34 ± 0.55) than sun-dried flour with WAC (190.75 ± 24.85), WS (14.43 ± 3.97) and SWP (3.29 ± 0.44).

The functional properties of flour are those that directly determine their end uses. The WAC of SP flour in this study is much higher than reported values (24–42) by Osundahunsi et al. (2003) for native and parboiled flour.

The authors noted that precious studies reported higher values and suggested that the varieties tested have unusually low WAC. Water absorption capacity is the ability of the starch or flour to absorb water and swell for improved consistency. It is desirable in food systems to improve yield and consistency and give body to food. Osundahunsi et al. (2003) reported that parboiling improved the WAC of the flour by 173–175% and suggested that it should be used for SP flour required as a thickener. Jangchud et al. (2003) also reported that blanching increased the SWP of flour at all temperatures investigated. In the present study however, blanched flour had the least mean value of WAC (191.67) while WS and SWP compared to native flour. The relatively lower values of WAC could be due to some genetic factors rather than the pretreatment.

Main and interactive effects of variety, pretreatment and drying methods on functional properties of sweet potato flour

Each of the Hunter color parameters (L*, a*, b*) were significantly ($P < 0.001$) affected by variety, as well as the combinations of variety and pretreatment, variety and drying, and variety, pretreatment, and drying (Table 5). The a* values were significantly ($P < 0.001$) affected by each of the main factors and all their combinations. The WAC, WS, and SWP were all significantly ($P < 0.001$) affected by variety, pretreatment, and the combinations of variety and pretreatment, variety and drying, as well as variety, pretreatment, and drying. WS was significantly affected by all the main factors and their interactions (Table 5). Drying, and the interaction between pretreatment and drying, affected SWP and WAC, respectively, only at $P < 0.05$. The implication of these interactive effects is that each of the main factors as well as the combinations are very important to the functional properties of flour and should therefore be selected appropriately during processing of SP flour.

Pasting properties of sweet potato flour

Table 6 shows the pasting properties of SP flour as affected by variety, pretreatment, and drying methods. The range of values were peak viscosity (PV) (24.50–260.92 RVU), trough (T) (7.08–145.83 RVU), breakdown viscosity (BV) (11.00–125.33 RVU), final viscosity (FV) (10.21–225.1), setback viscosity (SBV) (3.04–92.21 RVU), peak time (PTm) (6.07–9.76 min), and peak temperature (PTp) (69.78–81.25°C). Sun-dried flour had higher ranges for almost all the pasting properties PV (47.88–260.92 ± 53.12), T (17.21–145.83 ± 31.54), BV (13.25–125.33 ± 29.96), FV (26.04–222.50 ± 55.95), SBV (7.50–92.21 ± 25.82) than

Table 6. Range of values for pasting properties of sweet potato flour (n = 80) as affected by variety, pretreatment and drying method.

	Peak viscosity (RVU)	Trough (RVU)	Breakdown viscosity (RVU)	Final viscosity (RVU)	Setback viscosity (RVU)	Peak time (min)	Pasting temperature (^{0}C)
Minimum	24.50	7.08	11.00	10.21	3.04	6.07	69.78
Maximum	260.92	145.83	125.33	225.50	92.21	9.76	81.25
Mean	136.51	77.67	57.18	125.15	46.65	6.86	78.27
SD	52.61	34.48	25.72	60.55	27.13	0.52	2.96
CV (%)	38.54	44.40	44.98	48.38	58.16	7.51	3.78
Main effects							
Variety (V)	***	***	***	***	***	***	***
Pretreatment (P)	***	***	***	***	***	***	***
Drying (D)	***	***	***	***	***	***	***
Interactive effects							
V × P	***	***	***	***	***	***	***
V × D	***	***	***	***	***	***	***
P × D	***	***	***	***	***	***	***
V × P × D	***	***	***	***	***	***	***

CV, Coefficient of variation; NS, indicates not significant.

*, **, *** Indicate significant effects at $P < 0.05$, 0.01, 0.001, respectively.

oven-dried flour PV (24.50–197.00 ± 47.02), T (7.08–108.58 ± 33.63), BV (11.00–103.79 ± 25.70), FV (10.21–200.21 ± 59.14), and SBV (3.04–91.83 ± 26.33).

The most common objective method of determining the cooking property of starch-based food products is through an amylograph pasting profile. Such information have been used to correlate the functionality of starchy food ingredients in processes such as baking (Defloor et al. 1995; Rojas et al. 1999) and extrusion cooking (Ruales et al. 1993). In these studies, it was observed that the pasting properties (such as peak, trough, setback, breakdown, and final viscosities) of the cooked flour showed the widest variation amongst other pasting properties measured. This agrees with the results in this study in which the pasting viscosities varied between 39% and 58%, while the pasting time and temperature varied between 4–8% (Table 6). The values obtained for all the pasting properties except pasting time were much higher than those reported by Jangchud et al. (2003). The highest Peak viscosity (PV), Breakdown viscosity (BDV), and Final viscosity (FV) were found among the TIS 8250 flour in addition to having high T values. Flours from EX-OYUNGA had very low pasting viscosities, an indication of high enzymic activities in the flour (Osundahunsi et al. 2003; Collado and Corke, 1999).

The pasting temperature of the flour were higher than values reported by Osundahunsi et al. (2003) but lower than those reported by Aina et al. (2009) and Jangchud et al. (2003). Pasting temperature of flour can have energy-cost implications. It had earlier been reported that high pasting temperatures of flour may be due to higher amylose contents of the varieties (Aina et al. 2009). However, in this present study, there was a negative significant

($P < 0.001$) correlation between amylose content and pasting temperature, indicating that sweet potato flour with a higher amylose content is expected to have lower pasting temperature. Aina et al. (2009) also suggested that sugar content of the flour may have influenced the pasting temperature by competing with the starches for moisture, decreasing swelling, and viscosity while increasing the pasting temperature. The present study agrees with the low pasting viscosities due to high sugar, however the significant ($P < 0.01$) but low ($r = 0.322$–0.395) correlation shows that a high swelling power is expected. There was no significant ($P > 0.05$) correlation between sugars and pasting temperature. The setback viscosity indicates the retrogradation tendency of starch and flour. Low setback values showed a lower tendency for retrogradation. According to Garcia and Walter (1998), cultivars with the highest amount of amylose are expected to have stiffer pastes and high setback viscosities. There was no significant ($P > 0.05$) correlation between amylose and setback viscosities of flour in this study. Nevertheless, LYF flour which had the highest amylose content showed relatively high setback viscosities. Flour from 2532 OP-1-13 and EX-OYUNGA had the least setback values (3.04–43.42 RVU), hence suitable for pie fillings where retrogradation may cause syneresis and low quality product. Nevertheless, for all the varieties, except 2532 OP-1-13 and 199034.1, blanching resulted in a considerable decrease in values of all the pasting properties. This confirms the report of Jangchud et al. (2003) in which peak, final, and breakdown viscosities were lower than those of unblanched flour. In the same report, blanching significantly increased the pasting temperature by about 6–20°C for the two varieties studied. In the present study however, a difference of 1–2°C was observed which may

Table 7. Correlations between chemical composition and functional properties of sweet potato flour.

Chemical property	L*	a*	b*	WAC	WS	SWP
Moisture content	−.041	−.272**	−.328**	−.402**	−.120	.095
Protein	.380**	.168*	.426**	.458**	−.100	.051
Fat	.072	.249**	.517**	.431**	.391**	.118
Fiber	.134	.147	.478**	.493**	.226**	.241**
Ash	.157*	.206**	.291**	.341**	.051	−.255**
Carbohydrate	−.263**	−.108	−.423**	−.410**	−.084	−.150
Starch	−.241**	−.029	−.079	−.084	.109	−.420**
Amylose	−.218**	−.083	−.143	−.210**	.067	.028
Fructose	−.066	.514**	.323**	.068	.327**	−.059
Glucose	−.234**	.372**	.191*	.176*	.292**	−.144
Sucrose	.090	.241**	.512**	.167*	.801**	.395**
Total `sugar	.016	.416**	.588**	.211**	.882**	.322**
TTA	−.124	.323**	.357**	.380**	.455**	.045

L*, Lightness; a*, Redness; b*,Yellowness; WAC, Water absorption capacity; WS, Water solubility; SWP, Swelling power; TTA, Total Titratable acidity.
*Correlation is significant at $P < 0.05$.
**Correlation is significant at $P < 0.01$.

be an increase or decrease depending on variety and the interaction between pretreatment and drying method.

Main and interactive effects of variety, pretreatment, and drying methods on pasting properties of sweet potato flour

In this study, no particular trend was observed in pasting properties as a result of the treatments. This could imply that although the pasting properties were significantly ($P < 0.001$) affected by variety, pretreatment, and drying methods as well as the interactions among these treatments (Table 6), genetic variation may be a more dominant factor influencing the pasting properties of the SP flour.

Correlations among the properties of sweet potato flour

Table 7 shows the linear correlations between chemical composition and functional properties of sweet potato flour. Significant ($P < 0.01$) negative correlations were observed between L* values and carbohydrate, starch, amylose, and glucose while the significant ($P < 0.01$) correlations between b* values and protein, fat, fiber, ash, and sugars were positive. WAC was negatively correlated with moisture content, but showed positive correlations with protein, fat, fiber, and ash ($P < 0.01$). The WS showed positive correlations ($P < 0.01$) with all the sugars particularly sucrose ($r = 0.801$) and total sugar ($r = 0.882$). The SWP showed significant ($P < 0.01$) negative correlations with starch but positive correlations with sucrose and total sugar. Generally, the lower the moisture contents of the flour, the higher the WAC. On the other hand,

the higher the sugar contents, the higher the WS of the flour. Furthermore, lower starch content favors high SWP.

The pasting temperature showed significant ($P < 0.01$) correlations with more chemical properties than any other pasting property (Table 8). A strong positive correlation ($r = 0.728$) was particularly observed between pasting temperature and ash. Correlation showed that the lower the moisture content, the higher the pasting temperature. Fat showed negative but significant correlations ($P < 0.01$) with PV, T, BDV, FV, and SBV. Sucrose and total sugar showed similar trends as fat, except with SBV which was significant for sucrose only at $P < 0.05$. The lower the WAC and WS, the higher the peak, trough, breakdown, final, and setback viscosities as indicated by the significant negative correlations between these functional and pasting properties (Table 9). Peak time and peak temperature were positively correlated with WAC and this was significant ($P < 0.01$). SWP showed a significant negative correlation ($P < 0.01$) with breakdown viscosity and pasting temperature.

In order to determine attributes that could serve as quality indicators for sweet potato flour irrespective of variety, pretreatment and drying method, attributes that were significantly ($P < 0.001$) affected by the interaction among these treatments were selected. Correlations that were significant ($P < 0.01$) among the selected attributes were identified. The sugars show significant positive correlations with WAC, water solubility, and swelling power. Consequently, it may be deduced from these significant correlations that amylose, sucrose, or total sugar, and water absorption capacity could serve as a reliable quality indicators to predict the pasting properties of sweet potato flour and hence the functionality in specific food products.

Table 8. Correlations between chemical and pasting properties of sweet potato flour.

Chemical property	Peak viscosity (RVU)	Trough (RVU)	Break down viscosity (RVU)	Final viscosity (RVU)	Set back viscosity (RVU)	Peak time (min)	Pasting temperature (^{0}C)
Moisture content	.284**	.326**	.049	.337**	.292**	.038	−.401**
Protein	.072	.191*	−.026	.165*	.166*	.351**	.467**
Fat	−.375**	−.340**	−.260**	−.362**	−.352**	−.032	.240**
Fiber	−.268**	−.136	−.311**	−.173*	−.187*	.231**	.337**
Ash	−.047	−.062	.122	−.067	−.007	.418**	.728**
Carbohydrate	.042	−.099	.158*	−.069	−.057	−.375**	−.390**
Starch	−.111	−.224**	.157*	−.190*	−.099	.161*	.457**
Amylose	.077	.049	.027	.060	.041	−.244**	−.360**
Fructose	−.008	−.135	.196*	−.102	−.042	−.112	.028
Glucose	−.115	−.298**	.209**	−.297**	−.263**	−.182*	.056
Sucrose	−.334**	−.328**	−.240**	−.269**	−.183*	−.047	.065
Total `sugar	−.340**	−.403**	−.136	−.342**	−.241**	−.109	.080
TTA	.048	−.011	.210**	.008	.079	.185*	.385**

TTA, Total Titratable Acidity.
*Correlation is significant at $P < 0.05$.
**Correlation is significant at $P < 0.01$.

Table 9. Correlations between functional and pasting properties of sweet potato flour.

Functional property	Peak viscosity (RVU)	Trough (RVU)	Breakdown viscosity (RVU)	Final viscosity (RVU)	Setback viscosity (RVU)	Peak time (min)	Pasting temperature (^{0}C)
WAC	−.319**	−.206**	−.270**	−.266**	−.282**	.240**	.490**
WS	−.473**	−.505**	−.290**	−.437**	−.332**	−.146	−.003
SWP	−.026	.072	−.217**	.069	.029	−.037	−.258**

WAC, Water absorption capacity; WS, Water solubility; SWP, Swelling power; RVO, Rapid visco units.
*Correlation is significant at $P < 0.05$.
**Correlation is significant at $P < 0.01$.

Practical applications

Sensory acceptability of products such as 'cooked paste' ('amala') require sweet potato flour characterized by lower values of total sugar, water absorption capacity, swelling power amongst other properties (Fetuga et al. 2014). The correlations in this study indicates that sweet potato varieties that will give flour with low values of sugar, WAC and swelling power should be targeted for consumer acceptable 'amala'.

High paste viscosities are desirable in flour used as thickeners, whereas low paste viscosities are desirable for high-calorie food formulations such as weaning and speciality foods (Wiessenborn et al. 1994). The correlations in this study indicates that sweet potato varieties that will give flour with lower values of WAC and water solubility (WS) and consequently higher paste viscosities will be appropriate for thickening agents. On the other hand, varieties that will produce flour with high WAC and WS and consequently low paste viscosities, should be targeted at weaning and speciality foods.

Conclusions

This study has shown that variety had a significant effect on all the attributes of sweet potato flour (SPF). Pretreatment did not significantly affect moisture, fat, and lightness of SPF. The drying method did not significantly affect fiber and lightness of SPF. The interactive effect of variety, pretreatment, and the drying method had a significant effect on all the quality attributes of SPF except fat and fiber. The sweet potato flour had significantly different chemical, functional, and pasting properties. The flour showed a wider variability in pasting properties. Blanching resulted in considerable decrease in paste viscosities irrespective of variety and drying method. Varietal difference was a dominant factor in the differences observed in quality attributes of the flour studied, therefore, sweet potato varieties should be targeted at specific end uses. Sucrose or total sugar and water absorption capacity could serve as reliable quality indicators to predict the pasting properties of sweet potato flour and hence functionality in specific food products.

Acknowlegments

Funding for this study was provided by the Education Trust Fund of the Federal Republic of Nigeria. The corresponding author gratefully acknowledges the laboratory support provided at the Natural Resources Institute of the University of Greenwich at Medway, Kent, United Kingdom. The kind assistance provided by Aurelie Bechoff with the use of the Rapid Visco Analyzer is particularly acknowledged.

Conflict of Interest

None declared.

References

Afuape, S. O.. 2009. Sweet potato breeding program at National Root Crops research Institute (NRCRI), Umudike, Abia State. Pp. 73–74. in M. Akoroda, I. Egeonu, eds. Sweetpotato in Nigeria. Proceedings of the first National sweetpotato conference, 16 – 18 September, 2008, University of Ibadan, Nigeria.

Afuape, S. O.. 2013. Selected sweetpotato research in Nigeria. 6th SSP meeting, Accra, Ghana, 2013; Available at www.sweetpotatoknowledge.org (accessed April 24 2015).

Aina, A. J., K. O. Falade, J. O. Akingbala, and P. Titus. 2009. Physicochemical properties of twenty one Caribbean sweet potato cultivars. Int. J. Food Sci. Technol. 44:1696–1704.

Akubor, P. I. 1997. Proximate composition and selected functional properties of African breadfruit and sweet potato flour blends. Plant Foods Hum. Nutr. 51:53–60.

AOAC. 1984. Official methods of analysis, 14th ed. Association of Official Analytical Chemists, Washington, DC.

AOAC. 2000. Official methods of analysis, 17th ed. Association of Official Analytical Chemists, Washington, DC.

Bradbury, J. H., and U. Singh. 1986. Ascorbic and dehydroascorbic acid content of tropical root crops from the South Pacific. J. Food Sci. 51(4):975–978.

Bradbury, J. H., B. Hammer, T. Nguyen, M. Anders, and J. S. Millar. 1985. Protein quantity and quality and trypsin inhibitor content of sweet potato cultivars from the highlands of Papua New Guinea. J. Agric. Food Chem. 33(2):281–285.

Centro Internacional De La Papa (CIP). 2013. Sweet potato facts: production, utilization, consumption, feed use. Available at www.cipotato.org/cipotato/publications/pdf/005448.pdf (accessed 28 February 2013).

Collado, L. S., L. B. Mabesa, and H. Corke. 1997. Genetic variation in color of sweet potato flour related to its use in wheat-based composite flour products. Cereal Chem. 74:183–187.

Collado, L. S., R. C. Mabesa, and H. Corke. 1999. Genetic variation in the physical properties of sweet potato starch. J. Agric. Food Chem. 47:4195–4201.

Collado, L. S., and H. Corke. 1999. Accurate estimation of sweet potato amylase activity by flour viscosity analysis. J. Agric. Food Chem. 49:832–835.

Defloor, I., R. Leijskens, M. Bokanga, and J. A. Declour. 1995. Impact of variety, crop age and planting season on the breadmaking and gelatinization properties of flour produced from cassava (Manihot esculenta Crantz). J. Sci. Food Agric. 68:167–174.

Dubois, M., K. A. Gilles, J. K. Hamilton, P. A. Rebers, and E. Smith. 1956. Colorimetric methods for determination of sugars and related substances. Anal. Chem. 28:350–356.

Egeonu, I. N., and M. O. Akoroda. 2009. Assemblage, characterization and germplasm management of sweetpotato clones at Ibadan, Nigeria. Pp. 79–89. in M. Akoroda, I. Egeonu eds. Sweetpotato in Nigeria. Proceedings of the first National sweet potato conference, 16 – 18 September, 2008, University of Ibadan, Nigeria.

Falade, K. O., and J. O. Solademi. 2010. Modelling of air drying of fresh and blanched sweet potato slices. Int. J. Food Sci. Technol. 45:278–288.

FAOSTAT. 2013. Food and Agricultural Organization of the United Nations, Statistics Division.Avaliable at www.faostat3.fao.org/browse/Q/QC/E (Accessed April 24 2015).

Fernandez, A., J. Wenham, D. Dufour, C. C. Wheatley. 1996. The influence of variety and processing on the physicochemical and functional properties of cassava starch and flour. Pp. 263–269 in D. Dufour, G. M. O'Brien, R. Best, eds. Cassava starch and flour: progress in research and development. CIAT, Montpellier, France.

Fetuga, G., K. Tomlins, F. Henshaw, and M. Idowu. 2014. Effect of variety and processing method on functional properties of traditional sweet potato flour ('elubo') and sensory acceptability of cooked paste ('amala'). Food Sci. Nutr. 2:682–691.

Garcia, A. M., and W. M. Walter. 1998. Physicochemical characterization of starch from Peruvian sweet potato selections. Starch/Starke 50:331–337.

van Hal, M. 2000. Quality of sweet potato flour during processing and storage. Food Rev. Int. 16:1–37.

Horton, D., G. Prain, and P. Gregory. 1989. High-level investment returns for global sweet potato research and development. CIP Cir. 17:1–11.

International Life Sciences Institute (ILSI). 2008. Nutritionally improved sweet potato. in: Assessment of foods and feeds. Comprehensive Reviews in Food science and Food safety 7:81–91.

James, C. S. 1995. Analytical chemistry of foods. Chapman and Hall, London.

Jangchud, K., Y. Phimolsiripol, and V. Haruthaithanasan. 2003. Physicochemical properties of sweet potato flour and starch as affected by blanching and processing. Starch/Starke 55:258–264.

Mais, A., and C. S. Brennan. 2008. Characterization of flour, starch and fibre obtained from sweet potato (Kumara) tubers and their utilization in biscuit production. Int. J. Food Sci. Technol. 43:373–379.

Maruf, A., S. Akter Mst, and E. Jong-Bang. 2010a. Effect of pretreatments and drying temperatures on sweet potato flour. Int. J. Food Sci. Technol. 45:726–732.

Maruf, A., S. Akter Mst, and E. Jong-Bang. 2010b. Peeling, drying temperatures and sulphite-treatment affect physicochemical properties and nutritional quality of sweet potato flour. Food Chem. 121:112–118.

Noda, T., Y. Takahata, M. Hisamatsu, and T. Yamada. 1995. Physicochemical properties of starches extracted from sweet potato roots differing in physiological age. J. Agric. Food Chem. 43:3016–3020.

Oduro, I., W. O. Ellis, L. Nyarko, G. Koomson, and J. A. Otoo. 2003. Physicochemical and pasting properties of flour from four sweet potato varieties in Ghana. Proc. 8th ISTRC-AB Symp., Ibadan, Nigeria. Pp. 142–146.

Osundahunsi, O. F., T. N. Fagbemi, E. Kesselman, and E. Shimoni. 2003. Comparison of the physicochemical properties and pasting characteristics of flour and starch from red and white sweet potato cultivars. J. Agric. Food Chem. 51:2232–2236.

Owori, C., and A. Agona. 2003. Assesment of sweet potato cultivars for suitability for different forms of processing. Pp. 103–111 in D. Rees, Q. E. A. Oirschot and R. Kapinga, eds. Sweet potato post-harvest assessment: experiences from East Africa Natural Resources Institute. Chatham, U.K. (ISBn 0 85954 5482).

Owori, C., and V. Hagenimana. 2000. Quality evaluation of sweet potato flour processed in different agro-ecological sites using small scale processing technologies. African Potato Association Conference Proceedings. 5:483–490.

Pearson, D. A. 1976. The chemical analysis of foods, 7th ed. Churchill Livingstone, Edingburgh.

Peters, D., and Wheatley. 1997. Quart. J. Int. Agric. 36(4):331 in Van Hal, M. 2000. Quality of sweet potato flour during processing and storage. Food Rev. Int. 16(1), 1–37.

Picha, D. H. 1985. HPLC determination of sugars in raw and baked sweet potatoes. J. Food Sci. 50:1189–1190.

Ravindran, V., G. Ravindran, R. Sivakanesan, and S. B. Rajaguru. 1995. Biochemical and nutritional assessment of tubers from 16 cultivars of sweet potato (Ipomoea batatas L.). J. Agric. Food Chem. 43:2646–2651.

Rojas, J. A., C. M. Rosell, and B. C. de Barber. 1999. Pasting properties of wheat flour-hydrocolloid systems. Food Hydrocolloids 13:27–33.

Ruales, J., S. Valencia, and B. Nair. 1993. Effect of processing on the physico-chemical characteristics of quinoa flour. Starch/Starke 45:13–19.

Shittu, T. A., L. O. Sanni, S. O. Awonorin, B. Maziya-Dixon, and A. Dixon. 2007. Use of multivariate techniques in studying the flour making properties of some CMD resistant cassava clones. Food Chem. 101:1634–1643.

Sosulki, F. W., M. O. Garratt, and A. E. Slinkard. 1976. Functional properties of ten legume flours. Can. Inst. Food Sci. Technol. J. 9:66–69.

Tagodoe, A., and W. K. Nip. 1994. Functional properties of raw and precooked taro (Colocasia esculenta) flours. Int. J. Food Sci. Technol. 29:457–462.

Wiessenborn, D. P., P. H. Orr, H. H. Casper, and B. K. Tacke. 1994. Potato starch paste behavior as related to some physical/chemical properties. J. Food Sci. 59:644–648.

Williams, P. C., F. D. Kuzina, and I. Hylinka. 1970. A rapid colorimetric method for estimating amylose content of starches and flours. Cereal Chem. 47:411–421.

Woolfe, J. 1992. Sweet potato: an untapped food resource. Cambridge Univ. Press, Cambridge, U.K.

Yadav, A. R., M. Guha, R. N. Tharanathan, and R. S. Ramteke. 2006. Changes in characteristics of sweet potato flour prepared by different drying techniques. LWT 39:20–26.

10

Effects of fermentation time on the functional and pasting properties of defatted *Moringa oleifera* seed flour

Omobolanle O. Oloyede, Samaila James, Ocheme B. Ocheme, Chiemela E. Chinma & V. Eleojo Akpa

Department of Food Science and Technology, Federal University of Technology, PMB 65, Minna, Niger State, Nigeria

Keywords

Fermentation, functional, Moringa seed, pasting, solvent extractionsolvent extraction

Correspondence

Samaila James, Department of Food Science and Technology, Federal University of Technology, PMB 65, Minna, Niger State, Nigeria.

E-mail: samaila.james@futminna.edu.ng

Funding Information

No funding information provided.

Abstract

Effects of fermentation time on the functional and pasting properties of defatted Moringa oleifera seed flour was examined. Moringa seeds were fermented naturally at 0, 12, 24, 48 and 72 h; oven dried at 60°C for 12 h; milled into five different flour samples for each fermentation time and defatted. The functional and pasting properties of the samples were determined. The result shows significant increase in the water absorption capacity, oil absorption capacity, foaming capacity and emulsifying capacity with increase in fermentation time. However, there was a significant decrease in bulk density (0.53–0.32 g/cm^3) and dispersibility (36.00–20.50%) with an increase in fermentation time. There were significant increase in peak viscosity, trough, breakdown, final viscosity, and set back with increasing fermentation time. The swelling power and solubility of fermented Moringa seed flour was significantly affected.

Introduction

The insufficiency of animal protein and costliness of few available plant sources have prompted intense research in the area of local and underutilized vegetable plant proteins and legume seeds. The efforts would significantly improve nutrition and house hold food security (National Research Council, 2006).

Moringa oleifera is an important underutilized traditional vegetable tree widely cultivated in India and many countries in tropical Africa (Morton 1991; Anjorin et al. 2010; Ogunsina et al.,2010). It is commonly known as drum stick tree or horseradish tree in English and locally known as *Zogale* among the Hausa-speaking people of Nigeria (Anjorin et al. 2010). *Moringa* possesses many valuable properties, which is of great scientific interest. This includes the high-protein content (36.18%) in the seed which is not only abundant in good essential amino acids, but can be used to supplement cereal and tubers (Foidl et al. 2001). It is also rich in fat and oil, provitamins, and minerals as compared to most fruits. (Makkah and Becker 1997; Ogunsina et al., 2011).

Moringa is regarded as a versatile plant due to its multiple uses (Anwar and Rashid 2007). The incorporation of Moringa seed flour into maize flour for making cookies has been reported (Aluko et al. 2011). Different parts of the plant are edible and used as food. They include the leaves which are cooked and eaten like spinach, salad, or to make soup. The young green pods are boiled and eaten like green beans and the dry seeds are roasted and eaten like peas. Moringa seed is ground in to flour and used domestically in soup seasoning and industrially used as a flocculating agent for water purification(Gassenchmidt et al., 2005). Oil from the seed is used for cooking and as a solidifying agent in margarine production and other foodstuff that contain solid and semisolid fats thereby eliminating the hydrogenation process (FDA 2001).

Fermentation is an effective processing method used to improve nutritional quality of plant food as well as eliminating antinutritional factors (Steinkraus 1995). Fermentation improves digestibility, extends shelf life, and enhances the flavor and taste of raw seed (Achi and Okereke 1999).

Solvent extraction is a unit operation that involves separation of specific component of food, such as the removal of a desired component (the solvent) which is able to dissolve the solute (Clarke 1990). The solvent n-hexane is commonly used due to its extraction efficiency and ease of availability. It is preferred due to its low boiling point, low greasy residual, low corrosiveness, and high stability (Niosh 2007).

Processing methods tend to affect the characteristic of protein, carbohydrate, lipid, and their behavior in food systems as well as sensory properties of foods. The application of food flour in food systems depends largely on the knowledge of their functional properties. This is because of their influence on textural properties, sensory attributes, and consumer acceptability on their finished products (Adebowale et al. 2005; James and Nwabueze 2014). It is therefore necessary to study the effect of fermentation on the functional and pasting properties of defatted Moringa seed flour. The expected results would suggest the food applicability of the treated flour thereby expanding its usage.

Materials and Methods

Source of materials

Moringa seeds used for the study were purchased from Kure Ultra-Modern Market, Minna, Niger State, Nigeria.

Raw material preparation

Preparation of fermented *Moringa* seed flour

The method described by Oluwafemi and Ikeowa (2005) was adopted for the preparation of fermented *Moringa* flour with slight modification which was drying the fermented seeds before milling. About 100 grams of fresh *Moringa* seeds was put into four different 1000 mL beakers, and 500 mL of distilled water was added to each beaker. Moringa seeds were fermented for 12, 24, 48, and 72 h, respectively, at a room temperature 29 ± 02°C. At the end of the fermentation period, the seeds were washed with tap water, drained, and dried in a Gallenkamp oven at 60°C for 12 h. The dried seeds were then milled to

flour and passed through 100 μm mesh size. The resultant flour from each fermentation time was coded and stored in airtight containers at 4°C for further processing. The untreated *Moringa* flour was used as control.

Solvent extraction

The method described by Nwabueze and Iwe (2010) was adopted. Each flour sample was placed in a food grade hexane at 1:3 (flour: n-hexane) ratio for 3 h at room temperature and centrifuged at 3000 rpm for 10 min. The flour was separated from the supernatant and spread under the fan in order to remove the residual solvent (hexane) in the sample. The defatted flour was dried in an air convection oven (Gallenkamp, England) at 60°C for 12 h to further reduce the moisture level. The resultant defatted flour were milled using a hammer mill in order to break flour lumps and stored in an airtight container in the refrigerator at 4°C for further analysis.

Methods

Functional properties

The bulk density, oil, and water absorption capacity were determined by the method described by Adebowale et al. (2005). The dispersibility of the flour samples was measured by the method described by Kulkarin et al. (1991). The method described by Yatsumatsu et al. (1992) was used to determine the emulsion capacity. Foam capacity was determined by the method of Adebowale et al. (2005).

Determination of pasting properties of *Moringa* seed flour

Pasting parameters were determined using rapid visco analyzer (Newport Scientific Pty Ltd., Warrie-Wood, NSW, Australia). About 2.5 g of the flour sample was weighed into a dried empty canister; then 25 mL of distilled water was dispensed into the canister containing the sample. The suspension was thoroughly mixed and the canister was fitted into the rapid visco analyzer. Each suspension was kept at 50°C for 1 min and then heated up to 95°C at 12.2°C/min and held for 2.5 min at 95°C. It was then cooled to 50°C at 11.8°C/min and kept for 2 min at 50°C.

Determination of swelling power and solubility

The method of Sathe and Salunkle (1981) was used to determine the swelling power and solubility of the flour samples with slight modification in temperature. One gram of the flour sample was weighed into a previously tarred 50 mL centrifuge tube and 40 mL of 1% starch suspension (w/v) was added. The slurry was heated in a water bath at 60, 80, and 100°C, respectively for 15 min. During heating, the slurry was stirred gently to prevent clumping of the starch. After 15 min, the tubes containing the slurry were centrifuged at 3000 rpm for 10 min. The supernatant was decanted and the swollen granules weighed. A 10 mL sample was taken from the supernatant, placed in a crucible, and dried in an air convention oven at 120°C for 4 h to constant weight. The average value of the triplicate determinations was recorded.

Statistical analysis

Data were analysed using analysis of variance (Steel and Torrie 1980). The difference between mean values was determined by the least significant different test. Significance was accepted at the 5% probability level.

Results and Discussion

Functional properties of fermented seed *Moringa* flours

The functional properties of defatted *Moringa* flour samples are shown in Table 1. The result shows that water absorption capacity of the flour at different fermentation times was significantly ($P < 0.05$) higher than the control. This implies that there was an increase in water absorption capacity with increasing fermentation time. Gomez and Aguilera (1983) explained that low water absorption capacity value for raw sample is an indication of intact starch granules in the raw flour. The result of this study compares favorably with 1.16 g/mL and 1.31 g/mL for full fat and defatted Moringa seed flour, respectively (Ogunsina et al. 2014). Furthermore, the result shows similar trend reported by Alfaro et al. (2004); Filli et al. (2010) and James and Nwabueze (2014) for soya bean flour, extruded millet-soybean flour, and extruded African breadfruit flour mix, respectively. The increase in water absorption capacity of

Moringa seed flour could be due to the modification of macromolecules during fermentation. The modification exposes the hydrophilic domains of macromolecules which have high affinity for water. The significantly high value (2.31 g/mL) at the 48 h fermentation time represents the peak of bio modification. The high value recorded in this study implies the suitability of the flour and its isolates for incorporation into aqueous food formulation especially those involving dough handling.

The oil absorption capacity of the flour samples ranged from 0.87 to 1.91 g/mL. There was a significant ($P < 0.05$) increase in the oil absorption capacity of the flour samples with increase in fermentation time. Sample D had the highest value (1.91 g/mL). The oil-binding capacity reported in this study compares favorably with 1.30 and 2.08 g/mL for full fat and defatted Moringa seed flour, respectively (Ogunsina et al. 2014). The result is also in line with Periago et al. (1998) who reported that pea flour showed an increase in oil absorption capacity with fermentation time. According to Fagbemi (1999), good oil absorption capacity of flour samples suggest that they may be useful in food preparations that involve oil mixing as in bakery products, where oil is an important ingredient. Therefore, Moringa seed flour may be used to replace some legumes and oil seeds as thickeners in some liquid and semiliquid food reparations. It can also act as a flavor retainer, which implies that it can be incorporated in food for improved taste.

The emulsion capacity of fermented Moringa flour ranged from 50.71% to 68.75%. The result in this study agrees with Ogunsina et al. (2014) for raw Moringa seed flour however, low compared with 97.2% for defatted sample. Sample E was significantly ($P < 0.05$) high in the emulsion capacity while the control was significantly ($P < 0.05$) low. The emulsion capacity increased with increase in fermentation time. Emulsion capacity indicates the maximum amount of oil that can be emulsified by protein dispersion, whereas emulsion stability indicates the ability of an emulsion with a certain composition to remain unchanged (Enujiugha et al. 2003). Emulsion

Table 1. Effect of fermentation time on the functional properties of Moringa seed flour.

Functional properties	A	B	C	D	E
Water-holding capacity (g/mL)	0.86[c] ± 0.05	1.54[b] ± 0.01	1.59[b] ± 0.12	2.31[a] ± 0.33	1.81[b] ± 0.12
Oil-binding capacity (g/mL)	1.02[c] ± 0.02	0.87[d] ± 0.01	1.69[b] ± 0.02	1.91[a] ± 0.02	1.68[b] ± 0.014
Emulsifying capacity (%)	50.71[e] ± 0.01	60.85[d] ± 0.21	65.16[c] ± 0.31	65.96[b] ± 0.06	68.75[a] ± 0.01
Foaming capacity (%)	9.90[c] ± 0.14	16.31[a] ± 0.51	9.76[c] ± 0.06	13.87[b] ± 0.19	9.84[c] ± 0.05
Dispersibility (%)	36.00[a] ± 1.414	21.50[d] ± 0.71	24.00[c] ± 0.00	29.00[b] ± 0.00	20.50[d] ± 0.71
Bulk density (g/cm³)	0.60[a] ± 0.01	0.32[b] ± 0.01	0.37[b] ± 0.04	0.39[b] ± 0.02	0.54[a] ± 0.08

Values are means and standard deviation of two determinations.
Value followed by the same superscript letters in a column are not significantly ($P > 0.05$) different.
Keys: A = Control (unfermented Moringa flour); B = 12 h fermented Moringa flour; C = 24 h fermented Moringa flour; D = 48 h fermented Moringa flour; E = 72 h fermented Moringa flour.

capacity is an important consideration in the production of pastries and frozen desserts. Therefore, Moringa flour may thus be useful in such food formulations. The increasing trend in emulsion capacity is an indication that the flour is suitable in making cake batter, mayonnaise, and salad dressing among others.

The foaming capacity of fermented Moringa flour significantly ($P < 0.05$) increased during the 12 h fermentation time. However, during the 24 and 72 h fermentation periods, the value significantly ($P < 0.05$) reduced. Foaming capacity has been reported to improve the texture, consistency, and appearance of foods (Akubor and Chukwu 1999). Foam formation and stability are dependent on pH, viscosity, surface tension, and processing methods. The foaming properties recorded in this study is higher than those recorded for pumpkin by Oshodi and Fagbemi (1992) and germinated tiger nut varieties reported by Chinma et al. (2009). The trend in the foam capacity in this study agrees with James and Nwabueze (2014) for extruded African breadfruit flour mix. The foam capacity at 12 h fermentation time (16.31%) compares favorably with 20.60% for raw Moringa seed flour, however, the value was found to be high compared to 9.90% for the same raw Moringa seed flour reported in this study. The difference could be attributed to species variation and climatic difference among others. High foaming capacity by samples B and D implies that the flour samples could be used for leavening food products such as baked food, cakes, and biscuits.

The bulk density ranged from 0.32 to 0.59 g/cm³. There was significant ($P < 0.05$) decrease in the bulk density of the flour samples with respect to fermentation time. However, at 72 h fermentation period (sample E), the value significantly increased. The bulk density of a flour sample influences the amount and the strength of packaging material; texture, and mouth feel (Wilhelm et al. 2004). Therefore, the low value of bulk density obtained from this study makes the samples desirable for packaging.

Pasting properties of fermented *Moringa* seed flours

Pasting properties of flours are parameters used in determining the suitability of its application as functional ingredient in food and other industrial products. The effect of fermentation on the pasting properties of *Moringa* flour is presented in Table 2. Peak viscosity value ranged from 15.00 to 34.00 RVU. Peak viscosity is the ability of starch to swell freely before their physical break down. Peak viscosity indicates the water-binding capacity of starch. It generally depends on solubility and water-holding capacity as well as the structure of components in a food system (Leszek 2011). Sample C had the highest value while sample D had the least value. The value increased with increasing fermentation time at 12 and 24 h fermentation periods however, significantly ($P < 0.05$) reduced during the 48 h fermentation time. Studies have shown that flour with lower peak viscosity have a lower thickening power than flour with high peak viscosity which is attributed to many factors. Egouletey and Aworh (1991) reported that protein and fat interaction in the blend of African yam, beans, and cassava starch lowers the peak viscosity. Furthermore, Belitz and Grosch (1999) and James and Nwabueze (2014) reported that extrusion cooking lowers peak viscosity of extrudates. Therefore, to increase the pasting characteristics of Moringa flour, there is need for the flour to be blended with other high pasting flours.

The final viscosity of the flour samples ranged from 16.00 to 36.00 RVU. Sample E had the highest value while sample D the lowest value. Shimel et al. (2006) reported that final viscosity of flour sample is the ability of starch to form paste and gel during cooking and cooling. The final viscosity of starch paste is related to the amylose content. This implies that the flour with a higher amylose content gives a higher viscosity and flour with a lower amylose content gives a lower viscosity. The final viscosity significantly ($p < 0.05$) increased at the 24 h

Table 2. Pasting properties of fermented Moringa seed flour.

Pasting properties (RVU)	A	B	C	D	E
Trough	17.00[b] ± 1.41	17.00[b] ± 1.41	21.50[a] ± 0.71	11.00[c] ± 0.00	18.50[b] ± 0.71
Breakdown	7.00[b] ± 1.41	13.50[a] ± 0.71	12.50[a] ± 0.71	4.00[c] ± 0.00	14.50[a] ± 0.71
Final viscosity	27.50[b] ± 2.12	30.00[b] ± 1.41	34.50[a] ± 0.71	16.00[c] ± 0.00	36.00[a] ± 0.00
Setback	10.50[c] ± 0.71	13.00[b] ± 0.00	13.00[b] ± 0.00	5.00[d] ± 0.00	17.50[a] ± 0.00
Peak time	4.93[a] ± 0.94	4.37[a] ± 0.05	4.90[a] ± 0.05	4.57[a] ± 0.24	4.23[a] ± 0.05
Peak	24.00[b] ± 2.83	30.50[a] ± 0.71	34.00[a] ± 1.41	15.00[c] ± 0.00	33.00[a] ± 1.41

Values are means and standard deviation of two determinations.
Value followed by the same superscript letters in a column are not significantly ($p > 0.05$) different.
Keys: A = Control (unfermented Moringa flour); B = 12 h fermented Moringa flour; C = 24 h fermented Moringa flour; D = 48 h fermented Moringa flour; E = 72 h fermented Moringa flour.

Table 3. Effect of fermentation time on the swelling power of Moringa seed flour.

Temperature (°C)	Swelling power				
	A	B	C	D	E
60	$0.88^e \pm 0.06$	$80.78^a \pm 0.08$	$3.62^c \pm 0.03$	$2.12^d \pm 0.02$	$3.95^b \pm 0.08$
80	$6.33^a \pm 7.00$	$10.18^a \pm 0.10$	$5.61^a \pm 0.00$	$6.34^a \pm 0.15$	$6.10^a \pm 0.01$
100	$11.13^b \pm 1.59$	$17.25^a \pm 0.00$	$6.69^c \pm 0.04$	$7.07^c \pm 0.60$	$7.59^c \pm 0.02$

Values are means and standard deviation of two determinations.
Value followed by the same superscript letters in a column are not significantly ($P > 0.05$) different.
Keys: A = Control (unfermented Moringa flour); B = 12 h fermented Moringa flour; C = 24 h fermented Moringa flour; D = 48 h fermented Moringa flour; E = 72 h fermented Moringa flour.

fermentation period but, significantly reduced at the 48 h and latter increased at the 72 h fermentation period. The increase could be attributed to the breakdown of complex carbohydrates to lower sugars during fermentation.

Setback is related to the amylose content and reflects the retrogradation of starch. The setback values of the flour samples ranged from 5.00 to 17.50 RVU. Sample E (72 h fermentation time) was significantly ($P < 0.05$) high while, sample D (48 h fermentation time) was significantly ($P < 0.05$) low. The higher the setback value, the lower the retrogradation during cooling of the product made from the flour (James and Nwabueze 2014). This implies that Moringa seed flour fermented for 3 days would have high gel stability compared with others. Trough viscosity is the maximum viscosity value at the constant temperature phase of the RVU profile and measures the ability to withstand breakdown during cooling (Chinma et al. 2009). The trough value ranged from 11.00 to 21.50 RVU. Sample C was found to be significantly ($P < 0.05$) high in trough viscosity. Fermentation time significantly ($P < 0.05$) increased the trough at 24 h while at 48 h the fermentation time significantly ($P < 0.05$) reduced.

The breakdown value ranged from 4.00 to 14.50 RVU. The higher the break down viscosity, the lower the ability of the sample to withstand heating and shear stress during cooking (Adebowale et al. 2005). Fermentation time significantly ($P < 0.05$) increased the breakdown values of the flour samples except in sample D. The result

suggests that increased breakdown value with increase in fermentation time would enable easy cooking but, susceptible to stress when processed into the solid form. Breakdown viscosity value is a measure of ease with which the swollen granules can be disintegrated and hence an indicator of the stability of the flour product (Kaur and Singh 2005).

The pasting time of the flour samples ranged from 4.23 to 4.93 min. Fermentation time did not significantly ($P > 0.05$) affect the pasting time of the flour samples. Pasting time is a measure of the minimum time and temperature required to cook flour (Chinma et al. 2009). The pasting time was statistically similar for all the samples. This implies that the fermentation time has no influence on the pasting time of raw and fermented Moringa seed flour.

Swelling power of defatted Moringa seed flour at different temperatures

The swelling power of defatted Moringa seed flour at different temperatures is shown in Table 3. The value increased with an increase in fermentation time. Sample B (12 h) was significantly ($P < 0.05$) high while sample D was significantly ($P < 0.05$) low. Increased swelling power in fermented samples could be due to modification of starch granules during fermentation, resulting in higher water uptake by the granules. Claver et al. (2010) reported that temperature increase and vigorous

Table 4. Effect of fermentation time on the solubility of Moringa seed flour.

Temperature (°C)	Solubility C				
	A	B	C	D	E
60	$16.00^d \pm 0.00$	$107.00^a \pm 7.07$	$73.50^b \pm 2.12$	$41.50^c \pm 2.12$	$66.00^b \pm 2.83$
80	$45.00^c \pm 1.41$	$74.50^a \pm 2.12$	$65.00^b \pm 1.14$	$67.00^b \pm 4.24$	$67.00^b \pm 1.41$
100	$91.00^b \pm 1.41$	$105.00^a \pm 4.24$	$68.00^c \pm 0.00$	$67.00^c \pm 1.41$	$70.00^c \pm 2.83$

Values are means and standard deviation of two determinations.
Value followed by the same superscript letters in a column are not significantly ($P > 0.05$) different.
Keys: A = Control (unfermented Moringa flour); B = 12 h fermented Moringa flour; C = 24 h fermented Moringa flour; D = 48 h fermented Moringa flour; E = 72 h fermented Moringa flour.

starch vibration break intermolecular bonds, thereby allowing hydrogen bonding sites to accommodate more water molecules. At 80°C, the swelling power of the flour samples has no significant ($P > 0.05$) difference. This implies that fermentation time had no influence on swelling power at different temperatures. At 100°C, the swelling power decreased with increase in fermentation time with the exception of sample B which is significantly higher.

Solubility of defatted Moringa seed flour at different temperatures

The solubility of fermented Moringa seed flour at different temperatures is shown in Table 4. The solubility at 60 and 80°C increased with increase in the fermentation time. Sample B (12 h fermentation) was found to be significantly higher in solubility at both 60 and 80°C. At 100°C, the solubility significantly decreased in value with increase in fermentation time except at 12 h fermentation time. The decrease could be due to denaturation of protein present in the flour as a result of high temperature (100°C). Fermentation time shows increase in the solubility of Moringa flour samples, especially at 60 and 80°C. Solubility of flour is an indication of its quality and digestibility therefore, Moringa flour would be suitable for incorporation to low soluble flour to improve solubility.

Conclusion

Fermentation time showed beneficial effects on the functional and pasting properties of treated Moringa seed flour. Fermentation time significantly increased the water and oil absorption capacity as well as the emulsion capacity. However, dispersibility and bulk density were significantly reduced with fermentation time. These aspects imply the potential use as ingredient in food such as sauces, infant food, cakes, and bread products. The decreased in bulk density is desirable during flour packaging. Fermentation significantly increased trough viscosity, breakdown, final viscosity, setback, peak viscosity, and reduced the pasting time. The swelling power of the fermented flour showed significant increase at 60 and 100°C. The solubility of the flour was significantly affected at all temperatures, however, at 60°C, the samples had a significantly high solubility.

Conflict of Interest

None declared.

References

Achi, O. K., and E. G. Okereke. 1999. Proximate composition and functional properties of *Prosopis Africana* seed flour. J. Manage. Technol. 1:7–13.

Adebowale, Y. A., I. A. Adeyemi, and A. A. Oshodi. 2005. Functional and physiochemical properties of flours of six *Mucuna species*. Afr. J. Biotechnol. 4:1461–1468.

Akubor, P. I., and J. K. Chukwu. 1999. Proximate composition and selected functional properties of fermented and unfermented African oil bean (*Pentaclethra macrophylla*) seed flour. Plant Food Hum. Nutr. 54:227–238.

Alfaro, M. J., I. Alvarez, E. I. Khor, and F. C. Padika. 2004. Functional properties of protein products from *Barinus* nut. Am. J. Nutr. 54:223–228.

Aluko, O., M. R. Brai, and A. O. Adelore. 2011. Evaluation of sensory attributes of snack from maize-Moringa flour blends. Int. J. Biol. Food Vet. Agric. Eng. 7:597–599.

Anjorin, T. S., P. I. Kokoh, and S. Okolo. 2010. Mineral composition of *Moringa Oleifera* leaves, pods and seed from two regions in Abuja, Nigeria. Int. J. Agric. and Biol. 12:431–434.

Anwar, F., and U. Rashid. 2007. Physiochemical characteristics of *Moringa Oleifera* seeds and seed oil from a wild provenance of Pakistan. Pak. J. Bot. 39:1443–1453.

Belitz, H. D., and W. Grosch. 1999. Pp. 992. Food chemistry. Springer, Berlin Heidelberg, New York.

Chinma, C. E., O. Adewuyi, and O. J. Abu. 2009. Effect of germination on the chemical functional and pasting properties of flour from brown and yellow varieties of tiger nut (*Cyperus esculentus*). Food Res. Int. 42:1104–1109.

Clarke, R. J.. 1990. Instant coffee technology in a tunner (ED) Food Technology International Europe. Pp. 137–139. Sterling Publication International London, Great Britain.

Claver, I. P. H., Q. L. I. Zhang, Z. Kexue, and H. Zhou. (2010). Optimization of ultrasonic extraction of polysaccharides from Chinese malted sorghum using a response surface methodology. Pakistan Journal of Nutrition, 9:336–342.

Egouletey, M., and O. C. Aworh (1991). Production and Physioco-chemical properties of tempe-fortified maize based weaning food. Pp. 5–7 *in* S. Sefe Dedeh, ed. Proceedings of seminar on development of the protein energy foods from grain legumes. University of Lagos, Nigeria.

Enujiugha, V. N., A. A. Badejo, S. O. Iyiola, and M. O. Oluwamukomi. 2003. Effect of germination on the functional properties of African oil bean (*Penthaclethra macrophylla Benth*) seed flour. J. Food Agric. Environ. 1:72–75.

Fagbemi, T. N. 1999. Effect of blanching and ripening on functional properties of plantain (*Musa* spp) flour. Foods Hum. Nutr. 54:261–269.

Filli, K. B., I. Nkama, U. M. Abubakar, and V. A. Jideani. 2010. Influence of extrusion variables on some functional properties of extruded millet soybean for the manufacture of 'Fura': a Nigerian Traditional Food. Afr. J. Food Sci. 4:342–352.

Foidl, N. H., P. S. Makkah, and K. Becker. 2001Potential of *Moringa oleifera* in agriculture and industry. Pp. 20. Potential of Moringa Products Development, Dar Es Salaam, Tanzania.

Food and Drug Administration Agency Response Letter 2001. G.R.A Notice. 000069. Washington (DC).

Gassenschmidt, U., K. D. Janny, B. Tauscher, and H. Niebergall. 2005. Isolation and characterization of a flocculating protein from *Moringa oleifera*. Biochem. Biophys. Acta. 1243:477–481.

Gomez, M. H., and J. M. Aguilera. 1983. Changes in the starch fraction during extrusion-cooking of corn. J. Food Sci. 48:378–381.

James, S., and T. U. Nwabueze. 2014. Influence of extrusion condition and defatted soybean inclusion on the functional and pasting characteristics of extruded African breadfruit (*Treculia africana*) flour blends. Food Sci. Qual. Manage. 34:26–33.

Kaur, M., and N. Singh. 2005. Studies on functional, thermal and pasting properties of flour from different chick per cultivars. Food Chem. 91:403–411.

Kulkarin, K. D., D. N. Kulkarin, and U. M. Ingle. 1991. Sorghum malt-based weaning formulations, preparation, functional properties and nutritive value. Food Nutr. Bull. 13:327–332.

Leszek, M. 2011. Extrusion-cooking techniques applications, theory and sustainability. Wiley-VCH Verlag & Co, KGaA, Weinheim, Germany.

Makkah, H. P. S., and K. Becker. 1997. Nutrient and anti-quality factors in different morphological parts of the *Moringa Oleifera* tree. J. Agric. Sci. 128:311–322.

Morton, J. F. 1991. The horseradish tree, Moringa pterigosperma (*Moringaceae*). A boon to arid lands. Econ. Bot. 45:318–333.

National Research Council. 2006. Pp. 247. "Moringa" lost crops of African. Vegetable lost crops of Africa. National Academies Press, Abuja, Nigeria.

NIOSH. 2007. Pocket guide to chemical hazards. National Institute for Occupation Safety and Health, Washington DC, USA.

Nwabueze, T. U., and M. O. Iwe. 2010. Residence time and distribution (RTD) in a single screw extrusion of African breadfruit mixtures. Food Bioprocess Technol. 3:135–145.

Ogunsina, B. S., T. N. Indira, A. S. Bhatnagar, C. Radha, D. Sukumar, and A. G. Gopalakrishna. 2014. Quality characteristics and stability of *Moringa oleifera* seed oil of Indian origin. J. Food Sci. Technol. 51:503–510.

Ogunsina, B. S., C. Radha, and D. Indra. 2011. Quality characteristics of bread and cookies enriched with debittered *Moringa Oleifera* seed flour. International Journal of Food Sciences and Nutrition, 62:185–194.

Ogunsina, B. S., C. Radha, and R. S. V. Sign. 2010. Physicochemical and functional properties of full-fat and defatted Moringa oleifera kernel flour. International Journal of Food Science and Technology, 45:2433–2439.

Oluwafemi, F., and M. C. Ikeowa. 2005. Fate of aflatoxin B_1 during fermentation of *Ogi*. Nigeria Food J. 23:243–247.

Oshodi, A. A., and T. N. Fagbemi. 1992. Functional properties fluted pumpkin seed flour of defatted flour and protein isolates. Ghana J. Chem. 1:216–226.

Periago, M. J., M. L. Vidal, G. Ros, F. Rincon, C. Martinez, G. Lopez, et al. 1998. Influence of enzymatic treatment on the nutritional and functional properties of pea flour. Food Chem. 63:71–78.

Sathe, S. K., D. K. Salunkhe. 1981. Functional properties of great northern bean proteins: Emulsion, foaming, viscosity, and gelation properties. Journal of Food Science, 46:71–75.

Shimel, A. E., M. Meaza, and S. Rakshit. 2006. Physic-chemical properties, pasting behaviour and characteristics of flour and starch from improved Bean (*Phaseoluus vulgaris* L.) Varieties Grown in East Africa. CIGRE 8:1–18.

Steel, R. D. G., and J. H. Torrie, 1980. Principle and procedures of statistics. A biometrical approach. 2nd ed. Pp. 623. McGraw Hill Co., New York.

Steinkraus, K. H. 1995. Indigenous fermented foods involving alkaline. Fermented Food Res. Int. 54:234–237.

Wilhelm, L. R., A. S. Dwayna, and H. B. Gerand. 2004. Introduction to problem solving skills in food and process engineering technology. ASAE, Canada.

Yatsumatsu, K., K. A. Sawada, S. Moritaka, M. Misaki, J. Toda, and T. Wada. 1992. Whipping and emulsifying properties of Soya bean products. J. Agric. Biol. Chem. 36:717–725.

11

Optimization of some processing parameters and quality attributes of fried snacks from blends of wheat flour and brewers' spent cassava flour

Adebukola T. Omidiran[1], Olajide P. Sobukola[1], Ajoke Sanni[2], Abdul-Rasaq A. Adebowale[1], Olusegun A. Obadina[1], Lateef O. Sanni[1], Keith Tomlins[3] & Tosch Wolfgang[4]

[1]Departments of Food Science and Technology, Federal University of Agriculture, Abeokuta, Nigeria
[2]Nutrition and Dietetics, Federal University of Agriculture, Abeokuta, Nigeria
[3]Natural Resources Institute, University of Greenwich, Greenwich, U.K.
[4]SABMiller Plc, Surrey, U.K.

Keywords
Brewers' spent cassava flour, color, fried snack, microstructure, optimization, sensory

Correspondence
Adebukola T. Omidiran, Department of Food Science and Technology, Federal University of Agriculture, Pmb 2240, Abeokuta, Nigeria.

Funding Information
EU ACP-FP7 GRATITUDE (Gains from Losses of Root and Tuber crops

Abstract

The effect of some processing parameters (frying temperature [140–160°C], frying time [2–4 min], level of brewers' spent cassava flour (BSCF) [20–40%], and thickness [2–4 mm]) on some quality attributes of wheat-BSCF fried snack was investigated. Response surface methodology based on Box–Behnken design was used to optimize the effect of process parameters on product quality. Sensory evaluation of the optimized sample to determine its level of acceptability was carried out as well as the comparison with fried snack from 100% wheat flour. Increasing temperature had significant ($P < 0.05$) negative effect on the texture. Based on the desirability (0.771) concept, a frying temperature of 140 °C, frying time of 4 min, 32% level of BSCF, and 2 mm thickness was obtained as the optimized conditions. Sensory analyses showed that the optimized sample was preferred in terms of texture and its oiliness to fried snack prepared from 100% wheat flour, but, the aroma, taste and appearance of the wheat snack were preferred.

Introduction

Many formulated products are based on wheat flour (among other components) and its popularity is largely determined by the ability of the wheat flour to be processed into different products for example, a snack, which is mainly given by the unique properties of wheat-flour gluten proteins (Anjum et al. 2007). Snack foods are an integral part of the diet which constitute an important part of many consumers' daily nutrient and calorie intake (Amudha et al. 2002) and are typically produced to be durable, accessible, inexpensive, and easy to eat out of a bag or package without further preparation. Some common ones include biscuit, cake, chin chin among others. 'Chin chin' is a fried product made from 100% wheat flour, egg, baking powder, and sugar.

It is eaten as a snack by all classes of individuals in Nigeria. Cassava is one of the most important food security crops for approximately 700 million people. Postharvest losses are significant and come in three forms which are physical, economic through discounting or processing into low value products, and from bio-wastes. To reduce these losses, the role that cassava play in food and income security must be enhanced and new market opportunities for new products and added value products must be generated. In the manufacture of beer, various residues and by-products are generated. The most common ones are spent grains, spent hops, and surplus yeast, which are generated from the main raw materials (Mussatto 2009). The use of high quality cassava flour in brewing generates by-products called brewers' spent cassava mash which could be useful for food and industrial purposes. Brewers' spent cassava when converted to flour can be a rich source of fiber and protein as well as a functional marketable material. In fact, brewer spent grains obtained from brewing process using cereals has sparked interest in its usage as an adjunct in human food (Mussatto et al. 2006; Sobukola et al. 2013) especially due to an increase in the dietary fiber contents of foods. Spent cassava flour may also be a low-cost ingredient that can be used for snack production. Frying is an established process of food preparation worldwide. It is a simultaneous heat and mass transfer process where moisture leaves the food in the form of vapor bubbles, while oil is absorbed simultaneously. During the frying process of the snack, the physical, chemical, and sensory characteristics were modified. Various researches have been carried out on the use of brewers' spent grains, by-products from the use of barley and other grains for food and feed purposes (Awoyale et al. 2011; Sobukola et al. 2013). However, there is no information on the utilization of brewers spent cassava flour. Hence, the objective of this study is to evaluate some quality parameters of fried snacks produced from blends of wheat and brewers spent cassava flour.

Materials and Methods

The Brewers' spent cassava was obtained from SABMiller, U.K. Wheat flour, vegetable oil, sugar, baking powder, and butter were bought at a supermarket in Abeokuta, Ogun State, Nigeria.

Preparation of wheat-brewers spent cassava flour

The brewers' spent cassava was dried in a cabinet dryer at 70°C for 16 h to 7.78% moisture content; milled and sieved using 250 micron to obtain brewers' spent cassava flour (BSCF).Wheat flour and BSCF were weighed and mixed in the ratios as follows: 80:20%, 70:30%, and 60:40%. The various mixes were thoroughly blended and packed in low density polyethylene.

Product preparation

Dough sample was prepared using the method described by Gazmuri and Bouchon (2009). The dry ingredient proportion was modified to ensure that they all contained the specified amount of water added depending on the initial water content of the ingredients. To every 100 g of flour, 2 g of baking powder, 10 g of butter, and 20 g of sugar were added. The ingredients were mixed; distilled water was added to the dry mixture blend until it reached 40% water content (wb) to form the dough. Half of the water was heated at 100°C and added while mixing at room temperature. After mixing for 2 min, the rest of the water

was added. The dough was sheeted to get a final thickness of 2–4 mm which was then cut into cubes and fried.

Frying experiment

The dough (cubes) were placed inside the frying basket and covered with a grid to prevent them from floating. Frying was carried out in an electrically heated deep fryer (Bush Domestic FCO300, U.K.) containing 3 L of vegetable oil which was preheated to the test temperature for 1 h prior to frying. The product was fried by immersing the product in a wire basket in the oil for 2, 3, or 4 min for the 2 mm, 3 mm, and 4 mm thicknesses, at frying temperature of 140, 150, or 160°C accordingly. After frying, the samples were removed from the fryer and held on a stainless steel grid for 10 min to allow excess oil to drain from the fried products.

Optimization procedure

A four factor experimental set up was used with frying temperature (X_1), frying time (X_2), BSCF: wheat flour levels (X_3) and sample thickness (X_4) as the independent factors at three levels each as shown in Table 1. The data obtained was analyzed by response surface methodology (RSM) based on Box–Behnken design (Table 2) to optimize process variables. Twenty-nine combinations including four replicates of the center point was performed in random order according to the design.

Proximate composition

The composite flour (WF/BSCF) and fried snacks from it were analyzed for moisture, ash, and oil according to AOAC (2003). Protein was determined using Kjeldahl method (AACC, 46-12.01). The carbohydrate content was obtained by difference.

Color measurement of fried snacks

Color measurement was done using the technique explained by Papadakis et al. (2000). This was carried out by setting up a lightning system, using a high-resolution camera

Table 1. Coded values of the independent variables.

Variables	Codes		
	−1	0	+1
Frying temperature (°C)	140	150	160
Frying time (min)	2	3	4
Level of BSCF (%)	20	30	40
Thickness (mm)	2	3	4

Table 2. Experimental runs showing different combinations of the independent variables.

Experimental runs	X_1	X_2	X_3	X_4
1	150	4	20	3
2	140	3	40	3
3	140	3	30	2
4	150	2	30	4
5	150	3	30	3
6	150	2	40	3
7	150	3	30	3
8	160	4	30	3
9	160	3	30	4
10	150	3	40	4
11	150	3	40	2
12	150	3	30	3
13	150	2	20	3
14	150	3	30	3
15	150	4	30	4
16	150	4	30	2
17	150	4	40	3
18	150	3	20	2
19	140	3	20	3
20	140	2	30	3
21	140	4	30	3
22	140	3	30	4
23	160	2	30	3
24	150	2	30	2
25	160	3	30	2
26	150	3	30	3
27	150	3	20	4
28	160	3	40	3
29	160	3	20	3

Where X_1 = Temperature, X_2 = Frying time, X_3 = Level of BSCF, and X_4 = Thickness.

to capture images and Photoshop software to obtain color parameters. The image acquisition system consists of a color digital camera, Samsung HD 5X model which was used alongside a large box impervious to light with internal black surfaces. L, a, b coordinates was obtained using Adobe Photoshop 6.0 software, which was normalized to L^*, a^*, b^* coordinates, according to equations 1–3 (Yam and Papadakis 2004).

$$L^* = \frac{L}{255} \times 100 \tag{1}$$

$$a^* = a \times \frac{240}{255} - 120 \tag{2}$$

$$b^* = b \times \frac{240}{255} - 120 \tag{3}$$

The color difference between the raw (L_o^*, a_o^*, b_o^*) and fried (L^*, a^*, b^*) snack was determined by taking the Euclidean distance between them, according to Mariscal and Bouchon (2008) shown in equation (4):

$$\Delta E^* = ((L_o^* - L^*)^2 + (a_o^* - a^*)^2 + (b_o^* - b^*)^2)^{\frac{1}{2}} \tag{4}$$

Expansion analysis of fried snacks

Expansion was determined using a micrometer screw gage and was defined as the maximum height developed during frying (Gazmuri and Bouchon 2009). Reported values represent the mean of six measurements for each frying condition.

Texture measurement of fried snacks

Hardness of fried snacks was measured using the Texture analyser as described by Da Silva and Moreira (2008) using a three-point bending test where the sample is supported at two parallel edges and the load is applied centrally. The force (N) at the fracture point (highest value in the plot) was used as the resistance to breakage. The mean of three measurements for each frying condition is reported.

Sensory analysis of fried snacks

Acceptance test

The acceptance test was determined using the method described by Ihekoronye and Ngoddy (1985). Fifty consumer panellists made up of students of Federal University of Agriculture, Abeokuta, Ogun State, Nigeria evaluated the appearance, color, texture, oiliness, taste, and overall acceptability of fried snack prepared using the optimized frying conditions on a seven-point hedonic scale ranking seven for like extremely and one for dislike extremely. The average and mean values of scores for each of attributes was computed and analyzed statistically.

Preference test

A preference test was conducted to evaluate the sensory properties of fried snack from 100% wheat flour and optimized conditions. Thirty panelists made up of students of Federal University of Agriculture, Abeokuta, Ogun State, Nigeria were asked to compare each coded sample on basis of some specified characteristics (taste, aroma, texture, and overall appearance). Responses of the panellists were then analyzed statistically (Da Silva and Moreira 2008).

Scanning electron microscopy

The fried snacks were superficially defatted by immersing them in petroleum ether 35–60 for 2 h after frying. The samples were then coated with a thin gold layer (20 nm) using a Varian Vacuum Evaporator PS 10E (Evey

Engineering's Warehouse, Hoboken, NJ, USA) and analyzed using a variable pressure scanning electron microscope LEO 1420VP (LEO Electron Microscopy Ltd., Cambridge, U.K.) at an acceleration potential of 25 kV. An Oxford 7424 solid-state detector (Oxford Instruments, Oxford, U.K.) was used to obtain the electron microphotographs (Sobukola et al. 2013).

Statistical analysis

A second-order polynomial model for the dependent variables as shown in equation (5) was established to fit the experimental data. An analysis of variance (ANOVA) test was carried out using Design-Expert Version 6 (Stat-Ease, Inc., Minneapolis, MN) to determine level of significance at 5% level. The generalized regression model fitted was

$$Y = \beta_0 + \sum_{i=1}^{4} \beta_i X_i + \sum_{i=1}^{4} \beta_{ii} X_i^2 + \sum_{i<j=1}^{4} \beta_{ij} X_i X_j + \epsilon \qquad (5)$$

where Y is the response; β_0 is a constant; while β_i, β_{ii} and β_{iii} are linear, quadratic, and interaction coefficients, respectively; and ϵ is error.

Results and Discussion

Proximate composition of the flour and fried snacks

Proximate composition of the blends of brewers' spent cassava flour (BSCF) and wheat flour (WF) ranged as follows; 7.78–13.16% moisture, 13.14–16.01% protein, 2.50–3.65% ash, 67.74–69.11% carbohydrate, and 2.75–4.61% fat as presented in Table 3. The regression coefficients of the fried snacks vary between 0.42 and 0.76 as shown in Table 7. There were significant ($P < 0.05$) differences in the proximate composition of the blends. The higher protein content of the whole BSCF flour could be attributed to the addition of enzymes which are proteins to the cassava flour during brewing process to aid the conversion of starch into sugars. The proximate composition of the fried snack from BSCF

and wheat blends have the protein, moisture, ash, oil, carbohydrate content, and total dietary fiber, values ranging between 7.97–9.22%, 3–12.5%, 2–3%, 12.82–46.14%, 33.96–67.24%, and 3.43–3.87, respectively, as shown in Table 4. The shelf life of the fried products is mostly determined by the moisture content after frying and the values observed in this work suggests that the frying process reduces the final moisture contents of some of the fried snack products to a level that might be shelf stable. Ashworth and Draper

Table 4. Response surface analysis results of proximate composition of fried snack for the experimental runs.

Runs	Protein (%)	Moisture (%)	Ash (%)	Oil content (%)	Carbohydrate (%)	Total dietary fiber (%)
1	8.44	4.50	3.00	19.06	61.27	3.73
2	7.97	8.20	2.50	25.16	52.60	3.57
3	8.54	4.70	2.50	22.12	58.60	3.54
4	8.63	12.50	2.00	22.43	50.68	3.76
5	8.34	6.90	2.00	20.39	58.81	3.56
6	8.40	6.60	2.00	46.14	33.14	3.72
7	8.69	9.70	3.00	44.65	30.35	3.61
8	8.65	9.50	2.50	19.32	56.27	3.76
9	8.68	8.80	2.00	22.85	53.80	3.87
10	8.25	11.10	3.00	24.40	49.48	3.77
11	8.39	3.90	2.00	25.45	56.62	3.64
12	8.13	6.60	2.00	19.33	60.51	3.43
13	8.79	8.20	2.00	17.89	59.47	3.65
14	8.82	7.90	2.00	19.26	58.46	3.56
15	8.87	6.00	3.00	19.24	59.37	3.52
16	8.51	3.00	2.50	21.78	60.38	3.83
17	8.24	4.10	2.00	23.81	58.11	3.74
18	8.61	4.10	2.00	18.29	63.25	3.75
19	8.43	9.20	2.00	27.34	49.39	3.64
20	8.79	9.60	2.00	20.69	55.20	3.72
21	9.22	5.30	2.00	16.94	62.78	3.76
22	8.84	8.60	2.50	12.82	63.45	3.79
23	8.74	9.50	2.00	19.77	56.38	3.61
24	7.99	7.30	2.50	23.94	54.51	3.76
25	8.29	4.90	2.00	20.20	61.07	3.54
26	8.76	7.40	2.00	20.12	58.29	3.43
27	8.21	7.60	2.50	18.53	59.53	3.63
28	8.46	4.80	2.00	23.21	57.72	3.81
29	8.80	8.00	2.00	21.00	56.57	3.63

Values reported are means of duplicates.

Table 3. Proximate composition of Brewers' spent cassava – wheat flour blends.

	A	B	C	D	E
Protein (%)	16.01^d ± 0.92	14.09^c ± 0.14	14.0^c ± 0.18	13.7^b ± 0.50	13.14^a ± 0.14
Fat (%)	4.61^d ± 0.02	2.94^b ± 0.05	2.75^a ± 0.05	3.02^{bc} ± 0.02	3.06^c ± 0.04
Ash (%)	2.50^a ± 0.08	2.88^a ± 0.18	3.65^b ± 0.26	3.58^b ± 0.18	2.88^a ± 0.07
Moisture (%)	7.78^a ± 0.13	10.83^b ± 0.07	11.87^c ± 0.11	11.98^c ± 0.18	13.16^d ± 0.23
Carbohydrate (%)	69.11^b ± 0.14	69.27^b ± 0.04	67.74^a ± 0.02	67.73^a ± 0.39	67.77^a ± 0.20

Mean values followed by different superscript within the same row are significantly different ($P < 0.05$). Values are means of duplicates; A = 100% BSCF, B = 60:40 (W:BSCF), C = 70:30 (W:BSCF), D = 80:20 (W:BSCF), and E = 100WF.

(1992) reported that high-moisture products (>12%) usually have shorter shelf stability compared with low-moisture products (<12%). The lower the initial moisture content of a product, the better the storage stability of the product (Akubo 1997). The moisture level of the snacks decreased during frying as water vaporized, oil penetrated into the food. Ash content is similar in all products at 2–3%. The ash content of the fried snacks was noted to decrease significantly compared with the level in the BSCF itself. This could be attributed to the high level of wheat flour used (60, 70, and 80%) in all cases as these were sufficient enough to reduce its level in all samples. Oil content of the snacks reduced with increased frying time. The results may be explained by the formation of a crust, which acts as a barrier to reduce the oil uptake. The crust formation prevents the inside water from escaping to the outside and consequently preventing further oil uptake. Oil absorption is affected by the porosity of the product. Porosity increases during frying and longer frying times resulted in more uniform pore size distribution (Kawas and Moreira 2001). There were no significant ($P < 0.05$) effects by frying temperature, frying time, level of BSCF, and thickness on carbohydrate content.

Expansion and color of fried snacks

Table 5 shows the results of expansion and color parameters (texture, lightness, redness, yellowness, and change in color) of the fried snacks which ranged from 3.85 to 6.98 mm, 86.7 to 95.64, −1.48 to 4.03, 14.63 to 20.92, and 2.90 to 91.20, respectively. Expansion of the product was reported as the maximum height attained under different experimental conditions. Expansion decreased due to the fact that BSCF does not contain gluten that will support maximum expansion. Expansion was significantly ($P < 0.05$) affected by the level of BSCF, as the level of BSCF increased, it reduced expansion. Gazmuri and Bouchon (2009) and Sobukola et al. (2012) while working on fabricated matrices from wheat starch and vital gluten reported that products containing high amount of gluten and water tend to expand during frying with the gluten content of the matrix developing an elastic structure that traps water vapor producing an expanded product.

Texture of fried snacks

One important quality parameter of desirable textural characteristic of fried foods is crispness because it signifies freshness and high quality. Breaking force reduces if the fried snack becomes crispier and this could be made possible by increasing the frying time and temperature. This is in agreement with Rossell (2001) who reported that at higher frying temperature and time, crust formation is enhanced. At higher frying temperature and time, texture

Table 5. Response surface analysis results of color parameters, expansion, and texture of fried snack for the experimental runs.

Runs	Expansion (mm)	Texture (N)	Lightness	Redness	Yellowness	Change in color
1	6.26	45.90	94.61	−1.17	16.10	41.62
2	3.85	13.60	93.15	−0.37	14.63	14.99
3	5.29	32.80	92.94	1.07	19.05	10.53
4	5.83	26.70	90.72	2.15	20.53	21.88
5	5.44	29.00	93.07	0.77	17.04	14.77
6	5.89	21.05	87.11	4.03	20.92	23.47
7	5.77	21.40	92.82	1.29	19.87	10.51
8	5.03	34.60	92.12	1.06	18.81	13.64
9	6.98	34.70	92.76	0.64	18.73	11.15
10	4.94	26.40	92.05	0.76	19.11	2.90
11	5.91	18.00	86.70	3.40	20.82	21.84
12	5.97	38.15	94.12	−0.65	16.04	13.67
13	5.79	42.30	94.05	−0.19	19.24	75.26
14	6.13	36.80	95.64	−0.70	16.27	11.11
15	6.90	43.50	93.70	0.63	17.25	12.09
16	6.01	13.20	93.46	0.37	17.30	11.78
17	5.98	23.60	91.66	1.47	19.67	4.28
18	5.89	19.10	92.08	1.20	19.25	91.27
19	5.88	60.10	94.89	−0.49	17.32	51.54
20	5.96	28.30	93.41	0.45	18.34	9.02
21	6.16	31.40	91.14	1.47	20.10	17.40
22	6.07	67.30	94.83	−0.17	16.43	11.73
23	4.88	17.50	94.15	−0.29	18.95	4.63
24	6.10	18.00	93.22	0.25	20.48	6.08
25	6.16	17.80	94.55	0.37	17.10	10.52
26	6.16	18.90	92.34	0.48	16.24	30.81
27	6.43	40.20	95.05	−0.88	18.51	61.71
28	5.10	23.20	92.55	0.51	18.62	3.36
29	6.67	30.10	93.89	−1.48	17.65	58.22

Values are means of duplicates.

also reduced with increased level of BSCF but increased with thickness.

Color measurement of fried snacks

The changes in the color of fried products are as a result of the Maillard reaction that depends on the content of reducing sugars and amino acids at the surface, as well as the temperature and frying time as reported by Marquez and Anon (1986). Color is considered as one of the most important quality parameters of deep fat fried snacks. As the frying temperature increased, the lightness parameter of the fried product decreased, whereas the redness and yellowness parameters increased for the same frying time (Krokida and Oreopoulou 2000; Moyano et al. 2002). These results were consistent with this study. Lightness value of fried snack decreased with increase in frying temperature, frying time, and level of BSCF, while redness and yellowness values increased. The addition of wheat flour could be said to cause an increase in the amount of amino acid in the flour blend used for the fried snacks allowing the

Maillard browning reaction to easily occur, with increase in level of BSCF resulting in the decrease in lightness value, but increased redness and yellowness values. This was similar to the report of Jirawan et al. (2009). The increase in redness and yellowness values could be attributed to the color of the BSCF which is almost light yellow. At increasing thickness, lightness increased and this could be attributed to increasing quantity of the dough that was fried while redness and yellowness decreased. The lightness and redness was observed to have a positive significant ($P < 0.05$) effect on the level of BSCF but was not significantly affected by frying time, frying temperature, and thickness. Furthermore, the yellowness of the fried snacks was significantly ($P < 0.05$) affected by the interaction between frying time.

The regression coefficients of the quality parameters of the fried snack are as shown in Tables 6 and 7 while Figures 1–3 are the response surface plots for the expansion, texture and change in color, respectively.

Optimization of process variables

Expansion, protein content, carbohydrate content, total dietary fiber, and yellowness were maximized (6.98, 9.22, 67.24, 3.43, and 20.92, respectively, while texture [13.2], ash content [2.0], oil content [12.82], lightness [86.7], redness [−1.48], and change in color [2.90] were minimized). Frying temperature of 140.11°C, frying time of 4 min, level of BSCF of 32.09%, and thickness of 2 mm with a desirability of 0.771 was selected and an optimized sample was prepared under these conditions.

Table 6. Regression coefficients of the response surface models and statistical results of the color parameters, expansion, and texture of the fried snacks.

Coefficients	Expansion (mm)	Texture (N)	Lightness	Redness	Yellowness	Change in color
B_o	5.89	28.85	93.6	0.24	17.09	16.17
X_1	0.13	−6.30*	−0.03	−0.1	0.33	−1.14
X_2	0.16	3.2	0.34	−0.21	−0.77	−3.29
X_3	−0.44*	−9.32*	1.78*	1.07*	0.48	−25.73*
X_4	0.15	9.99*	0.51	−0.29	−0.29	−2.55
X_1^2	−0.27	4.11	0.55	−0.39	−0.12	−5.62
X_2^2	0.07	−1.56	−0.83	0.54	1.38*	−2.56
X_3^2	−0.2	0.62	−1.02	0.11	0.65	23.60*
X_4^2	0.29	−0.1	−0.49	0.49	0.99	1.48
X_{12}	−0.01	3.5	0.06	0.08	−0.48	0.16
X_{13}	0.11	9.90*	0.1	0.47	0.92	−4.58
X_{14}	0.01	−4.4	−0.92	0.38	1.06	−0.14
X_{23}	−0.09	−0.26	1	−0.4	0.47	3.61
X_{24}	0.29	5.4	0.69	−0.41	−0.03	−3.87
X_{34}	−0.38	−3.18	0.6	−0.14	−0.24	2.65
R^2	0.49	0.76	0.59	0.55	0.53	0.873
PRESS	32.59	5315.3	245.32	95.42	164.32	9638.83
P	0.53	0.02	0.25	0.25	0.42	0.0005

*Significant values at 5% level; B_o is intercept, $X_1 – X_{14}$ are regression coefficients.

Table 7. Regression coefficients of the response surface models and statistical results of the proximate composition of the fried snacks.

Coefficients	Protein	Moisture	Ash	Oil content	Carbohydrate	Total dietary fiber
B_o	8.55	7.70	2.20	24.75	53.28	3.52
X_1	−0.01	−0.01	−0.08	0.11	−0.02	0.02
X_2	0.05	−1.78*	0.21	−2.56	4.07	0.01
X_3	−0.13	−0.24	0.00	3.84	−3.48	0.02
X_4	0.10	2.23*	0.13	−0.96	−1.51	0.02
X_1^2	0.12	0.45	−0.12	−3.13	2.60	0.07
X_2^2	0.10	−0.17	0.07	−0.68	0.57	0.11*
X_3^2	−0.18	−0.89	0.00	1.73	−0.73	0.08
X_4^2	−0.08	−0.62	0.19	−3.05	3.47	0.09*
X_{12}	−0.13	1.08	0.13	0.83	−1.92	0.03
X_{13}	0.03	−0.55	−0.13	1.10	−0.52	0.06
X_{14}	0.02	0.00	0.00	2.99	−3.03	0.02
X_{23}	0.04	0.30	−0.25	−5.88	5.79	−0.02
X_{24}	−0.07	−0.55	0.25	−0.26	0.71	−0.08
X_{34}	0.07	0.93	0.13	−0.32	−0.86	0.06
R^2	0.42	0.76	0.50	0.44	0.42	0.57
PRESS	6.32	189.67	7.61	2262.59	2664.63	0.82
P	0.72	0.02	0.50	0.66	0.72	0.30

*Significant values at 5% level; B_o is intercept, $X_1 – X_{14}$ are regression coefficients where X_1, X_2, X_3 and X_4 are frying temperature, frying time, level of BSCF and thickness, respectively BSCF- Brewers' Spent high quality Cassava Flour

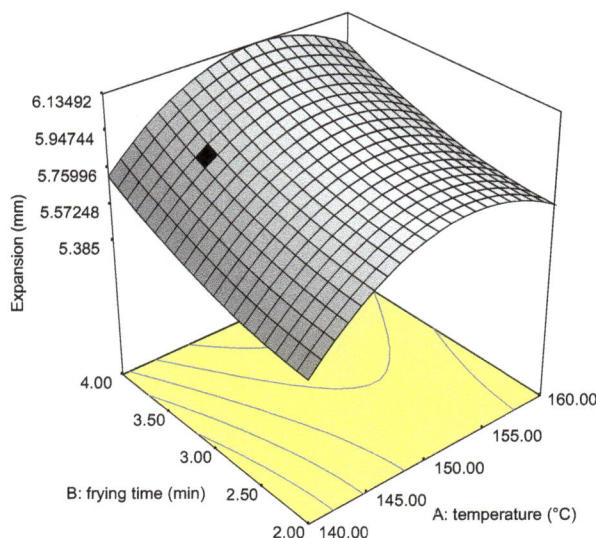

Figure 1. Response surface plot showing effect of independent variables on the expansion (mm) of fried snacks from Wheat-Brewers' Spent Cassava Flour.

Sensory evaluation of optimized sample

The result of the sensory evaluation of the fried snacks is presented in Figure 4 which shows the degree of likeness of the optimized sample based on the appearance,

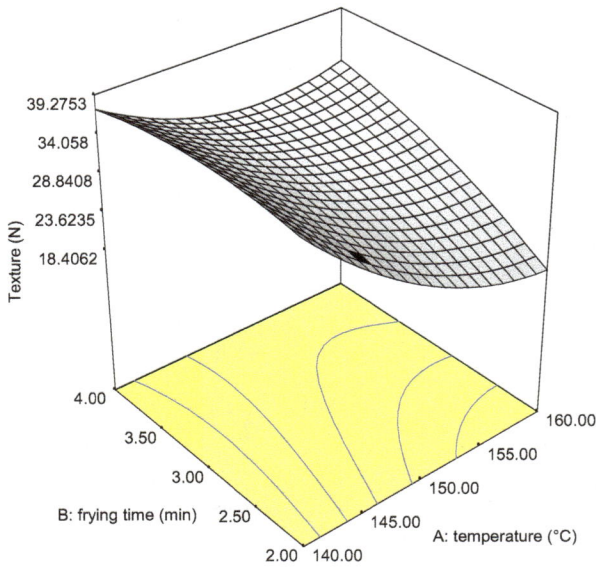

Figure 2. Response surface plot showing effect of independent variables on the texture (N) of fried snacks from Wheat-Brewers' Spent Cassava Flour.

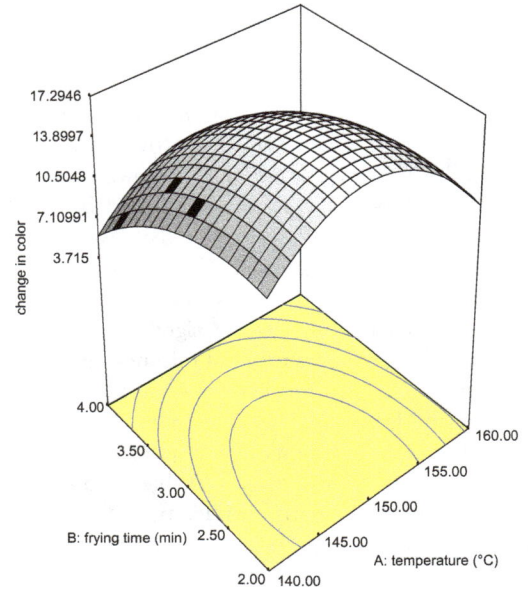

Figure 3. Response surface plot showing effect of independent variables on change in color of fried snacks from Wheat-Brewers' Spent Cassava Flour.

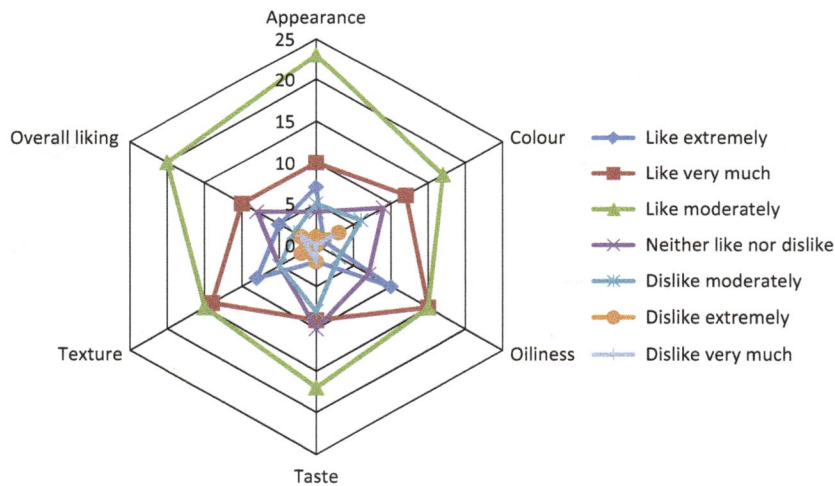

Figure 4. Radar chart showing the level of acceptability of the optimized fried snack based on its attributes.

color, oiliness, taste, texture, and overall liking. Figure 5 shows the comparison of the optimized sample with 100% wheat flour. The panellists preferred the optimized sample more in texture and greasiness.

Scanning electron micrographs of fried snacks

According to Hoseney (1994), wheat starch granules consist of large granules (over 10 μm) and smaller ones (less

than 10 μm). The micrograph of fried snack from 100% wheat flour showed the presence of larger air cells formed during frying with less continuity of network which could enhance the migration of oil into the fried snacks, thereby increasing the oil content while the micrograph of the optimized snack had air cells that are smaller in size, more continuity, and less porosity which might help in minimizing oil migration into the fried snack. Though most snacks require a porous texture for their desired

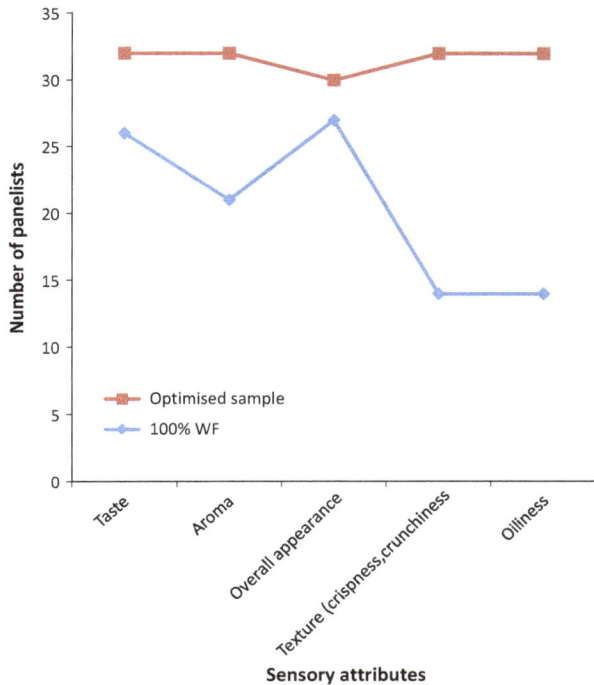

Figure 5. Bar chart showing the results of the preference sensory test.

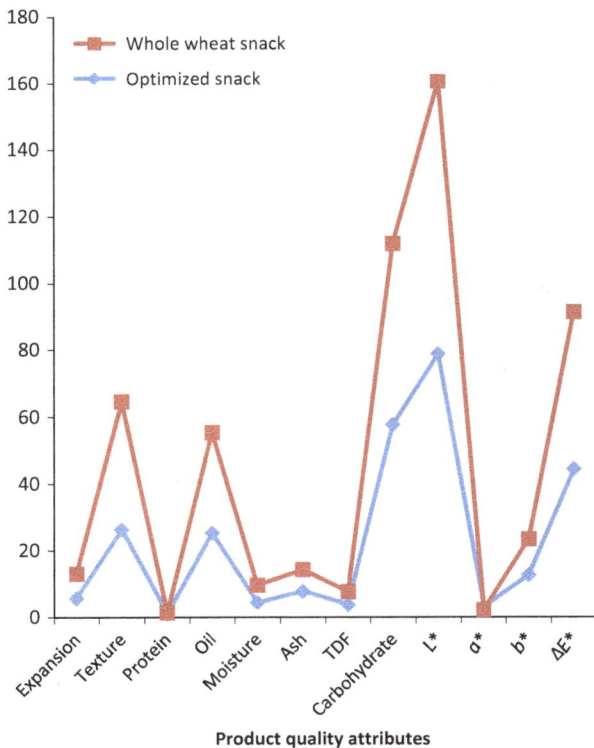

Figure 6. Bar chart showing the results of some quality attributes of the fried snacks.

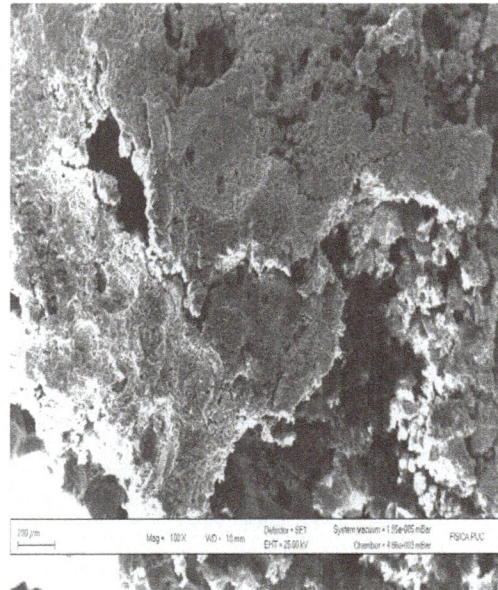

Figure 7. Scanning electron micrographs of fried snacks from 100% wheat flour.

Figure 8. Scanning electron micrographs of fried snacks from wheat plus 32.09% BSCF.

sensory properties, the pores should not be large enough to encourage oil migration. Result showed that fried snack from 100% WF had higher oil uptake than the optimized snack as shown in Figure 6 and this is supported by the scanning electron micrographs of the fried snacks (Figs. 7 and 8).

Conclusion

It can be inferred from this study that an acceptable fried snack can be developed from the inclusion of BSCF to wheat flour. The frying temperature of 140 °C, frying time of 4 min, level of BSCF of 32%, and 2 mm thickness was selected to give maximum values for the various responses with the highest desirability value of 0.771. The optimized fried snack (containing 32% BSCF) had a higher protein content and lower oil uptake than the snack from 100% WF. Also, the optimized fried snack was crispier and requires less breaking force than the snack from 100% WF.

Acknowledgment

Sincere appreciation goes to the GRATITUDE (EU-ACP FP7) for funding the research work.

Conflict of Interest

None declared.

References

Akubo, P. I. 1997. Proximate composition and selected functional properties of African breadfruit and sweet potato flour blends. J. Plant Foods Hum. Nut. 51:53–60.

Amudha, S., R. Ravi, K. K. Bhat, and M. K. Seethalaksmi. 2002. Studies on the quality of fried snacks based on blends of wheat flour and soya flour. Food Qual. Prefer. 13:267–273.

Anjum, F. M., M. R. Khan, and A. Din. 2007. Wheat gluten: High molecular weight glutenin subunits-structure, genetics and relation to dough elasticity. J. Food Sci. 72:R56–R61.

AOAC Method. 2003. in P. Cunnif, ed. Official methods of Association of Official Analytical Chemists International, 17th edn. AOAC Method, Arlington, VA, USA.

Ashworth, A., and A. Draper. 1992. The potential of traditional technologies for increasing the energy density of weaning foods. A critical review of existing knowledge with particular reference to malting and fermentation. WHO/CBD EDP/92.4.

Awoyale, W., B. Maxiya-Dixon, L. O. Sanni, and T. A. Shittu. 2011. Nutritional and sensory properties of a maize – based snack food (kokoro) supplemented with treated Distiller' spent grain. Int. J. Food Sci. Technol. 46:1609–1620.

Da Silva, P., and R. Moreira. 2008. Vacuum frying of high-quality fruit and vegetable-based snacks. Lebensmittel Wissenchraft Technologie 41:1758–1767.

Gazmuri, A., and P. Bouchon. 2009. Analysis of wheat gluten and starch matrixes during deep-fat frying. Food Chem. 115:999–1005.

Hoseney, R. C. 1994. Pp. 250–263Yeast leavened products: principles of cereal science and technology. American Association of Cereal Chemist, St. Paul, MN.

Ihekoronye, A. I., and P. O. Ngoddy. 1985. Integrated food science and technology for the tropics. Macmillan Publishers Ltd., London and Basingstoke 386 pp.

Jirawan, M., N. Athapol, and S. T. Pawan. 2009. Optimization of processing conditions to reduce oil uptake and enhance physico-chemical properties of deep fried rice crackers. Food Sci. Technol. 42:805–812.

Kawas, M., and R. Moreira. 2001. Effect of degree of starch gelatinization on quality attributes of fried tortilla chips. J. Food Sci. 66:300–306.

Krokida, M. K., and V. Oreopoulou. 2000. Water loss and oil uptake as a function of frying time. J. Food Eng. 44:39–46.

Mariscal, M., and P. Bouchon. 2008. Comparison between atmospheric and vacuum frying of apple slices. Food Chemistry, 107(4), 1561–1569.

Marquez, G., and M. Anon. 1986. Influence of reducing sugars and amino acids in the colour development of fried potatoes. J. Food Sci. 51:157–160.

Moyano, P. C., V. K. Vioseco, and P. A. Gonzalez. 2002. Kinetics of crust colour changes during deep fat frying of impregnated French fries. J. Food Eng. 54:249–255.

Mussatto, S. I. 2009. Biotechnological potential of brewing industry by products. Pp. 313–326 in P. Singh nee' Nigam and A. Pandey, eds. Biotechnology for agro-industrial residues utilization, Springer, New York, United States. doi:10.1007/978-1-4020-9942-7 16.

Mussatto, S. I., G. Dragone, and I. C. Roberto. 2006. Brewers spent grain: generation, characteristics and potential applications. J. Cereal Sci. 43:1–14.

Papadakis, S. E., S. A. Malek, R. E. Kamdem, and K. L. Yam. 2000. A versatile and inexpensive technique for measuring color of foods. Food Technol. 54:48–51.

Rossell. J. B. 2001. Factors affecting the quality of frying oils and fats. In: Rossell, J.B. (Ed.) Frying –Improving Quality. Woodhead Publishing, Cambridge, England. PP 115–164.

Sobukola, O. P., J. M. Babajide, and O. Ogunsade. 2012. Effect of brewers spent grain addition and extrusion parameters on some properties of extruded yam starch-based pasta. J. Food Process. Preserv. 37:734–742 ISSN 1745-4549.

Sobukola, O. P., V. Dueik, L. Munoz, and P. Bouchon. 2013. Comparison of vacuum and atmospheric deep-fat frying of wheat starch and gluten based snacks. Food Sci. Biotechnol., 22: 177–182.

Yam, K. L., and S. E. Papadakis. 2004. A simple digital imaging method for measuring and analyzing color of food surfaces. J. Food Eng. 61:137–142.

12

Effect of selected spices on chemical and sensory markers in fortified rye-buckwheat cakes

Małgorzata Przygodzka[1], Henryk Zieliński[1], Zuzana Ciesarová[2], Kristina Kukurová[2] & Grzegorz Lamparski[1]

[1]Division of Food Science, Institute of Animal Reproduction and Food Research of the Polish Academy of Sciences, Tuwima 10, P.O. Box 55, 10-748 Olsztyn 5, Poland
[2]National Agriculture and Food Centre – Food Research Institute, Priemyselná 4, P.O. Box 25, 824 75 Bratislava 26, Slovak Republic

Keywords
Antioxidant activity, buckwheat flour, cakes, Maillard reaction, sensory evaluation, spices

Correspondence
Małgorzata Przygodzka, Division of Food Science, Institute of Animal Reproduction and Food Research of the Polish Academy of Sciences, Tuwima 10, P.O. Box 55, 10-748 Olsztyn 5, Poland.

E-mail: m.przygodzka@pan.olsztyn.pl

Funding Information
This research was funded by grant No. 2012/07/N/NZ9/02250 from National Science Centre.

Abstract

The aim of this study was to find out the effect of selected spices on chemical and sensorial markers in cakes formulated on rye and light buckwheat flour fortified with spices. Among collection of spices, rye-buckwheat cakes fortified individually with cloves, nutmeg, allspice, cinnamon, vanilla, and spice mix revealed the highest sensory characteristics and overall quality. Cakes fortified with cloves, allspice, and spice mix showed the highest antioxidant capacity, total phenolics, rutin, and almost threefold higher available lysine contents. The reduced furosine content as well as free and total fluorescent intermediatory compounds were observed as compared to nonfortified cakes. The FAST index was significantly lowered in all cakes enriched with spices, especially with cloves, allspice, and mix. In contrast, browning index increased in compare to cakes without spices. It can be suggested that clove, allspice, vanilla, and spice mix should be used for production of safety and good quality cakes.

Introduction

Spices have been used by human since ancient times. According to the U.S. Food and Drug Administration (US FDA) spice is an "aromatic vegetable substance in the whole, broken, or ground form, the significant function of which in food is seasoning rather than nutrition" and from which "no portion of any volatile oil or other flavoring principle has been removed" (Sung et al. 2012). The compilation of current trends in bakery technology to enhance antioxidant activity of bakery products was widely described by Dziki et al. (2014). At the top of the list, spices have been suggested as a well-recognized source of compounds with antioxidant potential (Hinneburg et al. 2006; Wojdyło et al. 2007; Charles 2013). Recently, the inquisitive studies on

antioxidant capacities of spices employing updated analytical methods were reported by Przygodzka et al. (2014). According to data collected in Food Frequency Questionnaire, the average of spices/herbs intake was estimated as 1.1 grams per day for one person what revealed that spices are important contributor of antioxidants to our diet (Carlsen et al. 2011). Spices are mainly employed as flavoring and color agents, whereas potential use to preservation food and disease prevention has been already studied (Kaefer and Milner 2008; Cazzola and Cestaro 2014; Embuscado 2015). Spice application was demonstrated by Illupapalayam et al. (2014). Their probiotic-yogurt with cardamom, cinnamon, and nutmeg has increased sensorial acceptability among consumers, besides spice addition increased overall antioxidant activity of this functional product.

Presently, consumers seeking for new food products are focused on joining two aspects: a taste and functional properties (Wójtowicz et al. 2013). The functional properties of innovative products in prevention or therapy support in selected diseases are desirable. Anticancer, antiallergic, antiviral, cholesterol-reducing, blood pressure-reducing, and arteriosclerosis-reducing were ascribed as buckwheat's healing effects (Krkošková and Mrázová 2005). In this trend, buckwheat-based product with spices addition can be a good alternative to inclusion in varied and balanced diet. Moreover, several studies approved consumer acceptability of buckwheat-based products (Wronkowska et al. 2008; Filipčev et al. 2011; Sedej et al. 2011; Chlopicka et al. 2012). The high sensorial acceptability of 30% buckwheat flour incorporation in baked products was reached.

In this study, the recipe of rye-buckwheat cakes (RBC) was enriched with one spice form the list including: anise, allspice, cardamom, cinnamon, cloves, coriander, fennel, ginger, nutmeg, star anise, vanilla, white pepper, and commercial spice mix for ginger cakes. The sensory evaluation of cakes was used as a tool for selection cakes accepted by sensory panel. It seems to be rationale to use Maillard reaction (MR) products as markers for description quality of RBC fortified with spices. It is well-known that MR products are responsible for the development of color, taste, and aroma as well as the nutrients loss of thermally treated food (Markowicz Bastos and Gugliucci 2015). Virág et al. (2013) stated that remaining lysine after baking process is a good indicator of the progress of MR and important to monitor its content as essential amino acids. Several unfavorable food contaminants are simultaneously formed in thermal processing. During early step of MR, the nutritionally valued available lysine can be converted into furosine, a heat-treatment marker (Gökmen et al. 2008; Giannetti et al. 2014). The advanced stage of MR is characterized by the formation of fluorescence compounds with regard to advanced glycation end-products formation and monitoring protein degradation by FAST index (Delgado-Andrade et al. 2007; Liogier de Sereys et al. 2014). Positively, melanoidins formed in the final stage of MR are responsible for the color formation and possess the ability to scavenge free radicals (Langner and Rzeski 2014). It was concerned that MR products formation in a model systems and food products can be reduced/increased by an application of substances having a high antioxidant potential (Marková et al. 2012; Oral et al. 2014; Cheng et al. 2015).

The aim of this study was to find out an impact of selected spices on Maillard reaction progress and sensory quality of RBC fortified with spices. Therefore, analysis of selected chemical and sensorial markers such as quercetin 3-rhamnosylglucoside (rutin)—the main buckwheat flavonoid, available lysine, total phenolics contents (TPC),

antioxidant capacity (AC) of cakes using extracts scavenging activity against $ABTS^{\cdot+}$ radical cation and against superoxide anion radicals ($O_2^{\cdot-}$) measured by the photochemiluminescence method (PCL), and furosine, fluorescent compounds, and melanoidins, were addressed in this study. To determine the impact of thermal treatment on protein damage, FAST index was calculated.

Materials and Methods

Chemicals and reagents

2,2′-Azinobis(3-ethylbenzothiazoline-6-sulphonic acid) diammonium salt (ABTS), 6-hydroxy-2,5,7,8-tetramethylchroman-2-carboxylic acid (Trolox), rutin (quercetin-3-rutinoside), lysine (N^{α}-acetyl-L-lysine), and pronase E (*Streptomyces griseus lyoph.*) were purchased from Sigma (Sigma Chemical Co., St. Louis, MO). PCL ACW (Antioxidant Capacity of Water-soluble substances) kit for PCL assay was from Analytik Jena AG (Jena, Germany). o-phtaldialdehyde for fluorescence (OPA) and sodium dodecylsulfonate (SDS) were supplied by Fluka (Buchs, Switzerland). Furosine (2-furoylmethyl-lysine) was purchased from PolyPeptide (Strasbourg, France). Acetonitrile and methanol (HPLC purity) were provided by POCh (Gliwice, Poland). Water was purified with Mili-Q-system (Millipore, Bedford, MA).

Formulation of rye-buckwheat ginger cakes enhanced with spices

The cakes were baked using rye flour blended with light buckwheat flour in ratio 70:30 (w/w). The making process involved dough preparation by mixing flours, honey, and sugar. Each one of selected spices (2% on flour mixture basis; w/w) from the list: anise, allspice, cardamom, cinnamon, cloves, coriander, fennel, ginger, nutmeg, star anise, vanilla, white pepper, and commercial spice mix for ginger cake, was used in RBC recipe. According to the producer's declaration, commercial spice mix contained cinnamon, pepper, clove, anise, coriander, fennel, and nutmeg. The amounts of ingredients added to make each type of cake are presented in Table 1. The dough was cut into 0.5-cm-thick disks of 5.5 cm diameter and baked at 180°C for 18 min in a DC-32E electric oven (Sveba-Dahlen, Fristad, Sweden). Finally, the cakes were freeze-dried and grounded into powder. The powdered samples were sieved through a 60-mesh screen and then stored at −20°C until analyzed.

Sensory evaluation

Twenty-four attributes related to the appearance, odor, taste, and texture of rye-buckwheat ginger cakes with

Table 1. Formula of rye-buckwheat cakes fortified with selected spices: anise, allspice, cardamom, cinnamon, cloves, coriander, fennel, ginger, nutmeg, star anise, vanilla, white pepper, and commercial mix of spices for ginger cake.

Ingredients	Control cake	Rye-buckwheat cake with spice addition
Rye flour (T-720) (g)	70	70
Light buckwheat flour (g)	30	30
Buckwheat honey (g)	50	50
Sugar (g)	20	20
Baking powder (g)	3	3
Butter (g)	25	25
Selected spice (g)	0	2

spices were selected and thoroughly used to during profiling procedure. Sensory characteristics and overall quality of ginger cakes were evaluated according to international unified standards (ISO/DIS 1998). A six-member trained panel judged ginger cakes in a 10-point scale (0—for weak, 10—for very good) using quantitative descriptive analysis to determine differences between each type of ginger cakes (Stone et al. 2012). The description of sample preparation and standardized procedure of sensory evaluation were in details presented by Zieliński et al. (2012).

Overall acceptability of each sample was evaluated in relation to the sensory preferences on the basis of overall appearance, aroma, taste, and texture, in a 10-point hedonic scale, where: 0 = not accept, and 10 = fully accept. The profiling analysis of all samples was run in duplicate (two series) proceeded by introduction session. Ginger cakes were considered as acceptable if their mean scores for overall acceptability were above 6 (Kowalska et al. 2012).

Preparation of extracts from RBC

Rye-buckwheat cake powders (100 mg) were extracted with 1 mL of 65% (v/v) ethanol. After ultrasonic vibration for 30 sec, the solution was mixed and centrifuged for 5 min at 5000× g at 4°C. That step was repeated five times and the supernatants were collected into 5-mL flask. Final extracts concentration was 20 mg/mL. Ethanol extracts were prepared in triplicate. Next extracts were stored at −20°C until analysis of rutin content, total phenolic compounds (TPC), and AC by ABTS and PCL ACW assays.

Determination of total phenolic content (TPC) and rutin

The TPC was determined with Folic-Ciocalteu reagent as it was described in details by Przygodzka et al. (2014). TPC was standardized against gallic acid and expressed in terms of mg gallic acid equivalents (GAE)/g dry

matter. The content of rutin in ginger cakes was determined with HPLC (Shimadzu, Japan) with UV detector (SPD-10A) set up 330 nm as it was recently described by Zielińska et al. (2010). For quantitative analysis, rutin standard was prepared in triplicate at five concentrations within the range 1.0–40 μM. All solutions were filtered through a 0.45 μm nylon membrane before use. The results were expressed in μg per g of dry matter.

Antioxidant capacity determination

The AC of RBC enhanced with spices was determined by ABTS and photochemiluminescence (PCL ACW) assays as it was described in details by Przygodzka et al. (2014). The results provided by ABTS and PCL ACW methods were expressed as μmol of Trolox equivalents (TE)/g DM.

Available lysine content determination

The OPA assay as described by Michalska et al. (2008) was employed to determine available lysine content using the microplate reader (Infinite® M1000 PRO, Tecan, Switzerland). Exactly 50 μL of sample, 100 μL of OPA reagent, and 100 μL of water were added to well and incubated for 3 min (96-well microplate; Porvair Sciences, Norfolk, UK). Then the fluorescence reading was measured at extinction wavelength 340 nm and emission wavelength 455 nm. Quantitative analysis was performed by the external standard method, employing a calibration curve of N^{α}-acetyl-L-lysine ranged from 10 to 250 μM. Each result is a mean of three independent extractions.

Maillard reaction products determination

Furosine content determination

According to Delgado-Andrade et al. (2007), 30 mg of cake sample was hydrolyzed with 4 mL of 4.9 M HCl at 110°C for 23 h in a Pyrex screw-capped vial with PTFE-faced septa. Hydrolysis tubes must be sealed under nitrogen. After that the hydrolysates was centrifuged for 10 min. A 0.5 mL portion of the supernatant was applied to a Sep-pak C18 cartrigde (Millipore) conditioned with 5 mL of methanol and 10 mL of distilled water, then eluted with 3 mL of 3M HCL and evaporated under vacuum. The dried sample was dissolved in 1 mL of a mixture of water, acetonitrile, and formic acid (95:5:0.2) before HPLC analysis.

The furosine was quantified by HPLC system (Shimadzu, Japan) comprised of a controller (SCL-10AVP), a PDA detector (SPD-M10AVP). A Cadenza CD-C18 column (250 × 2 mm, 3 μm, Imtakt, Kyoto, Japan) at 35°C. The mobile phase consisted of a solution of 5 mM sodium heptanases sulfonate containing 20% of acetonitrile and 0.2% of formic acid. The elution was isocratic and the flow rate was 0.2 mL/min. The UV detector was set at 280 nm. Calibration curve was made by the external standard of furosine 0.2–9 μg/mL.

Measurement of MR fluorescence intermediatory compounds and FAST index calculation

The fluorescence of free, linked-to-protein, and total intermediary compounds (FIC) was determined after sample extraction and further enzymatic hydrolysis using pronase E according to Delgado-Andrade et al. (2006). Readings were recorded in a luminescent spectrofluorimeter (LS 50B; Perkin Elmer, Waltham, USA) setting at $\lambda_{ext.}$ = 347 nm and $\lambda_{em.}$ = 415 nm. Tryptophan fluorescence Trp_{FL} was measured at $\lambda_{ext.}$ = 290 nm and $\lambda_{em.}$ = 340 nm. Results are expressed in fluorescence intensity (FI) per mg of sample DM. The FAST index was calculated as recently reported by Zieliński et al. (2012) with a one novelty modification based on the use of fluorescent compounds linked-to-proteins for index calculation. The samples were analyzed in triplicate and FAST index data were expressed as a percentage (%).

Brown pigments assay

Formation of brown pigments was estimated as reported in details by Zieliński et al. (2012). All measurements

were performed in triplicate. Results were expressed as arbitrary absorbance units.

Statistical analysis

The results of the chemical analyses are given as the means and the standard deviation of three independent measurements. Statistical one-way analysis of variance (ANOVA) using Fischer test was performed. The significance level was set at $P < 0.05$. The correlation test between rutin content, antioxidant ability, and MRPs formation was performed and the Pearson correlation coefficients were calculated. Statistical analyses were performed using software package (StatSoft Inc., v. 7.1, Tulsa, OK).

Results and Discussion

Consumer acceptance

The overall quality of RBC made of light buckwheat flour incorporated with selected spices is presented on Figure 1. The overall acceptability for control cake was 6.4, in comparison its sensorial score was higher than for cakes made of buckwheat and wheat flour proposed by Kaur et al. (2014). The RBC fortified with spices showed following rank of acceptability: cakes with vanilla (7.9), with spice mix (7.5), with cinnamon (6.9), and with nutmeg (6.6). Also high acceptability showed cakes fortified with allspice (6.2) and cloves (6.1). Taking into account the overall acceptability rating, it was decided to use RBC

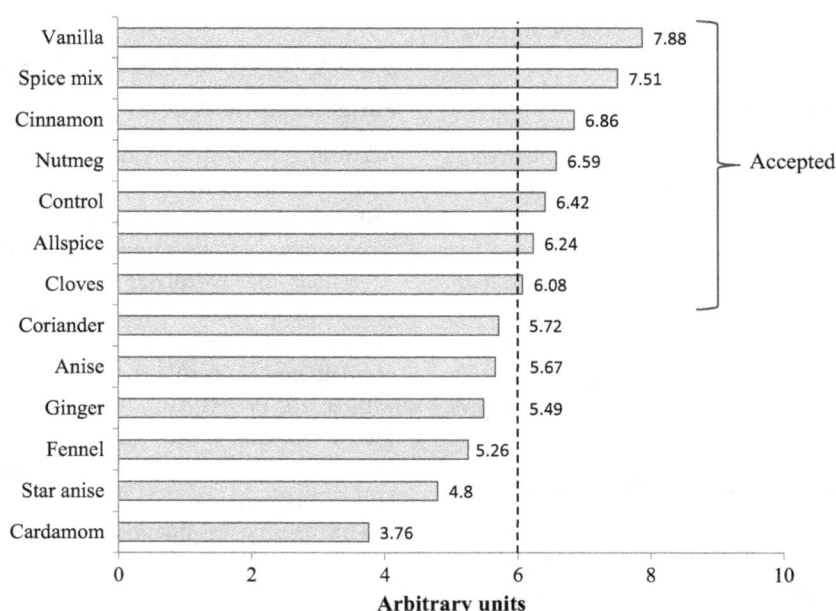

Figure 1. The overall quality of rye-buckwheat cakes fortified with spices addition.

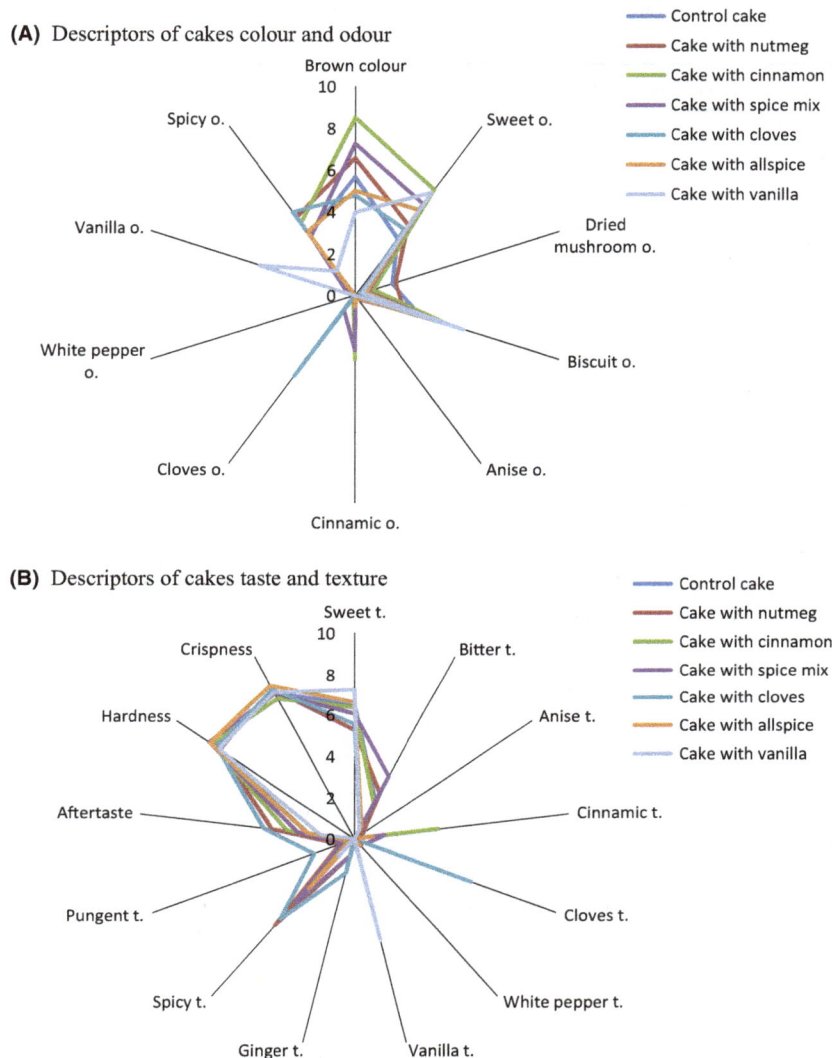

Figure 2. Sensory profiles of rye-buckwheat cake without spice (control cake) and rye-buckwheat cakes fortified with selected spices: nutmeg, cinnamon, spice mix, cloves, allspice, and vanilla. O, attributes of odor; t, attributes of taste.

fortified with allspice, cinnamon, cloves, spice mix, nutmeg, and vanilla for further chemical analysis.

Sensory evaluation

In order to observe the above differences in the analyzed samples more clearly, the sensory profiles of RBC as control and cakes enriched with clove, cinnamon, allspice, nutmeg, vanilla, and spice mix were displayed as spider diagrams in Figure 2A,B. The mean sensory ratings for the samples and the analysis of variance are presented in Table 2. The buckwheat honey, sugar as well as cinnamon and vanilla usage can contribute on high level of sweet taste and odor in presented cakes. The sweetness can be mitigated bitter taste and aftertaste of buckwheat flour. According to Pauly

et al. (2013), the high values of hardness can be linked to type of buckwheat flours used in recipe. ANOVA showed that there were significant differences in the intensity of attributes such as brown color, odor descriptors: "sweet," "biscuit," "cinnamic," "cloves," "vanilla," and "spicy"; taste descriptors: bitter, "cinnamic," "cloves," "vanilla," "spicy," pungent, and aftertaste. The high score of acceptability for cake with vanilla was involved with significantly low contribution of negative attributes as spicy, pungent taste, and aftertaste, masked by intensive biscuit and vanilla odor, and vanilla taste. The highest scores of cinnamic odor and cinnamic taste, vanilla odor and vanilla taste as well as cloves odor and cloves taste were characteristic for cakes with cinnamon, vanilla, and cloves, respectively. It can be concluded that 2% of spice addition was sufficiently for

Table 2. Descriptive analysis of results based on the analysis of variance (ANOVA) performed on rye-buckwheat cakes (RBC) fortified with selected spices.

Attribute		Rye-buckwheat cakes						
		Control	Cloves	Nutmeg	Allspice	Cinnamon	Vanilla	Spice mix
1	Brown color	5.67[cd]	4.81[de]	6.58[bc]	5.01[cde]	8.51[a]	3.98[e]	7.28[ab]
2	Sweet o.	3.38[c]	3.93[bc]	4.21[bc]	5.08[ab]	6.26[a]	6.08[a]	5.41[ab]
3	Dried mushroom o.	1.82[a]	0.25[a]	1.99[a]	0.54[a]	0.89[a]	0.36[a]	0.60[a]
4	Biscuit o.	3.23[bc]	0.97[d]	2.42[cd]	3.95[abc]	4.67[ab]	5.32[a]	4.31[ab]
5	Anise o.	0.01[a]	0.01[a]	0.20[a]	0.14[a]	0.03[a]	0.01[a]	0.02[a]
6	Cinnamic o.	0.09[b]	0.01[b]	0.00[b]	0.51[b]	3.10[a]	0.02[b]	2.66[a]
7	Cloves o.	0.00[c]	4.81[a]	0.02[c]	0.00[c]	0.13[c]	0.01[c]	0.88[b]
8	White pepper o.	0.07[a]	0.01[a]	0.01[a]	0.02[a]	0.02[a]	0.02[a]	0.02[a]
9	Vanilla o.	0.09[c]	0.01[c]	0.08[c]	0.02[c]	0.12[c]	4.66[a]	0.34[b]
10	Spicy o.	3.61[a]	4.92[a]	4.62[a]	3.76[a]	4.27[a]	1.45[b]	3.48[a]
11	Sweet t.	6.53[a]	5.53[a]	5.23[a]	6.61[a]	6.35[a]	7.25[a]	6.04[a]
12	Bitter t.	2.06[ab]	0.60[b]	2.51[a]	0.70[b]	1.95[ab]	0.51[b]	3.41[a]
13	Anise t.	0.51[a]	0.02[a]	0.46[a]	0.02[a]	0.02[a]	0.02[a]	0.02[a]
14	Cinnamic t.	0.02[b]	0.02[b]	0.02[b]	0.85[b]	3.92[a]	0.02[b]	1.40[b]
15	Cloves t.	0.02[c]	5.79[a]	0.02[c]	0.02[c]	0.00[c]	0.01[c]	0.62[b]
16	White pepper t.	0.01[a]	0.02[a]	0.44[a]	0.38[a]	0.02[a]	0.02[a]	0.02[a]
17	Vanilla t.	0.34[b]	0.03[b]	0.01[b]	0.02[b]	0.02[b]	5.01[a]	0.06[b]
18	Ginger t.	0.01[a]	1.68[a]	0.83[a]	0.02[a]	0.01[a]	0.03[a]	0.84[a]
19	Spicy t.	4.68[ab]	5.17[ab]	5.54[a]	3.18[c]	4.06[bc]	1.13[d]	4.08[bc]
20	Pungent t.	0.13[b]	1.99[a]	0.72[b]	0.26[b]	0.30[b]	0.02[b]	0.61[b]
21	Aftertaste	2.15[bc]	4.26[a]	3.85[ab]	2.17[bc]	3.13[abc]	1.57[c]	2.64[abc]
22	Hardness	7.98[a]	8.19[a]	8.19[a]	8.41[a]	7.66[a]	8.03[a]	7.96[a]
23	Crispness	6.42[a]	6.08[a]	6.59[a]	6.24[a]	6.86[a]	7.88[a]	7.51[a]

The cakes were marked in a 10-point scale (0—for weak, 10—for very good). Means in each row with the same letters do not have significant differences (Fisher test, $P < 0.05$). o, attributes of odor; t, attributes of taste.

Table 3. The content of rutin, total phenolic compounds, and antioxidant capacity of rye-buckwheat cakes enhanced with selected spices.

Type of cake	Rutin (μg/g DM)	TPC (mg GAE/g DM)	Antioxidant capacity (μmol TE/g DM)	
			PCL ACW	ABTS
Rye-buckwheat control cake	104.36 ± 3.75[cd]	1.12 ± 0.03[g]	6.15 ± 0.51[d]	21.13 ± 0.88[f]
Rye-buckwheat cake with vanilla	100.63 ± 2.33[d]	1.32 ± 0.08[f]	5.17 ± 0.83[d]	21.87 ± 1.21[f]
Rye-buckwheat cake with cinnamon	101.62 ± 6.17[d]	2.28 ± 0.05[b]	8.69 ± 0.87[c]	49.38 ± 0.19[c]
Rye-buckwheat cake with cloves	319.80 ± 3.51[a]	2.11 ± 0.04[c]	23.30 ± 1.00[b]	55.52 ± 2.73[b]
Rye-buckwheat cake with allspice	173.19 ± 7.52[b]	1.84 ± 0.17[d]	9.25 ± 0.22[c]	40.86 ± 2.28[d]
Rye-buckwheat cake with nutmeg	100.84 ± 3.53[d]	1.56 ± 0.10[e]	5.78 ± 0.13[d]	30.49 ± 0.84[e]
Rye-buckwheat cake with spice mix	111.16 ± 3.50[c]	2.70 ± 0.09[a]	31.56 ± 0.05[a]	63.24 ± 1.31[a]

TPC (total phenolic content) is expressed in mg of gallic acid equivalents/g of dry matter (mg GAE/g DM). Antioxidant capacity measured by ABTS and PCL ACW methods is expressed in μmol of Trolox equivalents (TE)/g DM. Values are means ± standard deviation ($n = 3$). Values in each column with different small superscript letters are significantly different (Fisher test, $P < 0.05$). DM, dry matter.

differentiation between samples. Therefore, the sensory evaluation proved that addition of spices to RBC formulation increased the sensory quality of products.

The total phenolic content (TPC) and rutin determination

The content of rutin and total phenolic compounds in RBC fortified with selected spices is compiled in Table 3.

The addition of spices to RBC formula resulted in increase in TPC in comparison to control cake. The significantly highest TPC values in cakes with spice mix, cinnamon and cloves were observed, 4.84-fold, 2.03-fold, and 1.97-fold, respectively ($P < 0.05$). Our results are in accordance to Przygodzka et al. (2014) who noted high TPC values for spice mix, cloves, and cinnamon. The level of rutin was significantly higher in cakes after cloves and allspice application in comparison to control cake (3.07 and 1.66

times, respectively). The novel RBC possess rutin content 5.5 times and 6 times higher than in gluten-free rice-light buckwheat and rice-wholegrain buckwheat (70:30, w/w) breads, respectively (Sakač et al. 2011), which might be related to differentiations of rutin content in buckwheat flours and application of buckwheat honey in the cakes recipe. Additionally, our TPC results for control cake are in agreement with the results of these gluten-free breads (Sakač et al. 2011).

A weak correlation between rutin and TPC was noted ($r = 0.23$). It can be suggested that other flavonoid compounds extracted from spices have higher contribution on the antioxidant properties of RBC. Moreover, the negative correlation was found between TPC and bitter taste ($r = -0.61$). It can be said that phenolic compounds increased bitterness, our findings are in accordance to information collected by Shahidi and Naczk (1995).

Antioxidant properties

The 2% spices substitution in the formulation of cakes made of rye and light buckwheat flour resulted in significant differences ($P < 0.05$) in the AC determined against scavenging ability of ABTS$^{\cdot+}$ and O$_2^{-\cdot}$ (PCL ACW method) radicals. The results for AC determination are presented in Table 3. Sorted by AC measured by ABTS method RBC supplemented with spice mix has the highest antioxidant value followed by cloves, cinnamon, then allspice, nutmeg and finally vanilla cakes. Significantly highest results were obtained for RBC with spice mix, cloves, cinnamon and allspice, 2.99, 2.63, 1.93 –times higher than in control cake. The antioxidant potential evaluation by PCL ACW method for RBC was listed as follows: spice mix> cloves> allspice ≈ cinnamon> nutmeg≈ vanilla. The addition of spice mix and cloves was more effective in enhancing antioxidant activity, as evaluated by means of PCL ACW, which increased 6.10-fold and 4.50-fold, respectively. These results are in agreement with findings of Hossain et al. (2008), which indicated that cloves and cinnamon have the highest AC among other spices.

Moreover, the TPC and rutin contribution on AC overall was expressed by correlation coefficient. The strong correlation between TPC/ABTS and TPC/PCL ACW data ($r = 0.97$ and $r = 0.81$, respectively) were observed. According to studies of Bi et al. (2015), the strong correlation between TPC and AC measured by ABTS was observed for cloves extracts. It may suggest that active compounds from cloves have high contribution to antioxidant overall capacity of RBC. However, the weaker correlations for rutin versus PCL ($r = 0.43$) and rutin versus ABTS ($r = 0.44$) were noted.

Available lysine determination

The results for available lysine amount after thermal processing are shown in Table 4. Available lysine values of 0.52 mg/g DM was found in control RBC without condiments supplementation. According to the obtained results for available lysine in rye-buckwheat cake with condiments addition a protective effect on lysine blockage was found. The statistically significantly most high lysine blockage content in cloves, allspice, and spice mix was noted (2.75, 2.64, 2.20 times higher). The observation of protective effect of spices on lysine blockage in cakes was confirmed by positive correlation between OPA values and rutin and TPC contents ($r = 0.74$ and 0.63, respectively), as well as AC measured by ABTS ($r = 0.74$) and PCL ACW ($r = 0.62$). On the basis of these results, it can be concluded that spices positively influenced the baking process and increased the nutritional value of the product.

Maillard reaction products evaluation

As shown in Table 4, furosine content decreased after spices addition from 4 up to 78% in comparison to cake without spices addition. The furosine contents in rye-buckwheat ginger cakes after spices addition were significantly lower than the maximum allowable tolerance of furosine in milk proposed by Martysiak-Żurowska and Stołyhwo (2007). Moreover the furosine content is even twice lower than determined in commercial breakfast cereals (Rada-Mendoza et al. 2004) and five times lower in cookies made of wheat flour (Gökmen et al. 2008). The observation of inhibition effect of spices on furosine formation in cakes with spices was confirmed by high correlation between furosine and rutin ($r = -0.80$) as well as AC measured by ABTS ($r = -0.81$) and PCL ACW ($r = -0.84$). Whereas the weaker correlation ($r = -0.68$) between furosine and TPC contents was calculated. The strong relationship between furosine and OPA values and melanoidins formation were observed. The negative correlation coefficients were calculated ($r = -0.91$, $r = -0.90$, respectively), that can suggest that available lysine is a dominant precursor of MR progress in early stage and furosine formation is competitive to melanoidins. Whereas the positive correlation between furosine and FAST index data was noted ($r = 0.81$). It can be said that, in a great percent of furosine is converted into fluorescent compounds.

Collected in Table 4, the total, free, and linked-to-protein FIC found in all cakes were within the range of 142.4–201.7, 54.0–116.1, and 73.1–113.5 FI/mg sample DM, respectively. The total FIC values significantly decreased after cloves, spice mix, and allspice supplementation (15%, 10%, and 8%, respectively), the same effect

Table 4. Data on Maillard reaction products in rye-buckwheat cakes fortified with spices.

Type of cake fortified with spices	Available lysine (mg/g DM)	Furosine (μg/g)	Free FIC (Fl/mg DM)	Total FIC (Fl/mg DM)	Linked-to-protein (Fl/mg DM)	Tryptophan (Fl/mg DM)	FAST (%)	Browning (AU)
Control	0.59 ± 0.04[d]	510.8 ± 12.0[a]	77.44 ± 1.44[d]	166.57 ± 2.20[c]	89.13 ± 1.76[bc]	19.30 ± 1.00[cd]	462 ± 8[b]	0.36 ± 0.02[e]
Vanilla	0.67 ± 0.09[d]	488.4 ± 6.5[b]	116.14 ± 0.97[a]	201.68 ± 6.12[a]	85.54 ± 5.15[c]	17.68 ± 0.73[d]	484 ± 2[a]	0.41 ± 0.01[d]
Cinnamon	0.95 ± 0.03[c]	450.2 ± 9.7[c]	69.06 ± 3.35[b]	182.60 ± 4.18[b]	113.54 ± 7.53[a]	30.87 ± 0.81[b]	368 ± 4[e]	0.49 ± 0.01[c]
Cloves	1.62 ± 0.06[a]	111.7 ± 7.2[f]	69.21 ± 1.04[b]	142.36 ± 6.23[e]	73.15 ± 7.27[d]	21.92 ± 1.21[c]	230 ± 9[g]	0.68 ± 0.01[a]
Allspice	1.56 ± 0.07[a]	292.9 ± 5.6[d]	65.63 ± 1.65[c]	153.26 ± 3.76[d]	87.63 ± 5.46[bc]	19.44 ± 0.22[cd]	451 ± 3[c]	0.52 ± 0.01[b]
Nutmeg	0.99 ± 0.02[c]	458.6 ± 10.8[c]	114.91 ± 1.11[a]	200.92 ± 8.34[a]	86.01 ± 7.23[c]	31.87 ± 3.61[b]	392 ± 4[d]	0.48 ± 0.02[c]
Spice mix	1.30 ± 0.16[b]	220.5 ± 10.1[e]	54.02 ± 2.05[e]	151.47 ± 1.23[d]	97.45 ± 3.28[b]	38.00 ± 2.26[a]	256 ± 2[f]	0.52 ± 0.01[b]

FIC is expressed in fluorescence intensity (Fl) per mg of sample DM. Browning is expressed as absorbance units (AU). Values are means ± standard deviation (n = 3). Values in each column with different small superscript letters are significantly different (Fisher test, $P \leq 0.05$). DM, dry matter.

was observed for free FIC data. It can be said that some spices promote fluorescence compounds formation, however between total FIC and rutin negative correlation was noted ($r = -0.66$). Between total FIC and OPA data according to calculated correlation coefficient ($r = -0.72$).

To describe protein nutritional loss, FAST index was calculated as a ratio between linked-to-protein fluorescence/tryptophan fluorescence and expressed in percentage. The strong correlation between tryptophan and TPC data and antioxidant ability measured by PCL ACW and ABTS assays ($r = 0.92, 0.87,$ and 0.93, respectively) were noted. According to tryptophan results, it can be said that compounds from RBC supplemented with spices with strong antioxidant ability, have a potential to increase nutritional value. Table 4 shows the FAST index values. The values ranged from 230% to 484%. FAST index cloves, spice mix, cinnamon, nutmeg, and allspice being 2.0, 1.92, 1.8, 1.2, and 1.0 times lower. The positive influence of spices on FAST index decreasing was proved by correlation coefficient calculated between FAST/TPC, FAST/PCL, and FAST/ABTS ($r = -0.80, r = -0.89,$ and $r = -0.81$, respectively). According to results presented by Zieliński et al. (2012), rye ginger cakes showed a higher FAST index values in comparison to rye-buckwheat ginger cakes. However the FAST index for rye ginger cakes before storage is twice lower than indexes for rye-buckwheat ginger cakes. This indicates that measuring the FAST index in rye-buckwheat ginger cakes before storage and then in a determined time intervals can be important to monitor ongoing changes.

Brown high molecular polymers of MRP pigments formation were determined and presented in Table 4. Addition of cloves, spice mix, allspice, cinnamon, nutmeg, and vanilla to the recipe significant increased ($P < 0.05$) browning index by 88%, 44% (both spice mix and allspice), 36%, 33%, and 14%, respectively. It has been proven that melanoidin formation is positively correlated with AC measured by ABTS test ($r = 0.76$) and PCL ACW assay ($r = 0.65$) and with TPC and rutin contents in the cakes ($r = 0.63$ and 0.86). These findings are in agreement with Zieliński et al. (2010), who also evaluated the positive correlation between the AC and melanoidin content in wheat-rye ginger cakes. Moreover, browning index and OPA data were positively correlated ($r = 0.89$). The correlation value between FAST and browning indexes suggests that there is no relationship between loss of nutritional and melanoidin formation ($r = -0.81$).

Conclusions

Among collection of spices, the RBC fortified individually with cloves, nutmeg, allspice, cinnamon, vanilla, and spice mix addition revealed the highest sensory characteristics and overall quality. Cakes fortified with cloves, allspice, and spice mix showed the highest AC, total phenolics, rutin, and almost threefold higher available lysine contents. The reduction in furosine content as well as free and total fluorescent intermediatory compounds was observed as compared to control nonfortified cakes. In contrast, browning index was increased as compared to cakes without spices. In this study, the chemical and sensorial markers were fully applicable for description of the quality of RBC fortified with spices. It can be suggested that cloves, allspice, vanilla, and spice mix should be used for production of safety and good quality cakes.

Acknowledgment

This research was funded by grant No. 2012/07/N/NZ9/02250 from National Science Centre. The article is a part of the Ph.D. thesis of Małgorzata Przygodzka.

Conflict of Interests

The authors declare that there is no conflict of interests regarding the publication of this paper.

References

Bi, X., Y. Y. Soong, S. W. Lim, and C. J. Henry. 2015. Evaluation of antioxidant capacity of Chinese five-spice ingredients. Int. J. Food Sci. Nutr. 66:289–292.

Carlsen, M.H., R. Blomhoff, and L.F. Andersen. 2011. Intakes of culinary herbs and spices from a food frequency questionnaire evaluated against 28-days estimated records. Nutr. J. doi: 10.1186/1475-2891-10-50.

Cazzola, R., and B. Cestaro. (2014). Antioxidant spices and herbs used in diabetes. Pp. 110–119 in V. Preedy, ed. Aging. Diabetes: oxidative stress and dietary antioxidants. Academic Press, London, UK.

Charles, D. J. 2013. Natural antioxidants. Pp. 39–64 in J.D. Charles, ed. Antioxidant properties of spices, herbs and other sources. Springer, New York, USA.

Cheng, J., X. Chen, S. Zhao, and Y. Zhan. 2015. Antioxidant-capacity-based models for the prediction of acrylamide reduction by flavonoids. Food Chem. 168:90–99.

Chlopicka, J., P. Pasko, S. Gorinstein, A. Jedryas, and P. Zagrodzki. 2012. Total phenolic and total flavonoid content, antioxidant activity and sensory evaluation of pseudocereals breads. LWT Food Sci. Technol. 46:548–555.

Delgado-Andrade, C., J. A. Rufián-Henares, and F. J. Morales. 2006. Study on fluorescence of Maillard reaction compounds in breakfast cereals. Mol. Nutr. Food Res. 50:799–804.

Delgado-Andrade, C., I. Seiquer, M. P. Navarro, and F. J. Morales. 2007. Maillard reaction indicators in diets usually consumed by adolescent population. Mol. Nutr. Food Res. 51:341–351.

Dziki, D., R. Różyło, U. Gawlik-Dziki, and M. Świeca. 2014. Current trends in the enhancement of antioxidant activity of wheat bread by the addition of plant materials rich in phenolic compounds. Trends Food Sci. Technol. 40:48–61.

Embuscado, M.E. (2015). Herbs and spices as antioxidants for food preservation. Pp. 251–283. in F. Shahidi, ed. Handbook of antioxidants for food preservation. Woodhead Publishing Series in Food Science, Technology and Nutrition, Cambridge, UK.

Filipčev, B., O. Šimurina, M. Sakač, I. Sedej, P. Jovanov, M. Pestorić, et al. 2011. Feasibility of use of buckwheat flour as an ingredient in ginger nut biscuit formulation. Food Chem. 125:164–170.

Giannetti, V., M. Boccacci Mariani, P. Mannino, and E. Testani. 2014. Furosine and flavor compounds in durum wheat pasta produced under different manufacturing conditions: multivariate chemometric characterization. LWT Food Sci. Technol. 56:15–20.

Gökmen, V., A. Serpen, Ö. Cetinkaya Acar, and F. J. Morales. 2008. Significance of furosine as heat-induced marker in cookies. J. Cereal Sci. 48:843–847.

Hinneburg, I., H. J. Damien Dorman, and R. Hiltunen. 2006. Antioxidant activities of extracts from selected culinary herbs and spices. Food Chem. 97:122–129.

Hossain, M. B., N. P. Brunton, C. Barry-Ryan, A. B. Martin-Diana, and M. Wilkinson. 2008. Antioxidant activity of spice extracts and phenolics in comparison to synthetic antioxidants. Rasayan J. Chem. 4:751–756.

Illupapalayam, V. V., S. C. Smith, and S. Gamlath. 2014. Consumer acceptability and antioxidant potential of probiotic-yogurt with spices. LWT Food Sci. Technol. 55:255–262.

ISO/DIS 13299. (1998). Sensory analysis – methodology – general guidance for establishing a sensory profile. International Organization for Standardization, Geneva, Switzerland.

Kaefer, C. M., and J. A. Milner. 2008. The role of herbs and spices in cancer prevention. J. Nutr. Biochem. 19:347–361.

Kaur, M., K. Singh Sandhu, A. Arora, and A Sharma. 2015. Gluten free biscuits prepared from buckwheat flour by incorporation of various gums: physicochemical and sensory properties. LWT Food Sci. Technol. 62:628–632.

Kowalska, H., A. Marzec, and M. Mucha. 2012. Sensory evaluation of some types of bread and functional consumer preferences among the bread. Zeszyty Problemowe Postępu Nauk Rolniczych 571:67–78 (abstract in english).

Krkošková, B., and Z. Mrázová. 2005. Prophylactic components of buckwheat. Food Res. Int. 38:561–568.

Langner, E., and W. Rzeski. 2014. Biological properties of melanoidins: a review. Int. J. Food Prop. 17:344–353.

Liogier de Sereys, A., S. Muller, S. D. Desic, A. D. Troise, V. Fogliano, A. Acharid. 2014. Potential of the FAST index to characterize infant formula quality. Pp. 457–475 in V.R. Preedy, R.R. Watson, S. Zibadi, eds. Handbook of dietary and nutritional aspects of bottle feeding, human health handbooks. Academic Publishers, Wageningen, the Netherlands.

Marková, L., Z. Ciesarová, K. Kukurová, et al. 2012. Influence of various spices on acrylamide content in buckwheat ginger cakes. Chem. Pap. 66:949–954.

Markowicz Bastos, D. H., and A. Gugliucci. 2015. Contemporary and controversial aspects of the Maillard reaction products. Curr. Opin. Food Sci. 1:13–20.

Martysiak-Żurowska, D., and A. Stołyhwo. 2007. Content of furosine in infant formulae and follow-on formulae. Polish J. Food Nutr. Sci. 57:185–190.

Michalska, A., M. Amigo-Benavent, H. Zielinski, and M. D. del Castillo. 2008. Effect of bread making on formation of Maillard reaction products contributing to the overall antioxidant activity of rye bread. J. Cereal Sci. 48:123–132.

Oral, R. A., M. Dogan, and K. Sarioglu. 2014. Effect of certain polyphenols and extracts on furans and acrylamide formation in model system and total furans during storage. Food Chem. 142:423–429.

Pauly, A., B. Pareyt, M. A. Lambrecht, E. Fierens, and J. A. Delcour. 2013. Flours from wheat cultivars of varying hardness produces semi-sweet biscuits with varying textural and structural properties. LWT Food Sci. Technol. 53:452–457.

Przygodzka, M., D. Zielińska, Z. Ciesarová, K. Kukurová, and H. Zieliński. 2014. Comparison of methods for evaluation of the antioxidant capacity and phenolic compounds in common spices. LWT Food Sci. Technol. 58:321–326.

Rada-Mendoza, M., J. L. García-Bañosa, M. Villamiela, and A. Olano. 2004. Study on nonenzymatic browning in cookies, crackers and breakfast cereals by maltulose and furosine determination. J. Cereal Sci. 39:167–173.

Sakač, M., A. Torbica, I. Sedej, and M. Hadnađev. 2011. Influence of breadmaking on antioxidant capacity of gluten free breads based on rice and buckwheat flours. Food Res. Int. 44:2806–2813.

Sedej, I., M. Sakać, A. Mandić, A. Mišan, M. Pestorić, O. Šimurina, et al. 2011. Quality assessment of gluten-free crackers based on buckwheat flour. LWT Food Sci. Technol. 44:694–699.

Shahidi, F., and M. Naczk. 1995. Contribution of phenolic compounds to sensory characteristics of food. Pp. 199–226 in F. Shahidi, ed. Food phenolics. Sources. Chemistry. Effects. Applications. Taylor and Francis, London, UK.

Stone, H., R. N. Bleibaum, and H. A. Thomas. 2012. Sensory evaluation practices. Academic Press, Orlando, FL.

Sung, B., S. Prasad, V. R. Yadav, and B. B. Aggarwal. 2012. Cancer cell signaling pathways targeted by spice-derived nutraceuticals. Nutr. Cancer 64:173–197.

Virág, D., A. Kiss, P. Forgó, C. Csutorás, and M. Szabolcs. 2013. Study on Maillard-reaction driven transformations and increase of antioxidant activity in lysine fortified biscuits. Microchem. J. 107:172–177.

Wojdyło, A., J. Oszmiański, and R. Czemerys. 2007. Antioxidant activity and phenolic compounds in 32 selected herbs. Food Chem. 105:940–949.

Wójtowicz, A., A. Kolasa, and L. Mościcki. 2013. Influence of buckwheat addition on physical properties, texture and sensory characteristics of extruded corn snacks. Polish J. Food Nutr. Sci. 63:239–244.

Wronkowska, M., A. Troszyńska, M. Soral-Śmietana, and A. Wołejszo. 2008. Effects of buckwheat flour (Fagopyrum esculentum Moench) on the quality of gluten-free bread. Polish J. Food Nutr. Sci. 58:211–216.

Zielińska, D., D. Szawara-Nowak, and H. Zieliński. 2010. Determination of the antioxidant activity of rutin and its contribution to the antioxidant capacity of diversified buckwheat origin material by updated analytical strategies. Polish J. Food Nutr. Sci. 60:315–321.

Zieliński, H., M. Amigo-Benavent, M. D. del Castillo, A. Horszwald, and D. Zielińska. 2010. Formulation and baking process affect Maillard reaction development and antioxidant capacity of ginger cakes. J. Food Nutr. Res. 49:140–148.

Zieliński, H., M. D. del Castillo, M. Przygodzka, Z. Ciesarova, K. Kukurova, and D. Zielińska. 2012. Changes in chemical composition and antioxidative properties of rye ginger cakes during their shelf-life. Food Chem. 135:2965–2973.

13

Quality changes and freezing time prediction during freezing and thawing of ginger

Poonam Singha & Kasiviswanathan Muthukumarappan

Department of Agricultural and Biosystems Engineering, South Dakota State University, Brookings, South Dakota 57007

Keywords

Essential oil, freezing, ginger, microstructure, quality, simulation

Correspondence

Poonam Singha, Department of Agricultural and Biosystems Engineering, South Dakota State University, Brookings, SD 57007.

E-mail: poonam.singha@sdstate.edu

Funding Information

Agricultural Experiment Station, South Dakota State University, Brookings, SD.

Abstract

Effects of different freezing rates and four different thawing methods on chemical composition, microstructure, and color of ginger were investigated. Computer simulation for predicting the freezing time of cylindrical ginger for two different freezing methods (slow and fast) was done using ANSYS® Multiphysics. Different freezing rates (slow and fast) and thawing methods significantly ($P < 0.05$) affected the color and composition of essential oil in ginger. Fresh ginger was found to contain 3.60% gingerol and 18.30% zingerone. A maximum yield of 7.43% gingerol was obtained when slow frozen gingers when thawed by infrared method. Maximum zingerone content of 38.30% was achieved by thawing slow frozen gingers using infrared-microwave method. Microscopic examination revealed that structural damage was more pronounced in slow frozen gingers than fast frozen gingers. Simulated freezing curves were in good agreement with experimental measurements ($r = 0.97$ for slow freezing and $r = 0.92$ for fast freezing). Slow freezing damaged ginger's cellular structure. Data obtained will be helpful in selecting appropriate thawing method to increase desirable essential oil components in ginger. Computer simulation for predicting freezing time may help in developing proper storage system of ginger.

Introduction

Ginger (*Zingiber officinale*) is regarded as one of the most important spices grown in the world. It is extensively grown in the tropical and subtropical regions of the world particularly in Bangladesh, India, Taiwan, Jamaica, Africa, Mexico, China, and Japan (Thompson et al. 1973; Baliga et al. 2011; Aziz et al. 2012; Mishra et al. 2013). It is not only used as a spice, but also as a traditional medicine against several diseases. It has been used as a remedy to dyspepsia, nausea, indigestion, colic, and diarrhea (Aziz et al. 2012). The health-promoting functionality of ginger is often attributed to its rich phytochemistry (Shukla and Singh 2007). The constituents of ginger are numerous and vary depending on the place of origin and form of rhizomes, for example, fresh or dry. The rhizomes contain highly valued aromatic, volatile, and pungent compounds. The nonvolatile components of the ginger are mainly responsible for imparting its pungency, whereas the volatile components are responsible for its aroma (Govindarajan

and Connell 1983). The chemical investigations carried out in the past showed that monoterpene hydrocarbons, oxygenated monoterpenes, sesquiterpene hydrocarbons, and nonterpenoid compounds were the main constituents in ginger oils (Onyenekwe and Hashimoto 1999; Kim and Lee 2006; Aziz et al. 2012). Among the many components, α-zingiberene is the most predominant component of ginger oil (Ravindran and Babu 2005). Gingerols are attributed for ginger-specific pungency and possess substantial antioxidant activity as determined by various antioxidant assays (Butt and Sultan 2011).

Like all fresh fruits and vegetables, ginger is perishable because of its high moisture content. The optimum conditions for storing fresh ginger roots are 13–15°C and 90–95% relative humidity (Enyama 1981; Choi and Kim 2001). For ginger growers it is costly and difficult to maintain the optimum storage conditions of ginger roots and hence, they store them in underground tunnels, where the optimum temperature and humidity control are impossible. This results in spoilage and the ginger roots sprout after a few months (Lee et al. 1994). Ginger roots when gamma irradiated with up to 80 Gy and stored at 25–28°C showed a deterioration in external appearance after 1 month, and shrinkage and discoloration after 2 months (Yusof 1990). Similar results were reported by (Gonzalez et al. 1969), Sirikulvadhana and Prompubesara (1979) and Queirol et al. (2002). Other pretreatments, such as a citric acid treatment (Brown and Lloyd 2007), wax coating (Okwuowulu and Nnodu 1988) and antimicrobial treatment (Subramanyam et al. 1962) exhibited limited effects on extending the storage life of ginger roots, and were not found to be suitable for long-term storage.

Developing methods for the long-term storage of harvested ginger roots without a loss in quality is very important to ginger growers as well as to people who process food, and who need a continuous supply of good quality ginger roots throughout the year. It appears that freezing could be a suitable method to preserve the quality of the ginger roots for a long period.

Freezing is an efficient process of preserving the quality of food because in frozen state, water is immobilized as ice and the rates of deterioration are much slower than at higher temperatures. Two major thermal events occur during freezing. At first there is formation of ice crystals (or nucleation) followed by increase in crystal size (crystal growth). The rate of crystal growth is determined by three factors: rate of reaction at the crystal surface, diffusion rate of water to the growing crystal, and rate of heat removal. Crystal size varies inversely with the number of nuclei. Freezing stands on two basic prerequisites to deliver high quality products: (1) rapid freezing rates; and (2) rapid thawing rates (Petzold and Aguilera 2009). Rapid freezing can be accomplished using cryogenic systems employing liquefied gases such as nitrogen or carbon dioxide as refrigerant. Liquid nitrogen has a boiling point of −196°C at atmospheric pressure and the cooling effect is almost instantaneous when sprayed on food stuff. Liquid carbon dioxide when released at atmospheric pressure creates 50% dry ice (solid CO_2) and 50% vapor, both at −70°C. Dry ice has an extremely high rate of heat removal from food product surface, which can often more than compensate for the higher temperatures compared with liquid nitrogen freezing. The use of CO_2 freezing depends on the individual application (George 1997).

Frozen storage should markedly enhance storage life. However, freezing process may cause severe changes to tissues, resulting in excessive softening (Delgado and Rubiolo 2005). Complications may also arise due to freezing process itself causing alterations to physicochemical characteristics, biochemical quality, and microbiological safety affected adversely by freeze–thaw cycle (Opoku-Nkoom 2015). The freezing rate is responsible for tissue damage (Fuchigami

et al. 1997) and can result in unacceptable or suboptimal product characteristics after thawing. It is generally accepted that high freezing rates retain the quality better than slow freezing rates (Partmann 1975). However, ultrarapid freezing result in mechanical cracking(Kalichevsky 1995) particularly in large samples with high moisture content and low porosity. (Kim and Hung 1994)

Thawing generally occurs more slowly than freezing. Theoretically, thawing is the inverse process of freezing; they are different not only in phase change direction, cooling and heating process, but also in food freezing time and internal temperature variations (Min 2001). The thawing process is to make the freezing ice melt into water inside the food, and get absorbed by the food to restore the freshness similar to that before frozen. Thawing process is much more complex than the freezing process. During thawing, foods are subject to damage due to chemical and physical changes. Therefore, optimum thawing procedures should be of concern to food technologists (Fennema et al. 1973; Kalichevsky 1995). Quick thawing of food is desirable to assure food quality of frozen vegetables, bread, pastries and so on. But slow thawing is better for thawing fish and meat (Ji et al. 2014) as it allows to reabsorb much of the moisture from the melting ice crystals so there's less "drip out". Li and Sun (2002) reviewed on application of novel thawing methods such as microwave thawing, acoustic thawing, and ohmic thawing.

Furthermore, it is essential to predict the temperature and freezing time of foods when designing and evaluating freezing equipment (Mannapperuma and Singh 1989). Freezing time and temperature profile within food can be determined experimentally or predicted approximately by analytical, numerical, or computational simulation methods. Experimental procedures are often too expensive, time consuming and may lack a generalized theoretical description of the process. By comparison, numerical and computational simulation methods based on finite differences and finite element techniques (Mannapperuma and Singh 1989) are more effective in analyzing actual situation. The physical changes of food during phase change have to be understood for proper prediction of its thermal behavior at different temperature conditions. Knowledge of the thermo-physical properties of food material and surrounding environment such as specific heat and enthalpy, density, thermal conductivity, etc., are required for the prediction of temperature within food (Matuda et al. 2011).

Adjusting the freezing–thawing process variables will help to preserve and retain the quality of the product. No information has been published in the literature on the effects of different thawing methods on the microstructure and chemical composition of ginger. In our study, we employed four different thawing techniques namely room temperature thawing which is a slow thawing process and is also a common household practice to thaw frozen foods, and three quick thawing processes viz. microwave thawing, infrared thawing and infrared-microwave thawing. Quick thawing maintains the quality of food. During microwave thawing electromagnetic radiation is transmitted through food product which is transferred into heat energy. Infrared thawing uses mechanism of radiation to supply heat which favors high rate of surface heat transfer (Venugopal 2005). The objectives of this study were: (1) to investigate and compare the effects of different freezing and thawing methods on the volatile and nonvolatile contents, the microstructure and appearance of ginger; and (2) to study the temperature profile during freezing and predict freezing time using computer simulation.

Materials and Methods

Experimental procedure

Freezing

Fresh ginger rhizomes were purchased from local store in Brookings, SD, USA. They were selected to be homogeneous in size and color. The moisture content as determined using AOCS methods (AOAC, 1990) was in the range 85–89% (wb). The gingers were cut into regular cylindrical shape of 40 ± 0.01 mm in length and 8 ± 0.003 mm in diameter. Ginger samples were frozen by two methods: slow freezing in a refrigerator maintained at −18°C and fast freezing by using liquid nitrogen (~203°C). In fast freezing, ginger samples were suspended in closed chamber filled with liquid nitrogen until desired temperature (−18°C) at the geometric center was reached. During freezing (both slow and fast freezing), the temperature profile were recorded at regular intervals, normally every 3–5 sec, using thermocouples inserted at the geometric center of the gingers and connected to a data acquisition system (Personal Daq/56). Freezing rate during slow and fast freezing was 0.24°C·min⁻¹ and 1.29°C·sec⁻¹ respectively. The core temperature of ginger samples reached −18°C within 28 sec during fast freezing and after 187 min during slow freezing.

Thawing

As the final freezing temperature was reached approximately −18°C, the frozen gingers were individually thawed under four different conditions viz. at room temperature (~23°C), in microwave, infrared, and infrared-microwave condition. Advantium™ 120 oven (SCA1001KSS02, Louisville, KY) was used for microwave and infrared thawing purposes. During thawing, temperature at the

center of the gingers were recorded. Thawing was considered complete when the final core temperature was ~19°C. All treatments and measurements were carried out in triplicate. Thawing time for slow frozen gingers was approximately 87 min when thawed at room temperature (~23°C), 1 min in microwave, 10 min in infrared and 3 min in infrared-microwave. Thawing time for fast frozen gingers was 27 min at room temperature (~23°C), 56 sec in microwave, 6 min in infrared, and 2 min in infrared-microwave.

Physico-chemical analyses

Analysis of essential oil composition

Extraction of essential oil fractions in ginger samples (fresh, frozen, and thawed) was done following a method mentioned elsewhere (Usman et al. 2013). Typically, all ginger samples were dried in oven at 60°C for 24 h and pulverized by using a coffee grinder (SmartGrind, Black & Decker, CH Annex Company, China). Approximately, 0.5 g of each pulverized samples was weighed and placed in a 25 mL volumetric flask with methanol as extracting solvent. The solvents were allowed to percolate the materials which were soaked in it for 48 h before collecting the extract. The extracts were centrifuged at 5000 g for 10 min at room temperature (~23°C) and filtered through a 0.2 μ filter prior to analysis.

Analysis was performed using a Gas Chromatography-Mass Spectrometry (GC-MS) (Agilent GC–7890A, MSD-5975C and auto-sampler-7693). The capillary columns were 30 m × 0.25 mm × 0.25 mm DB-5MS (J&W Scientific, Folsom, CA). Ultrapure hydrogen was used as the carrier gas at a flow rate of 1.5 mL·min⁻¹ and column head pressure of 52.4 kPa (7.6 psi). The auto-sampler introduced a 1 μL sample into an injection port with a total inlet flow of 54 mL·min⁻¹. The injection port was held at 250°C and contained an Agilent inlet liner of deactivated borosilicate single-taper with glass wool packing. The purge flow was initiated at 1 min with a flow of 50 mL·min⁻¹. The GC oven temperature was initially held at 80°C for 2 min, then elevated at a rate of 9°C·min⁻¹ up to 200°C and held for 4 min. The gradient was then increased to 10°C·min⁻¹ up to 280°C where it was held constant for 5 min. This gradient resulted in an overall run time of 13 min with ATCA-(TMS)₃ eluting at approximately 8.76 min. The GC was interfaced with a mass selective detector with the transfer line held at 265°C. Fragmentation of the sample was as accomplished through electron impact with selected ion monitoring (SIM) mode for monitoring abundant ions of ATCA (m/z 245, 347, and 362) and ATCA-d2 (m/z 349, and 364) with a dwell time of 100 ms each. The MS conditions were as follows: ion source pressure 2.0 Pa ($1.5 \times 10 -5$ Torr), source temperature 200°C, quadrupole temperature 150°C, electron energy 70 eV, electron emission current 34.6 A, and electron multiplier voltage +400 relative to the autotune setting. The major chemical compositions were identified through a NIST Mass Spectral library.

Color measurements

Color measurements of the fresh, frozen, and thawed ginger samples were carried out using Minolta Spectrophotometer (CM-2500d; Minolta Co. Ltd, Osaka, Japan). The spectrophotometer was first calibrated with a white plate and checked for recalibration between measurements, although no adjustments were necessary. Readings were reported in the L^*, a^*, b^* system. The color values, expressed as L^* (whiteness or brightness/darkness), a^* (redness/greenness), and b^* (yellowness/blueness), for respective samples were determined. Three readings were taken and average values were calculated for each data. Reference color values for the fresh samples (L_o^*, a_o^*, b_o^*) and color values from frozen and thawed samples were employed to determine the changes in each individual color parameter and were calculated as follows:

$$\Delta L^* = L^* - L_o^*. \tag{1}$$

$$\Delta a^* = a^* - a_o^*. \tag{2}$$

$$\Delta b^* = b^* - b_o^*. \tag{3}$$

The total color difference (ΔE^*) was determined using the following equation:

$$\Delta E^* = [\Delta L^{*2} + \Delta a^{*2} + \Delta b^{*2}]^{1/2}. \tag{4}$$

The chroma and hue angle ($H°$, hue angle; red = 0°; yellow = 90°; green = 180°; blue = 270°) were also calculated on the basis of the following equations:

$$\text{Chroma} = \sqrt{(a^{*2} + b^{*2})}. \tag{5}$$

$$H° = \tan^{-1}(b^*/a^*) \text{ when } a^* < 0 \text{ and } b^* \geq 0. \tag{6}$$

and

$$H° = 180 + \tan^{-1}(b^*/a^*) \text{ when } a^* < 0. \tag{7}$$

Microstructure analysis

Structural observation was carried using a Hitachi-S3400 N (Tokyo, Japan) scanning electron microscope (SEM) operated at 10 kV. The samples were freeze-dried to remove

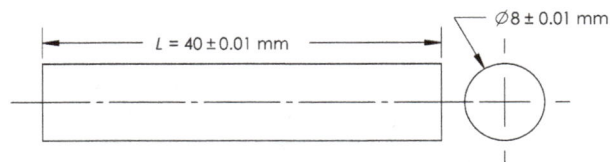

Figure 1. Geometrical dimensions of the ginger: Top view (Left) and Front view (Right).

water prior to the SEM observation. Freeze-dried samples approximately 8 mm in diameter and 0.2 mm thickness were mounted on stubs and 10 nm gold was coated using a CrC-150 sputtering system set to a pressure of 5–10 millitorr. Magnification of 160× was used in all the micrographs in the present study. At least two samples for each treatment showing similar images were used for the results.

Simulation of freezing process

The problem considered in this study involves the unsteady one-dimensional heat transfer in a food (cylindrical in shape) during freezing process. The temperature profile at the core of ginger during freezing (slow and fast) was determined using transient thermal analysis in ANSYS® Multiphysics (ANSYS, Inc., Canonsburg, PA, USA). The geometry of cylindrical ginger was developed using SolidWorks®14 (Dassault Systèmes SolidWorks Corporation, Waltham, MA, USA) and its geometrical dimensions are represented in Figure 1. The meshing of the ginger was conducted using ANSYS® Multiphysics. The mesh refinement process was repeated until further mesh refinements have insignificant effects on the results. This process reduces the uncertainties associated with the complexity of heat flow. The generated mesh (Fig. 2) consisted of 19,856 bricks (volume elements). Thermal boundary conditions were applied to the finite element model; (1) free convection from the surface of the ginger and (2) initial temperature of the cylindrical ginger set at 21°C. Following assumptions were made for simulation: (1) heat was transferred radially;

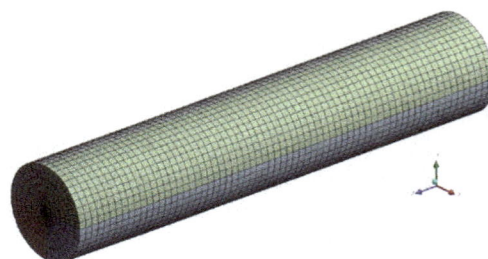

Figure 2. Finite element mesh of ginger model.

(2) the cylindrical ginger was at uniform temperature and was exposed suddenly at time zero to a cooling medium; (3) the cooling medium consisted of air for slow freezing and liquid nitrogen for fast freezing with constant temperature; (4) the food was isotropic; (5) mass transfer between the ginger and the environment was negligible. Inputs for computer simulation were selected to parallel the conditions of the actual experimental freezing trials. A detailed description of input parameters for simulation and thermophysical properties of air and liquid nitrogen are shown in Tables 1 and 2 respectively. The specific heat and thermal conductivity of ginger were measured using KD2 Pro thermal analyzer (Decagon devices, Inc., Pullman, WA, USA).

Statistical analysis

For data analysis, analysis of variance (ANOVA) was used. Post hoc Tukey's test was used to determine where significant differences ($P < 0.05$) occurred, unless otherwise mentioned. All statistical analysis was performed using SPSS version 16.0 for windows software (SPSS Inc., Chicago, IL).

Results and Discussion

Chemical composition of essential oil

Volatile and nonvolatile compounds in the essential oils of fresh, frozen, and thawed ginger samples were identified using GC-MS (Table 3, Fig. 3A). The major volatile and nonvolatile compounds identified were zingiberene ($C_{15}H_{24}$), zingerone ($C_{11}H_{14}O_3$), β-sesquiphellandrene ($C_{15}H_{24}$), β-bisabolene ($C_{15}H_{24}$), curcumene ($C_{15}H_{22}$), gingerol ($C_{17}H_{26}O_4$), and farnesene ($C_{15}H_{24}$). Fresh ginger essential oil contained 35.5% zingiberene, 18.3% zingerone, 11.3% β-sesquiphellandrene, 6.38% β-bisabolene, 3.6% gingerol, and 1.87% α-farnesene.

The nonvolatile compounds responsible for the pungency in ginger are gingerol, shogaol, and zingerone (Vasala 2004). Shogaol was not detected in fresh, frozen, and thawed ginger essential oils. Gingerol content in essential oil fractions was 4.2% for slow frozen (SF), 4.4% for fast frozen (FF), 6.6% for slow frozen-microwave thawed (SFMW), 7.43% for slow frozen-infrared thawed (SFIR), and 6.81% for slow frozen-infrared microwave thawed (SFIR-MW). Thawing resulted in increase in gingerone content when compared to fresh and frozen gingers. This increase in zingerone content may be due to transformation of gingerol to zingerone through retro-aldol reaction at the β-hydroxy ketone group during thawing (Zachariah 2008).

Compounds responsible for the aroma of ginger are β-sesquiphellandrene, zingiberene, β-bisabolene, and farnesene. Zingiberene is the major volatile compound in both fresh and slow frozen (36.27%) gingers. However, zingiberene content decreased in FF (28%), SFMW (31%), FFMW (29%), FFIR (23%), FFIR-MW (22%), FFRT (20%), SFRT (19%) with maximum decrease found to be in SFIR-MW (17.6%). β-bisabolene content increased in SF (9.4%), FF (8.9%), SFIR (7.13%), and FFMW (7.9%) compared to that in fresh ginger. β-sesquiphellandrene content in all sample of ginger essential oils decreased except in SF (12%). α-farnesene was not identified in most of the ginger samples. However, α-farnesene content was found to have increased in SF (3.3%), SFMW (2%) and FFIR (2.2%) ginger samples in comparison to that of fresh ginger. Traces of stereoisomers of α-farnesene; (E,Z)-α-farnesene and (Z,Z)-α-farnesene were identified in some thawed ginger samples. 2,3-dihydro-3,5-dihydroxy-6-methyl-4H-pyran one (DDMP; $C_6H_8O_4$) which has been reported to inhibit colon cancer cell growth (Ban et al. 2007) was found to have increased in the essential oil extract of FFIR-MW (2.5%). The increase in the volatile and nonvolatile compounds observed in some thawed gingers may be due to more disruption of cell walls and/or influence of heat.

Color alterations

Color properties of fresh, frozen, and thawed gingers were determined by using CIE (Commission international de l' éclairage) $L^*a^*b^*$ measurements as shown in Table 4. Color is an important attribute and undergoes significant changes during freezing and thawing (Fig. 4). The L^* value (lightness) of fresh ginger was 77.08 which decreased ($P < 0.05$) when gingers were frozen and thawed. The decrease in L^* value was more pronounced in all fast frozen and thawed gingers and for SFRT and SFMW. The a^* value, which is a measure of redness and greenness, was found to decrease ($P < 0.05$) more during slow freezing (-3.11) than during fast freezing (-1.94) when compared to fresh ginger (-1.54). After thawing at room temperature (~23°C), slow frozen gingers showed more yellowness (-1.32) than fresh ginger. However, all other thawing methods significantly decreased ($P < 0.05$) the a^* values of both slow and fast frozen gingers. Except for SFRT, all samples showed lower a^* value ($P < 0.05$) suggesting more greenness. All frozen and thawed samples had lower b^* ($P < 0.05$) value compared to that of fresh ginger (40.45) suggesting loss in yellow color. Hue angle values increased ($P < 0.05$) for all ginger samples except for FF and SFRT. Figure 5 shows the ΔE^* value, which was calculated based on the L^*, a^*, and b^* values of ginger according to different freezing and thawing methods. The total color

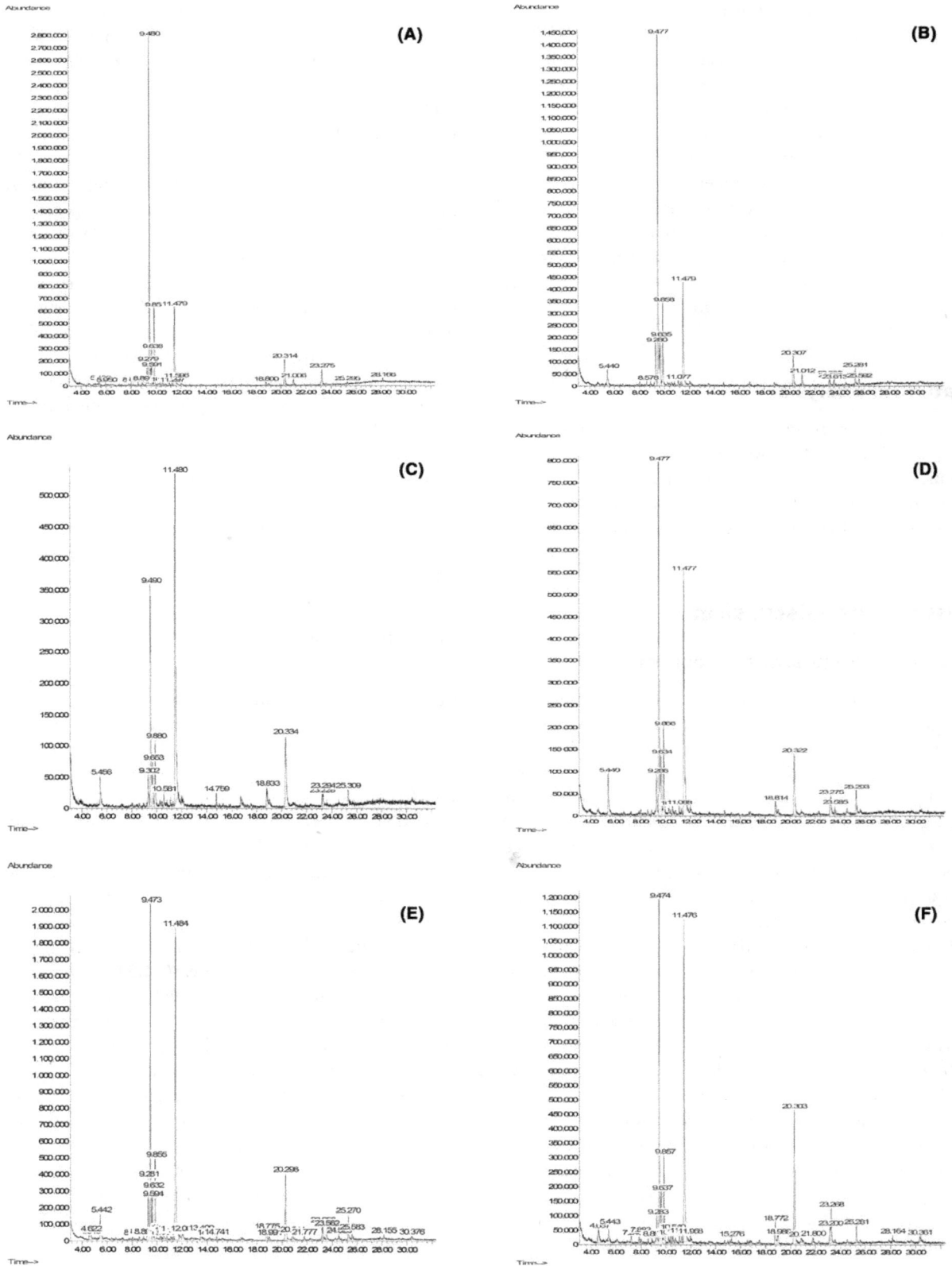

Figure 3.

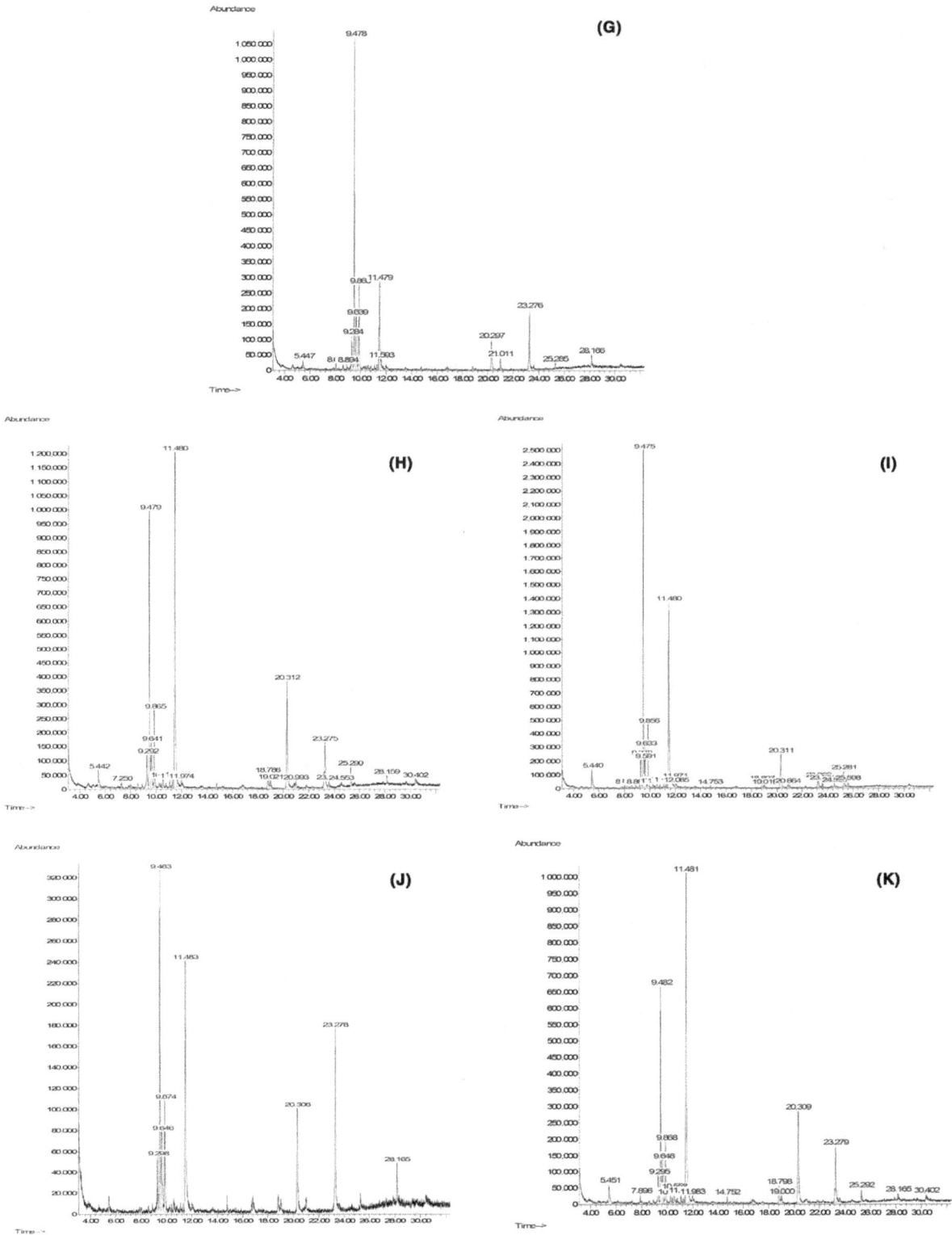

Figure 3. GC-MS analysis of (A) fresh ginger, (B) fast frozen, (C) fast frozen-room temperature thawed, (D) fast frozen-microwave thawed, (E) fast frozen-infrared thawed, (F) fast frozen-infrared microwave thawed, (G) slow frozen, (H) slow frozen-room temperature thawed, (I) slow frozen-microwave thawed, (J) slow frozen-infrared thawed, and (K) slow frozen-infrared microwave thawed.

difference (ΔE^*) was highest in fast frozen-microwave thaw (FFMW) and lowest in SF. The mean ΔE^* value of SF was significantly lower ($P < 0.05$) than that of other frozen and thawed samples, indicating that the

surface color of the SF ginger was closer to that of fresh ginger. The color difference may be explained by the fact that chroma values of all the frozen and thawed gingers was lower than fresh ginger. The chroma value indicates the saturation or color purity and is effected by a^* and b^* values.

Table 1. Input parameters for simulation of slow and fast freezing of cylindrical ginger.

Parameters	Values
Ginger initial temperature (°C)	21
Density of unfrozen ginger, ρ_g (kg·m^{-3})	1050
Specific heat capacity of unfrozen ginger, Cp_g (J·kg^{-1}·K^{-1})	3100
Thermal conductivity of air, k_{air} (W·m^{-1}·K^{-1})	0.8
Air temperature (°C)	−18
Liquid nitrogen temperature (°C)	−203

Table 2. Thermophysical properties of air and liquid nitrogen (Anzaldua-morales et al. 1999; Vasala 2004; Ban et al. 2007; Zachariah 2008; Usman et al. 2013).

	Air (T = 255 K)	Liquid nitrogen (T = 70 K)
Density, ρ (kg·m^{-3})	1.3835	840
Specific heat, Cp (J·kg^{-1}·K^{-1})	1003	2024
Viscosity, μ (Pa·s)	1.650×10^{-5}	2.20×10^{-4}
Thermal conductivity, k (W·m^{-1}·K^{-1})	0.0228	0.150
Prandtl number, Pr	0.715	2.97
Thermal expansion coefficient, β (K^{-1})	0.002	0.00504

Microstructure analysis

To gain insight into the effects of freezing and thawing on the structure of ginger, scanning electron microscopic (SEM) images were obtained to provide visual evidence of the changes in structure. Figure 6 shows microscopic image of fresh sample of ginger rhizome, which did not receive any other treatment other than preparation for SEM. The impacts of freezing on quality of food are directly related with the growth of ice crystals which can break cellular walls (Anzaldua-morales et al. 1999). Ginger rhizome typically contains 85–89% moisture (wb). When ginger was subjected to slow freezing large ice crystals were formed which disrupted the cells. Figure 7A shows the structural damage caused due to formation of large ice crystal during slow freezing. Contrary to this, fast or rapid freezing leads to formation of smaller ice crystals and hence causes minimum damage to cellular structure (Fig. 7B). Rapid freezing is appropriate to retain the tissue structure. This is in agreement with Delgado and Rubiolo (2005).

Thawing also plays an important role in regulating the cellular structure of food. In our study, we investigated the effects of different thawing process on the microstructure

Table 3. The chemical composition of essential oils of fresh (control), frozen, and thawed gingers (*Zingiber officinale*) analyzed by GC-MS.

Compound	Peak area %										
	FG	SF	FF	SFRT	SFMW	SFIR	SFIR-MW	FFRT	FFMW	FFIR	FFIR-MW
Gingerol	3.6^d	4.2^c	4.4^c	2.7^e	6.6^b	7.43^a	6.81^b	1.5^h	1.9^f	1.8^f	1.7^{fg}
Zingerone	18.3^f	3.6^h	14^g	33^b	27^d	21.5^e	38.3^a	32^{bc}	30^{cd}	29^{cd}	29^c
Zingiberene	35.5^a	36^a	28^c	19^f	31^b	19.7^{ef}	17.6^f	20^{ef}	29^{bc}	23^d	22^{de}
β-bisabolene	6.38^e	9.4^a	8.9^b	5.3^{gh}	6.1^f	7.13^d	5.01^i	5.4^g	7.9^c	5.1^{hi}	5.3^{gh}
β-sesquiphellandrene	11.3^a	12^a	11^a	6.6^{cd}	7.8^{bc}	8.19^{bc}	6.57^{cd}	6.7^d	8.6^b	6.3^{cd}	7.1^{bcd}
α-farnesene	1.87^c	3.3^a	–	–	2^{bc}	–	–	–	–	2.2^b	–
(E,Z)-α-farnesene	–	–	–	0.3	–						
(Z,Z)-α-farnesene	–	–	–	–	0.5						
2,3-dihydro-3,5-dihydroxy-6-methyl-4H-pyran-4-one (DDMP)	0.33^c	0.2^d	–	–	–	–	–	–	–	1.2^b	2.5^a
Curcumene	4.6^d	4.9^c	6.5^a	3.8^f	3.8^f	3.67^f	3.49^g	3.1^h	4.1^e	5.5^b	2.8^i
Copaene	0.45^a	–	–	–	0.4^a	–	–	–	–	–	–
Limonene	0.53	–	–	–	–	–	–	–	–	–	–
α-pinene	–	–	3	–	–	–	–	3.2	–	–	–

$^{a–i}$Mean values with different superscript letters are significantly different ($P < 0.05$, Tukey's test).FG, Fresh ginger; SF, Slow frozen; FF, Fast frozen; SFRT, Slow frozen and room temperature thawed; SFMW, Slow frozen and microwave thawed; SFIR, Slow frozen and infrared thawed; SFIR-MW, Slow frozen and infrared – microwave thawed; FFRT, Fast frozen and room temperature thawed; FFMW, Fast frozen and microwave thawed; FFIR, Fast frozen and infrared thawed; FFIR-MW, Fast frozen and infrared – microwave thawed.

Table 4. Color values of fresh (control), frozen, and thawed ginger (*Zingiber officinale*) samples.

Parameter	Ginger samples										
	FG	SF	FF	SFRT	SFMW	SFIR	SFIR-MW	FFRT	FFMW	FFIR	FFIR-MW
L* (Lightness)	77.08[a]	74.33[b]	58.42[h]	62.08[f]	61.38[fg]	71.51[c]	68.68[d]	59.29[h]	60.42[g]	66.12[e]	66.20[e]
	(0.05)	(0.07)	(0.30)	(0.56)	(0.50)	(0.30)	(0.64)	(0.16)	(0.10)	(0.13)	(0.54)
a* (Redness)	−1.54[b]	−3.11[f]	−1.94[c]	−1.32[a]	−3.69[g]	−2.56[e]	−2.54[e]	−3.95[i]	−2.06[d]	−3.06[f]	−3.84[h]
	(0.02)	(0.02)	(0.01)	(0.01)	(0.02)	(0.03)	(0.02)	(0.03)	(0.02)	(0.02)	(0.03)
b* (Yellowness)	40.45[a]	37.18[bc]	31.11[h]	34.37[e]	32.29[g]	36.69[c]	35.48[d]	35.18[d]	35.65[d]	37.72[b]	33.38[f]
	(0.17)	(0.06)	(0.11)	(0.34)	(0.16)	(0.14)	(0.33)	(0.23)	(0.29)	(0.37)	(0.15)
Chroma	40.48[a]	37.31[bc]	31.20[h]	34.39[e]	32.50[g]	36.78[c]	35.57[d]	35.39[d]	35.71[d]	37.76[b]	33.59[f]
	(0.17)	(0.06)	(0.11)	(0.34)	(0.16)	(0.14)	(0.33)	(0.24)	(0.29)	(0.48)	(0.15)
H° (Hue angle)	179.55[de]	181.41[b]	176.95[f]	179.10[e]	181.23[b]	180.17[cd]	179.84[de]	181.75[b]	180.02[de]	183.49[a]	181.14[bc]
	(0.44)	(0.28)	(0.19)	(0.60)	(0.02)	(0.09)	(0.11)	(0.09)	(0.19)	(0.95)	(0.22)

[a-i]Mean values with different superscript letters are significantly different (*P* < 0.05, Tukey's test). Parentheses indicate ± standard deviation (*n* = 3). FG, Fresh ginger; SF, Slow frozen; FF, Fast frozen; SFRT, Slow frozen and room temperature thawed; SFMW, Slow frozen and microwave thawed; SFIR, Slow frozen and infrared thawed; SFIR-MW, Slow frozen and infrared – microwave thawed; FFRT, Fast frozen and room temperature thawed; FFMW, Fast frozen and microwave thawed; FFIR, Fast frozen and infrared thawed; FFIR-MW, Fast frozen and infrared – microwave thawed.

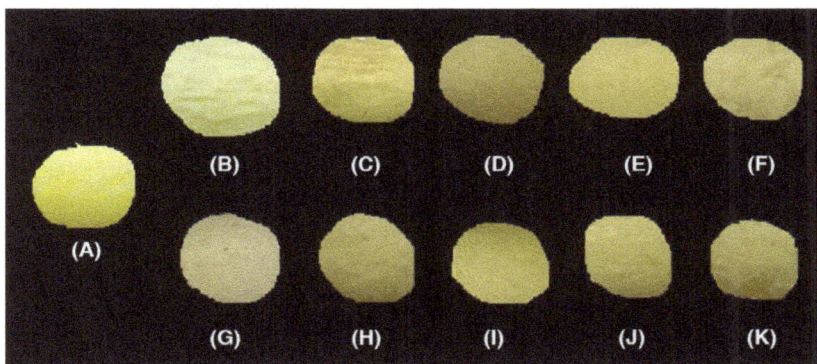

Figure 4. Changes in color of gingers during freezing and thawing: (A) fresh ginger, (B) slow frozen, (C) slow frozen-room temperature thawed, (D) slow frozen-microwave thawed, (E) slow frozen-infra red thawed, (F) Slow frozen-infrared microwave thawed, (G) fast frozen, (H) Fast frozen-room temperature thawed, (I) Fast frozen-microwave thawed, (J) Fast frozen-infrared thawed, and (K) Fast frozen-infrared microwave thawed.

of ginger. When gingers are thawed, the cells try to resume or regain their original shape. However, this greatly depends on the thawing method. During thawing reabsorption of water and other soluble substances by the cells takes place. From the SEM images (Fig. 7C–J), it can be said that after reabsorption, water and other products are better distributed within the cellular compartments of the fast frozen gingers than those of slow frozen ones. The reason behind such phenomenon could be due to less cell damage occurring during fast freezing than while slow freezing.

Freezing time prediction

Verification of the simulated results for the freezing time prediction was achieved by comparing with the experimental data obtained. Good correlation (*r* = 0.97

for slow freezing and *r* = 0.92 for fast freezing) was found between experimental measurements and the simulated results. The freezing curves for slow and fast freezing of ginger are shown in Figures 8 and 9, plotted over the simulated temperature history (solid line). From the figures it can be seen that the simulated curves closely predicts the actual freezing curves. Simulated curves slightly deviate from experimental data but are positioned quite closely to the observation for the practical-need accuracy. In case of slow freezing, at temperatures <−5°C the prediction method computed freezing rate ~1.06× faster than those determined experimentally. This amounts to freezing time difference of 9 min. On the other hand for fast freezing, at temperatures <−5°C the prediction method computed freezing rate ~1.39× faster than those determined experimentally. This amounts to freezing time

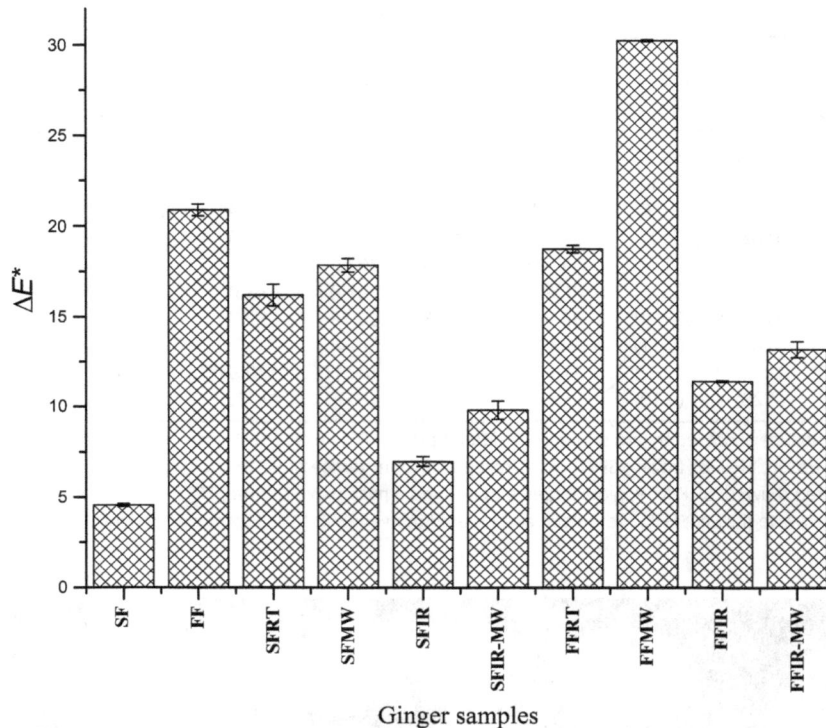

Figure 5. Effect of freezing and thawing on total color difference (ΔE*). SF, Slow frozen; FF, Fast frozen; SFRT, Slow frozen and room temperature thawed; SFMW, Slow frozen and microwave thawed; SFIR, Slow frozen and infrared thawed; SFIR-MW, Slow frozen and infrared – microwave thawed; FFRT, Fast frozen and room temperature thawed; FFMW, Fast frozen and microwave thawed; FFIR, Fast frozen and infrared thawed; FFIR-MW, Fast frozen and infrared – microwave thawed.

Figure 6. Electron micrograph showing cellular structure of fresh ginger rhizome. Scale bar = 300 μm.

difference of 2 sec. Variation in temperature measurements was attributed to uncertainty in the thermocouple position and limited control over freezing conditions. Precautions were taken during measurements to minimize possible influence of these factors. Taking all the uncertainties with the experimental results in to consideration the prediction method adequately simulates the freezing process.

Conclusions

All freezing and thawing methods to which the ginger samples were subjected caused changes in color and alteration in internal structure. Slow frozen gingers suffered more damage due to large ice crystal formation than fast frozen gingers. These structural changes affected the chemical compounds identified in the essential oils of frozen and thawed gingers. Rapid freezing is desirable in restoring the structure of food materials but slow freezing may contribute in increasing the chemical compounds of ginger essential oil. Simulated freezing curves closely modelled corresponding experimental freezing curves. Predicted freezing time for ginger during slow and fast freezing differed from experimental result by 9 min and 2 sec respectively. This study shows how different freezing and thawing methods affects the

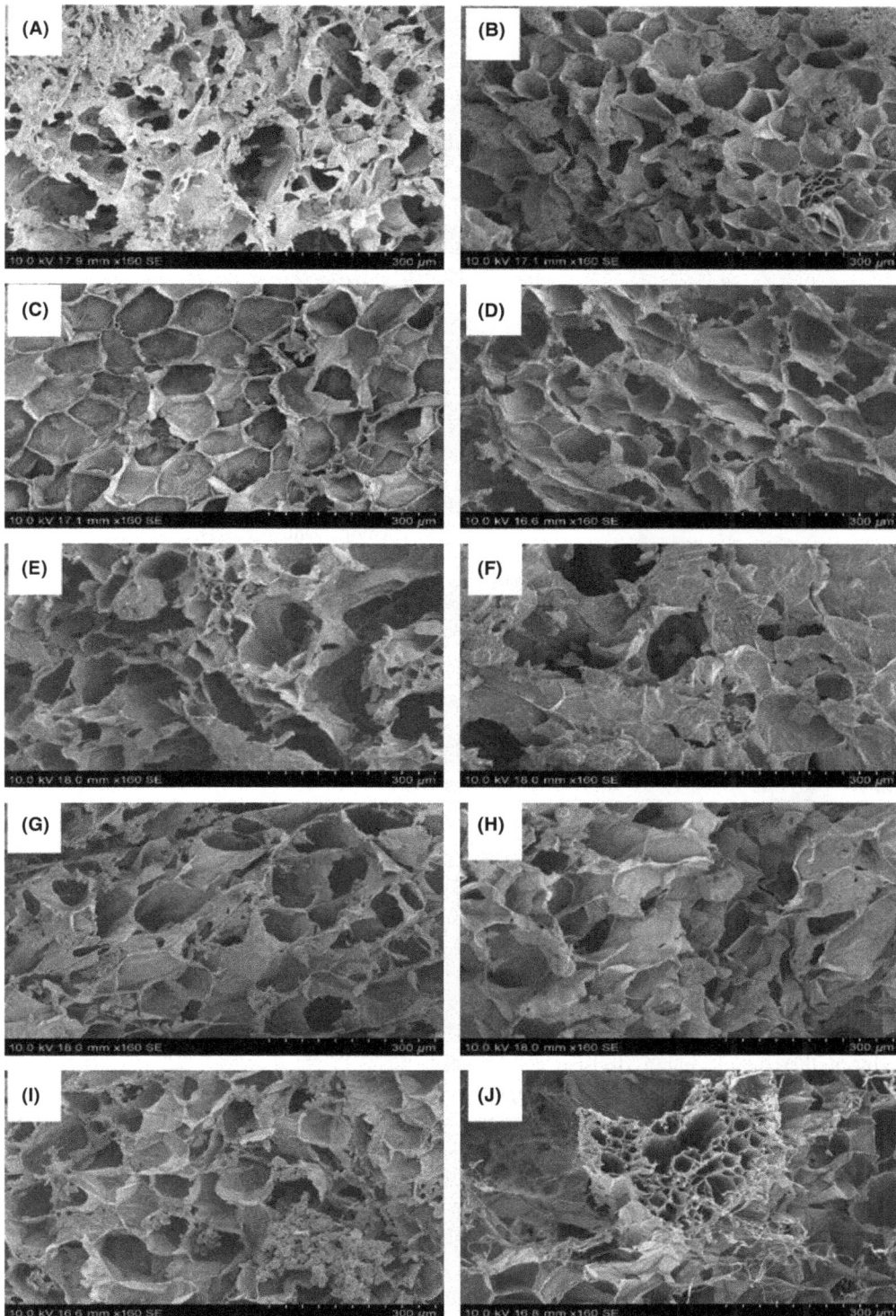

Figure 7. Electron micrograph of (A) slow frozen, (B) fast frozen, (C) slow frozen-room temperature thawed, (D) slow frozen-microwave thawed, (E) slow frozen-infra red thawed, (F) slow frozen-infrared microwave thawed, (G) fast frozen-room temperature thawed, (H) fast frozen-microwave thawed, (I) fast frozen-infrared thawed, and (J) fast frozen-infrared microwave thawed gingers. Scale bar = 300 μm.

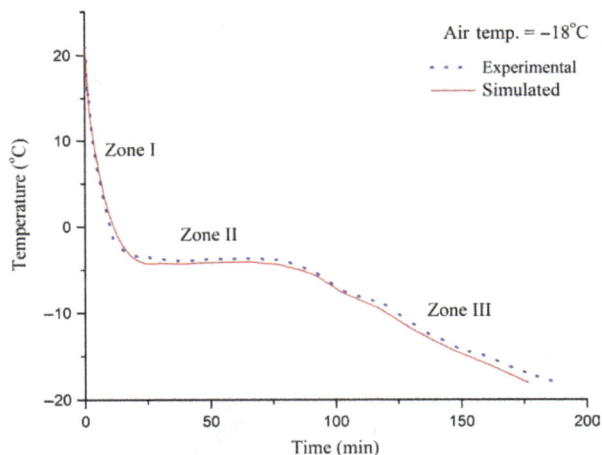

Figure 8. Comparison of simulated (solid line) and experimental (dotted line) results for slow freezing of ginger ($r = 0.97$).

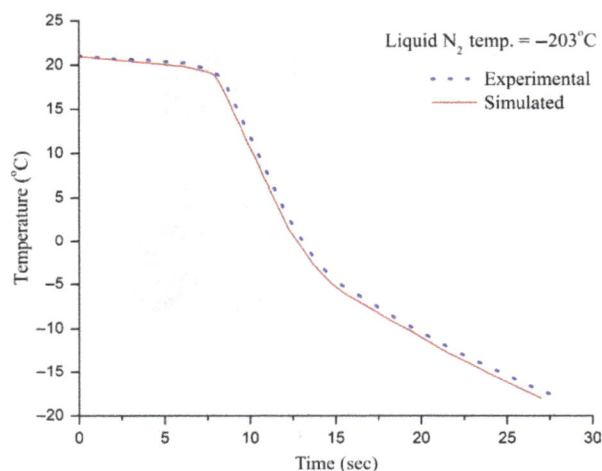

Figure 9. Comparison of simulated (solid line) and experimental (dotted line) results for fast freezing of ginger ($r = 0.92$).

structure of ginger thereby playing a decisive role in enhancing some aromatic and pharmacologically important components of ginger. Also, computer simulation for predicting the temperature and freezing time of ginger will be an important tool in designing and evaluating freezing equipment.

Acknowledgements

The authors acknowledge the financial support from Agricultural Experiment Station, South Dakota State University, Brookings, South Dakota. We thank Sushil K. Singh for his valuable input and suggestions in developing ginger model in SolidWorks and running simulation in ANSYS.

Conflict of Interest

None declared.

References

Anzaldúa-Morales, A., G. H. Brusewitz, and J. A. Anderson. 1999. Pecan texture as affected by freezing rates, storage temperature and thawing rates. J. Food Sci. 64:332–335.

AOAC. 1990. Official methods of analysis of the association of official analytical chemists. 15th edn. Association of Official Analytical Chemists, Inc., Arlington, VA.

Aziz, S., S. M. M. Hassan, S. Nandi, S. Naher, S. K. Roy, R. P. Sarkar, et al. 2012. Comparative studies on physicochemical properties and GC-MS analysis of essential oil of the two varieties of ginger (Zingiber officinale). Int. J. Pharm. Phytopharmacol. Res. 1:367–370.

Baliga, M. S., R. Haniadka, M. M. Pereira, J. J. D'Souza, P. L. Pallaty, H. P. Bhat, et al. 2011. Update on the chemopreventive effects of ginger and its phytochemicals. Crit. Rev. Food Sci. Nutr. 51:499–523.

Ban, J. O., I. G. Hwang, T. M. Kim, B. Y. Hwang, U. S. Lee, H.-S. Jeong, et al. 2007. Anti-proliferate and pro-apoptotic effects of 2,3-dihydro-3,5-dihydroxy-6-methyl-4H-pyranone through inactivation of NF-κB in human colon cancer cells. Arch. Pharm. Res. 30:1455–1463.

Brown, B. I., and A. C. Lloyd. 2007. Investigation of ginger storage in salt brine. Int. J. Food Sci. Technol. 7:309–321.

Butt, M. S., and M. T. Sultan. 2011. Ginger and its health claims: molecular aspects. Crit. Rev. Food Sci. Nutr. 51:383–393.

Choi, Y. H., and M. S. Kim. 2001. Effects of CO_2 absorbence in the PE film bag and styrofoam box during the ginger storage. Korean J. Postharvest Sci. Technol. 8:286–290.

Delgado, A. E., and A. C. Rubiolo. 2005. Microstructural changes in strawberry after freezing and thawing processes. LWT - Food Sci. Technol. 38:135–142.

Enyama, H. 1981. Kou Dan Shutsupan Ken Kyusha. Encyclopedia of food science. p. 300. Kou Dan Publishing Co., Tokyo, Japan.

Fennema, O. R., W. D. Powrie, and E. H. Marth. 1973. Low temperature preservation of foods and living matter. p. 598. Marcel Dekker, New York.

Fuchigami, M., N. Kato, and A. I. Teramoto. 1997. High-pressure-freezing effects on textural quality of carrots. J. Food Sci. 62:804–808.

George, R. M. 1997. Freezing systems. Pp. 3–9 in M. Erickson and Y.-C. Hung, eds. Quality in frozen food. International Thomson Publishing, New York, NY.

Gonzalez, O. N., L. B. Dimaunahan, L. M. Pilac, and V. O. Alabastro. 1969. Effect of gamma irradiation on peanuts, onions, and ginger. Philipp. J. Sci. 98:279–293.

Govindarajan, V. S., and D. W. Connell. 1983. Ginger – chemistry, technology, and quality evaluation: part 2. Crit. Rev. Food Sci. Nutr. 17:189–258.

Ji, A. M., X. C. Niu, B. N. Tang, and F. L. Li. 2014. Numerical simulation and experimental study on thawing time of cylindrical frozen food. Adv. Mater. Res. 989–994:3513–3517.

Kalichevsky, M. 1995. Potential food applications of high-pressure effects on ice-water transitions. Trends Food Sci. Technol. 6:253–259.

Kim, N. K., and Y. C. Hung. 1994. Freeze-cracking in foods as affected by physical properties. J. Food Sci. 59:669–674.

Kim, D.-h., and Y.-c. Lee. 2006. Changes in some quality factors of frozen ginger as affected by the freezing storage conditions. J. Sci. Food Agric. 1445:1439–1445.

Lee, S. E., M. C. Chung, and T. Y. Chung. 1994. Studies on the development of facilities for ginger storage. Korea Food Research Institute, Research Report, E1294-0538.

Li, B., and D.-W. Sun. 2002. Novel methods for rapid freezing and thawing of foods – a review. J. Food Eng. 54:175–182.

Mannapperuma, J. D., and R. P. Singh. 1989. A computer-aided method for the prediction of properties and freezing/thawing times of foods. J. Food Eng. 9:275–304.

Matuda, T. G., P. A. Pessôa Filho, and C. C. Tadini. 2011. Experimental data and modeling of the thermodynamic properties of bread dough at refrigeration and freezing temperatures. J. Cereal Sci. 53:126–132.

Min, Y. Y. 2001. Review on thawing technology of frozen foods. Food Sci. 22:87–90.

Mishra, A. P., S. Saklani, and S. Chandra. 2013. Estimation of gingerol content in different brand samples of ginger powder and their anti-oxidant activity: a comparative study. Recent Res. Sci. Technol. 5:54–59.

Okwuowulu, P. A., and E. C. Nnodu. 1988. Some effects of pre-storage chemical treatments and age at harvesting on the storability of fresh ginger rhizomes (*Zingiber officinale* Roscoe). Trop. Sci. 28:123–125.

Onyenekwe, P. C., and S. Hashimoto. 1999. The composition of the essential oil of dried Nigerian ginger (*Zingiber officinale* Roscoe). Eur. Food Res. Technol. 209:407–410.

Opoku-Nkoom, W. 2015. Safety and quality characteristics of freeze-defrost cycles in muscle foods. EC Nutr. 1:140–144.

Partmann, W. 1975. The effects of freezing and thawing on food quality. Pp. 505–537 in R. B. Duckworth, ed. Water relations of foods. Academic Press Inc., London.

Petzold, G., and J. M. Aguilera. 2009. Ice morphology: fundamentals and technological applications in foods. Food Biophys. 4:378–396.

Queirol, M. A. P., J. T. Neto, V. Arthur, F. M. Wiendl, and A. L. C. H. Villavicencio. 2002. Gamma radiation, cold and four different wrappings to preserve ginger rhizomes, *Zingiber officinallis* Roscoe. Radiat. Phys. Chem. 63:341–343.

Ravindran, P. N., and N. Babu. 2005. Introduction. Pp. 1–14. The Genus Zingiber, Ginger.

Shukla, Y., and M. Singh. 2007. Cancer preventive properties of ginger: a brief review. Food Chem. Toxicol. 45:683–690.

Sirikulvadhana, S., and C. Prompubesara. 1979. Effects of gamma irradiation and temperature on ginger sprout and weight. Food 11:55–61.

Subramanyam, H., S. Souza, and H. C. Srivastava. 1962. Storage behavior of ginger. Proc Symposium of Spices – Role National Economy, 1:5–10.

Thompson, E. H., I. D. Wolf, and C. E. Allen. 1973. Ginger rhizome: a new source of proteolytic enzyme. J. Food Sci. 38:652–655.

Usman, Y. O., S. E. Abechi, O. O. Benedict, O. Victor, U. U. Udiba, N. O. Ukwuije, et al. 2013. Effect of solvents on [6]-Gingerol content of ginger rhizome and alligator pepper seed. Ann. Biol. Res. 4:7–13.

Vasala, P. A. 2004. Handbook of herbs and spices. Woodhead Publishing, Abington, UK.

Venugopal, V. 2005. Quick freezing and individually quick frozen products. Pp. 95–139 in V. Venugopal, ed. Seafood processing: Adding Value Through Quick Freezing, Retortable Packaging and Cook-Chilling. CRC Press, Boca Raton, Florida.

Yusof, N. 1990. Sprout inhibition by gamma irradiation in fresh ginger (*Zingiber officinale* Roscoe). J. Food Process. Preserv. 14:113–122.

Zachariah, T. J. Ginger. 2008. Pp. 70–96. in V. A. Parthasarathy, B. Chempakam and T. J. Zachariah, eds. Chemistry of spices. CAB International, Wallingford, UK.

14

Procedure of brewing alcohol as a staple food: case study of the fermented cereal liquor *"Parshot"* as a staple food in Dirashe special woreda, southern Ethiopia

Yui Sunano

Graduate School of Bioagricultural Sciences, Nagoya University, Furocho, Chikusa-ku, Nagoya, Aichi 464-8601, Japan

Keywords

Africa, brewing process, diet, fermented food, food culture, lactic fermentation

Correspondence

Yui Sunano, Nagoya University Graduate School of Bioagricultural Sciences School of Agricultural Sciences, Furocho, Chikusa-ku, Nagoya, Aichi, 464-8601, Japan.

E-mail: sunano@nuagr1.agr.nagoya-u.ac.jp

Funding Information

This work was supported by JSPS KAKENHI Grant Number 15K16188 and 25300012, The Japan Science Society Sasagawa.

Abstract

For most brews, alcohol fermentation and lactic fermentation take place simultaneously during the brewing process, and alcohol fermentation can progress smoothly because the propagation of various microorganisms is prevented by lactic fermentation. It is not necessary to cause lactic fermentation with a thing generated naturally and intentionally. The people living in the Dirashe area in southern Ethiopia drink three types of alcoholic beverages that are prepared from cereals. From these alcoholic beverages, *parshot* is prepared by the addition of plant leaves for lactic fermentation and *nech chaka* by adding cereal powder for lactic fermentation before alcohol fermentation. People living in the Dirashe area partake of *parshot* as part of their staple diet. The brewing process used for *parshot* and a food culture with alcoholic beverages as parts of the staple diet are rare worldwide. This article discusses the significance of using lactic fermentation before alcoholic fermentation and focuses on lactic fermentation in the brewing methods used for the three kinds of alcoholic beverages consumed in the Dirashe area. We initially observed the brewing process and obtained information about the process from the people in that area. Next, we determined the pH and analyzed the lactic acid (g/100 g) and ethanol (g/100 g) content during lactic fermentation of *parshot* and *nech chaka*; the ethyl acetate (mg/100 g) and volatile base nitrogen (mg/100 g) content during this period was also analyzed. In addition, we compared the ethanol (g/100 g) content of all three kinds of alcoholic beverages after completion of brewing. The results showed that it was possible to consume large quantities of these alcoholic beverages because of the use of lactic fermentation before alcoholic fermentation, which improved the safety and preservation characteristics of the beverages by preventing the propagation of various microorganisms, improving flavor, and controlling the alcohol level.

Introduction

Food fermentation dates back to prehistoric times and humans have been making fermented food by exploiting the chemical action of bacteria and enzymes that exist naturally or are added deliberately (Harlander 1992). Fermentation can improve nutritional value (Simango 1997), dissolve unwanted components (Simango 1997; Sharma and Kapoor 1996), lessen the labor of food preparation (Simango 1997), and facilitate portability by reducing the volume. There are many types of fermented food in the world that are made of varied materials, using varied methods, and microorganisms. However, there are only four modes of fermentation in food processing; alcohol, lactic acid, acetic, and alkaline fermentation (Soni and Sandhu 1990). In alcohol fermentation, ethyl alcohol is produced from sugar using yeast. In lactic acid fermentation, lactic acid is produced mainly by lactic acid bacteria. In acetic fermentation, ethyl alcohol is dissolved into acetic acid and oxygen by acetobacter, and alkaline fermentation usually refers to the fermentation occurring to fish and seeds (Blandino et al. 2003). Many microorganisms are involved in these four modes of fermentation. Among these fermented foods, the most widely found are fermented beverages, a type of alcohol-fermented food, and lactic acid-fermented food.

Fermented alcoholic beverage has been produced since ancient times, and there are myriads of alcoholic drinks produced in this way. Roughly speaking, there are two types of fermented alcoholic beverages; a single-step fermented alcohol that is produced by a single alcohol fermentation process, and a multistep fermented alcohol that includes saccharification as well as alcohol fermentation processes. When these fermented materials are distilled, then the beverage becomes spirits. When fermented alcoholic beverage and spirits are mixed, then it turns into liquor. Single-step fermentation alcoholic beverages include fruit-based drinks such as wine and cider, popular in Europe, sap-based drinks such as palm wine, popular in Africa, Asia, and South America (Uzochukwu et al. 1994), honey-based drinks such as mead, popular in Africa and Europe (Bekele et al. 2006), and milk-based fermented drinks in Mongolia among others. Since these raw materials contain sugar, with naturally existing yeast, it turns into ethyl alcohol to make an alcoholic drink. On the other hand, in multistep fermentation beverages, starch in grains and tubers is saccharified first using enzymes in saliva, mold, and germinated seeds, followed by alcohol fermentation by yeast. Examples of such drinks include chicha, in which maize (Yamamoto 1995) or cassava (Mowat 1989) is saccharified by saliva, huangjiu in which grains such as rice, barley, wheat, millet and Japanese millet, or soya beans and peas are saccharified using accelerated growth of molds such as *rhizopus* and *mucor* (Yamamoto 1995), as well as beer in which a grain such as barley is saccharified using malt and germinated seeds. This type of alcoholic drink, especially bean sprout liquor, made using germinated seeds for saccharification is widespread around the world and in Africa, traditional local beverages using pearl millet (*pennisetum glaucum*) and sorghum are produced in many places (Oyewole 1997).

Another popular fermented food consumed worldwide is produced by lactic acid fermentation using mainly lactic acid bacteria. There are two types of lactic acid fermentation; homolactic fermentation produces only lactic acid while heterolactic fermentation produces carbon dioxide and ethanol in addition to lactic acid (Aguirre and Collins 1993). In both cases, any bacteria whose product is composed of over 50% lactic acid are called lactic acid bacteria (Ozaki 2009). There are several kinds of lactic acid bacteria in nature such as lactobacillus, leuconostoc, and lactococcus, and fermented foods including lactic acid-fermented foods exploit the interaction of such bacteria as well as

many other microbes (Blandino et al. 2003). Lactic acid fermentation enhances the flavor of the food (Chavan and Kadam 1989) and improves storage stability (Oyewole 1997). On the other hand, Western Asia is the birthplace of lactic acid-fermented food using animal milk as raw materials, such as yoghurt, butter and cheese (Ozaki 2009). Eastern Asia is the birthplace of lactic acid-fermented food using vegetables, such as pickles in salt and rice bran etc., sauerkraut and bread (Ozaki 2009).

Lactic acid bacteria are involved in the production process of almost all fermented food, not only in lactic acid-fermented food (Conway 1996). Examples include kimchi, vinegar, bread, miso, and sake. Take the example of a small local brewery. Since most of such operations are carried out in nonantiseptic conditions, many microbes find their way in during the fermentation, and thus it is important to control these elements to carry out a desirable alcohol fermentation process. In most of these types of operations, while ethyl alcohol is produced by yeast from sugar, lactic acid is also made simultaneously from sugar by lactic acid bacteria which are in the environment or growing on the surface of raw materials. When lactic acid is generated, the pH of the brew decreases to under 4, which inhibits the growth of microorganisms that cause food spoilage (Daly et al., 1990), and facilitates appropriate alcohol fermentation.

Among the Dirashe people, who live in Dirashe special woreda in southern Ethiopia, lactic acid fermentation is deliberately carried out before alcohol fermentation during the production process of an alcoholic drink made of germinated grains. As described earlier, the processes of alcohol fermentation and lactic acid fermentation in small-scale brewing operations are going naturally hand in hand, making it unnecessary to carry out deliberate lactic acid fermentation alone. The fermentation processes in this local area, both for lactic acid and alcohol, are different from any other processes in the world, in terms of fermented alcoholic drinks or preparation of lactic acid-fermented food. Further, an "alcoholic drink", *parshot* is given the longest lactic acid fermentation period, is considered as the "staple food", and is thus consumed in large quantities every day. The local people obtain almost all the calories and nutrients from *parshot*. This kind of food culture is rare in the world with other few cases found in Konso in southern Ethiopia (Shinohara 2000).

As overviewed above, though various form of fermented food and drinks have been consumed in the world for millennia, the mechanisms of fermentation were elucidated only in the 1980s (Caplice and Fitzgerald 1999), thus there are many areas that need to be investigated. Identifying the preparation method of fermented foods consumed locally is important in demonstrating the diversity of the world's food culture as well as in understanding human's dietary history. On this article, focuses on the lactic acid fermentation among the above mentioned three kinds of alcoholic drink made of sorghum, describing the unique production process and discussing the significance of the lactic acid fermentation of the raw materials. The study also discusses how such a unique fermentation method and food culture was born, focusing on the interaction social and ecological elements.

Methodology and Research Field Site

Methodology

A total of 13 months (December 2008 to March 2009, June to August 2009, January to March 2011, January to February 2012 and February to April 2013) were spent in a village called "W" in the lower land and "Y" in the higher land in Dirashe special woreda in southern Ethiopia. Participatory observation and interviews were conducted during the stay with regard to the fermentation processes of alcoholic drinks and the diet.

The author also brought home-frozen samples of *kalala*, *nech chaka*, *parshot*, *syuka*, which is the fermenting intermediate product of *parshot*, as well as *plota*, which is the fermenting intermediate product of *nech chaka*. As described in detail later, *syuka* is made of sorghum and maize flours by adding moringa (*moringa stenopetala*) or Ethiopian kale (*brassica carinata*) leaves. It is used after at least 2–3 months of preservation. In this study, 3-week-old and 3-months old *syukas* with moringa, which is often used in "W" village, are examined as the samples. *Plota*, on the other hand, is made of sorghum and maize flour mix only, and is used after 1 month of keeping. In this study, 1-week-old *plota* is brought home as the sample. After arriving back to Japan, quantities of lactic acid and ethanol in grams per 100 g were determined. High-performance liquid chromatography and gas chromatography were used for the measurements of lactic acid and ethanol, respectively. As for the *syukas*, ethyl acetate (in ppm) and volatile basic nitrogen (in mg) per 100 g were also determined. Microdiffusion analysis and absorptiometric analysis were used for determining volatile basic nitrogen and ethyl acetate, respectively.

While *kalala* was taken in its neat form among the locals, *nech chaka* was diluted twice with water and *parshot* with 1.3 times to twice with water. To prepare samples for the analyses, *kalala* was used as is, while the samples of *nech chaka* and *parshot* were diluted twice parts and 1.3 times of water, respectively. When making *nech chaka*, the yeast mash will be subjected to alcohol fermentation on the second day. It can be consumable for only 1 day after the fermentation process ends is done. On the other hand, the yeast mash will be kept 2–14 days before

starting alcohol fermentation for *parshot*. It can be consumable for 3 days after preparation is completed. *Parshot* samples used in this study were those made of 2-days old yeast mash and kept 1–3 days after completion, and made of 10-days old yeast mash and kept 1 day after completion. The samples were frozen after stopping the fermentation by applying a water bath of 60°C for 30 min in heat-proof containers. In order to assess alcohol concentration, ethanol per 100 g for three types of the samples was analyzed by gas chromatography after they had been brought back to Japan. All these analyses were outsourced to Japan Food Research Laboratories.

In the research field, in order to find out when lactic acid fermentation would start for *syuka* with added moringa and Ethiopian kale leaves and for *plota*, the pH of the samples were measured every day from the first day they were prepared to the 30th day of keeping, using pH meter at two villages, one in the higher land and another in the lower land. *Syuka* can be kept for 2–3 months and can be used for several months after that. So, in order to find out whether lactic acid fermentation would be still continuing during this 2–3 months period, the pH of *syuka* samples with added moringa and Ethiopian kale were measured on every month for 6 months, using pH meter. Further, in order to investigate the storage stability of the yeast mashes, the pH of yeast mashes with moringa and Ethiopian kale as well as yeast mash made only of grain flour were measured for 10 days on the research field site.

Overview of research field site

The field research was conducted in a village called "W" in Dirashe special woreda in Ethiopia, which is approximately 550 km southwest of Addis Ababa (Fig. 1). According to the Health Ministry of Dirashe special woreda, the area is approximately 1500 square kilometers with a population of approximately 130,000, of which most are ethnically Dirashe people, at the time of the survey in 2008. The highest point in the area is Mount Gardolla (2561 m) with a steep slope that leads to Segen valley plateau at 1100 m above sea level. Some parts of the slope and the valley plateau where their mainstay crops of sorghum and maize are cultivated are terraced and enclosed using stones and harvest wastes. The weather is cool/cold (annual temperature ranges between 13 and 25°C) at the higher ground near the top of the mountain with year-round rainfall. The average annual rainfall is 1300 mm. However, most areas in the woreda including "W" village are situated in semiarid land (annual temperature ranges between 17 and 31°C), with two rainy seasons in a year. With the average rainfall of only 800 mm.

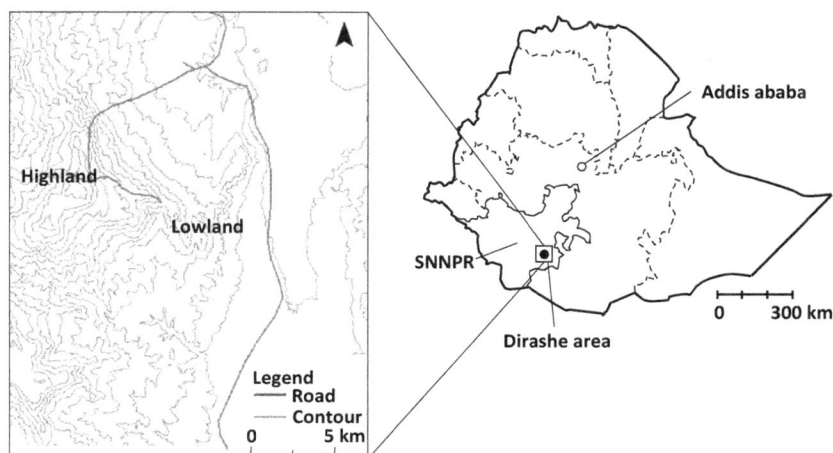

Figure 1. Location of research field.

The rainfall is unpredictable and unstable with large year-to-year differences. Good and very poor harvests alternate every few years. People in this area make storage holes in the ground called *polota* where sorghum crops can be stored for several years. They store the excess crop in a good year in preparation for a future lean year. Staple food of the people is *parshot*, a fermented alcoholic drink made of sorghum or sorghum with some maize mixed in, which have been stored in the storage holes.

Result

Brewing methods

Making *kalala* is easy; sorghum and maize flours are heated and gelatinized. Powdered germinated seeds are added to trigger saccharification and alcohol fermentation simultaneously (Fig. 2). Alcohol concentration of *kalala* is low at 1.8%. The taste is sweet and children drink it as a juice.

On the other hands, making *parshot* is more complicated and steps up to the lactic acid fermentation are different in the lower and the higher land (Fig. 2). In the lower land, a small amount of dried and powdered leaves of moringa or Ethiopian kale are added to sorghum and maize flours. The mixture is kept for 2–3 months during which lactic acid fermentation takes place until the smell of the mixture changes. Ethiopian kale is a brassica with large population of lactic acid bacteria. Moringa is also related to brassica family and can give the same, albeit inferior, effect. The mixture under lactic acid fermentation is called *syuka*. The brewer washes the surface, adds and mixes in more grain flour every day until the smell of the *syuka* changes. After 2–4 days, fruit flies (*drosophila melanogaster*) start to gather on the surface. After a further

4–6 days, *syuka* starts to emit the mixture of ammoniacal and rancid odors. If the *syuka* is left without handling, white mold starts to grow in few days. Once the smell changes, the *syuka* can be stored for several months with a monthly maintenance of washing the surface with water, adding and mixing in more grain flour, thus making it possible to make a large quantity at once and be used little by little. In the higher land, on the other hand, a small amount of Ethiopian kale powder is added to sorghum and maize flours, which will be wrapped in ensete (*ensete ventricosum*) leaves and kept in shallow holes dug near the roots of ensete plants in garden/field with several stones on top as lids and kept for 14–15 days. The steps after this stage are the same for both lower and higher lands. After adding and mixing more grain flour, *syuka* is heated to be gelatinized. Then, powdered germinated seeds of grains such as sorghum, maize, barley, and wheat are added to promote saccharification and growth of yeast. This yeast mash, with added germinated seeds, can keep up to 15 days. After that, brewing liquid made of grain flour and hot water is added to the yeast mash to start alcohol fermentation. *Parshot* is usually diluted 1.3–2 times with water to be taken and can be consumable for 3 days after the final product is made. The alcohol concentrations of *parshot*, which was made of the 2-days old yeast mash and diluted with 1.3 parts of water just before consumption were 2.97%, 3.63%, and 3.64% on the first, second, and third day, respectively.

During the short period of very dry season of January to February in which moringa leaves are not obtainable, the people in the lower land make *nech chaka* and *kalala* from grain only to substitute *parshot* for their staple food. Leaves of moringa can be harvestable even during dry seasons, however, if the rainfall is extremely poor, it stays dormant without leaves. The production process of *nech*

How to make *parshot*

1. Lactic acid fermentation

Lowland ①

Moringa

Highland

② To dry the plants and mix with flour of grain.

③ Work to perform every day

④ The ordor turns into a sweet aroma similar to that of a fermented food in 2 to 3 month time. The *syuka* will be stored for several months afterward.

Ethiopian kale

③

④ To wrap *syuka* in an ensete leave and bury it near the ensete tree. After 2 weeks .

How to make nech chaga

1. Lactic acid fermentation

① To wrap the mixture of grain flour and water in an ensete leave and stored in the anaerobic condition for a month.

② After a month, its smell change into a sweet aroma.

2. Pregelatinization and Sterilization

① Frist day

② Second day

③ To make *syuka* into a ball shape and boil for 30 minutes.

④

To add and mix in grain into *syuka* and leave.

3. Making Yeast Mash

⑤ To mash them.

⑥ To cool them down.

Third day ①

To saccharify *syuka* by adding the germination grains.

②

To store it for one to 14 days for *parshot* and for one day for *nech chaka* in order to increase yeast.

4. Alcohol Fermentation

To mix grain flour with boiled water to make fermentation base.

①

② To add a portion of yeast mash to the fermentation base, mix in water.

Forth day

Parshot is consumable 3 days while *nech chaka* for only day.

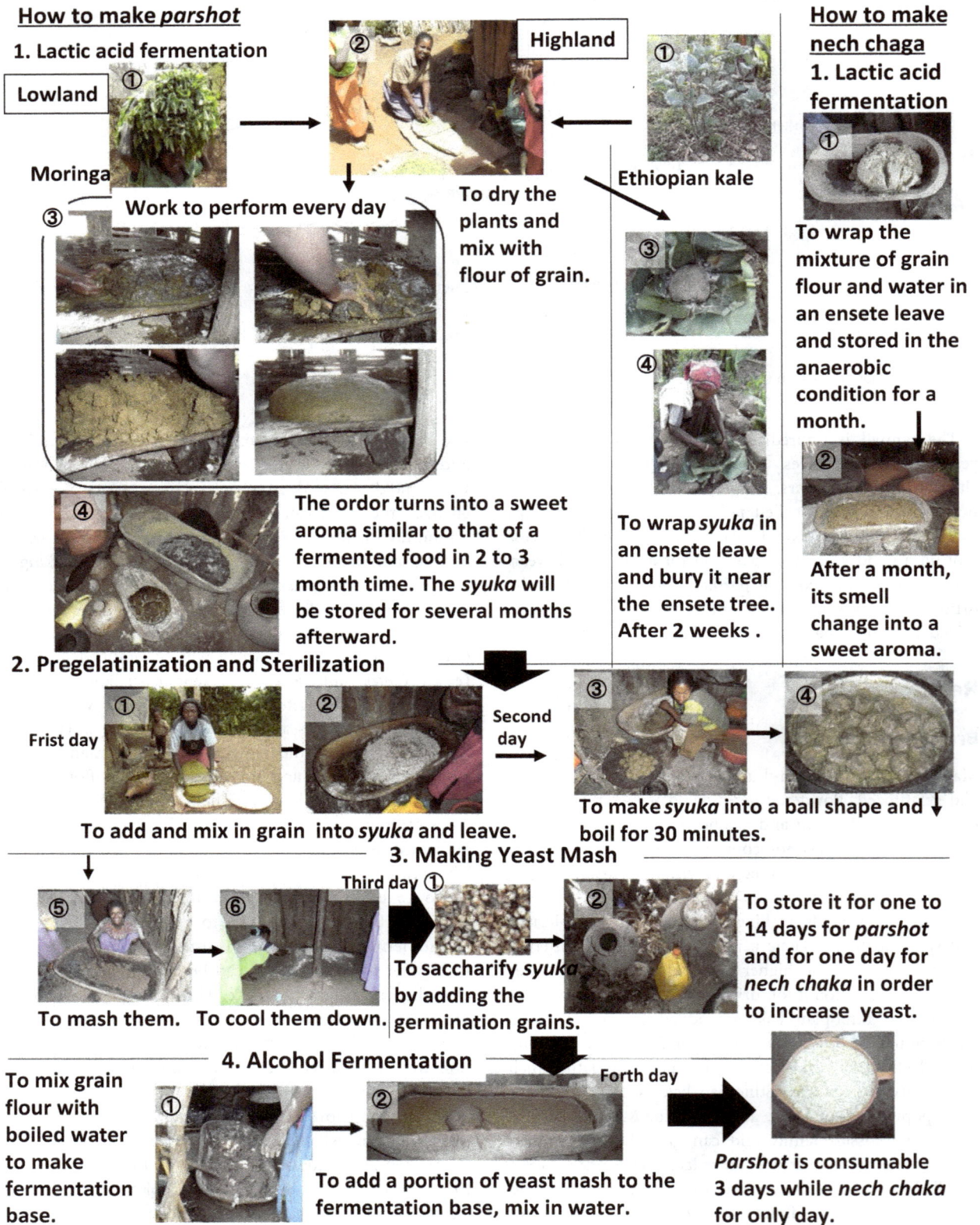

Figure 2. How to make alcoholic beverage.

chaka is almost the same as that of *parshot*: sorghum and maize flour mix is kept for 1 month to undergo lactic acid fermentation until the smell changes. The mixture under lactic acid fermentation is called *plota*. As with *syuka*, the brewer repeats the process of washing the surface and adds and mixes in more grain flour every day until the smell changes. After adding and mixing more grain flour, *plota* is heated to be gelatinized (Fig. 2). Powdered germinated seeds are then added to promote saccharification and growth of yeast. At this point, in *nech chaka* making, a greater amount of germinated seeds, 1.3–1.5 times more than parashot making, is added. The local people say that unless a large amount of germinated seeds is added, fermentation does not go well. Not like *parshot*, the yeast mash of *nech chaka* does not keep and goes rotten easily, thus it has to be used all at one time. Brewing liquid is then added to the yeast mash to start alcohol fermentation, but the finished product is consumable only for 1 day. *Nech chaka* is usually diluted twice with water when consumed and its alcohol concentration is rather high at 4.10%.

It can be said that the production processes of *parshot* and *nech chaka* are divided into two; production of vegetable material pickles by lactic acid fermentation and alcoholic drink production by saccharification and alcohol fermentation. While *nech chaka* utilizes lactic acid bacteria on the grains, and in the air, *parshot* utilizes the bacteria population on the vegetable leaves. And while parshot can keep well as an intermediate product during fermentation as well as a finished product, *nech chaka* does not. Local people say, "*parshot* smells better than *nech chaka* and has more rich taste". They also say, "*nech chaka* may sometimes upset the stomach". They also agree that the lactic acid fermentation stage is the most important in the whole process, with comments such as "without good *syuka* or *plota*, fermentation will not proceed well and the taste will be spoiled, this is the most important", "if you fail at the stage, the final product goes off quickly," and "if a strange smell is mixed in at this stage, the final product may upset your stomach". The difference in lactic acid fermentation process seems to be affecting the differences between *parshot* and *nech chaka* in terms of alcohol concentration, storage period, taste, and flavor as well as food safety. Therefore, the article hereafter focuses on the process of lactic acid fermentation.

Maintaining low pH

In the higher land, the author carried out measurement of the pH of two types of syuka, one with moringa and the other with Ethiopian kale, as well as *plota* made only with grain flours, for a month. The result is shown in Figure 3. As described earlier, it takes only 14–15 days

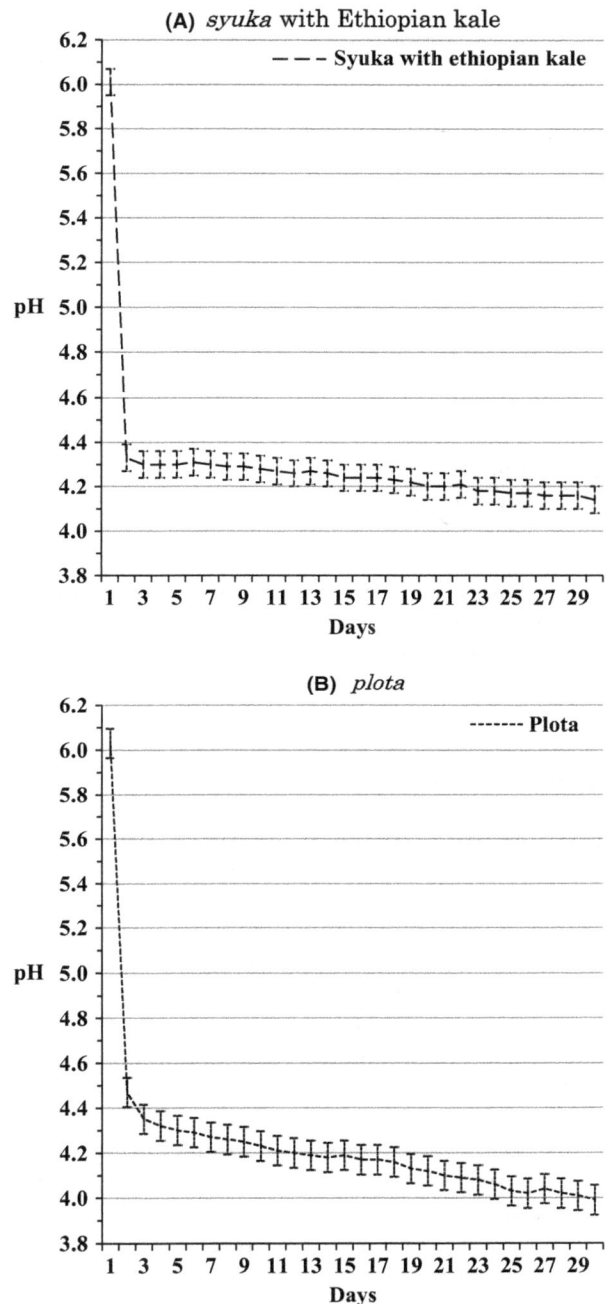

Figure 3. (A) pH level of syuka with Ethiopian kale produced in the highland over a month (Average ± SE Average = 4.30, SE = 0.011). (B) pH level of plota produced in the highland over a month (Average ± SE Average = 4.23, SE = 0.012).

for *syuka* to be produced in the higher land. Also, in the higher land, *plota* is traditionally very rarely made. However, the people there confirm that *plota* can be generated in 14–15 days as with the syuka. The pHs of the *syukas* with moringa and Ethiopian kale went down to pH 4.33 on

Table 1. pH profile of syuka and plota over 6 months.

Duration of lactic acid fermentation	Syuka		Plota
	With Moringa	With Ethiopian kale	
1 month	3.91	3.88	3.89
2 months	3.90	3.87	–
3 months	3.88	3.78	–
4 months	3.79	3.90	–
5 months	3.85	3.88	–
6 months	3.91	3.85	–

the second day and stayed at that level afterwards. By the 14th day, when it is said to become usable for the next step of fermentation, the pH was further reduced to pH 4.26 and pH 4.24 on the 15th day. Continuous measurement show that the reduction in the pH value continued slowly, down to pH 4.14 after 1 month. Since it has kept the pH level of 4 for a month, syuka can be kept for a month, however, it is usually used after 2 weeks for the next process. The pH reduction speed of plota is slower than syuka and it was pH 4.47 on the 2nd day and pH 4.35 on the 3rd day. However, the pH level afterwards did not stabilize, showing pH 4.18 on the 14th day, pH 4.19 on the 15th day, and pH 3.99 after 1 month.

The other conducted the same pH measurement of two types of syuka and plota, which were in the same stage. The results are shown in Figure 4. The pH level for the plota did not stabilize at first. It took 4 days to be reduced to pH 4.3 and continued to decrease slowly afterwards. The syuka, on the other hand, reached pH 4.3 on the 2nd day and stayed at that level afterwards. Further measurements of both syukas, one with moringa and another with Ethiopian kale, were conducted every month for the next 6 months. The results show that both kept acidity with the pH of between 3.78 and 3.91 during that time (Table 1).

The plota, on the other hand, reached pH 3.89 after 1 month and became suitable for the next fermentation process. However, it started to rot immediately afterwards (Fig. 4). As an experiment, the author continued to wash the surface off as well as adding and mixing grain flour into the sample of plota, as is the usual practice for syuka, even after 1 month had passed. Then, the plota started to emit abnormal order, gave an irritating sensation on the hand when mixing on the 34th day and started to change its color on the 36th day. At that point, the

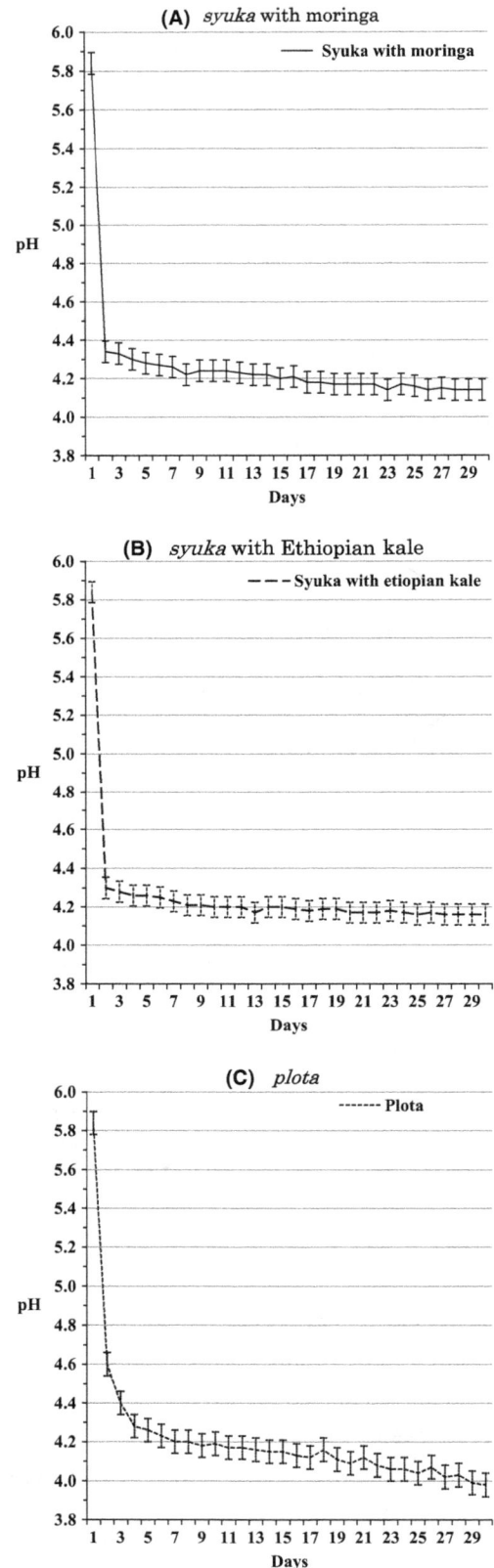

Figure 4. (A) pH level of syuka with moringa produced in the lowland over a month (Average ± SE Average = 4.26, SE = 0.010). (B) pH level of syuka with Ethiopian kale produced in the lowland over a month (Average ± SE Average = 4.25, SE = 0.010) (C) pH level of plota produced in the lowland over a month (Average ± SE Average = 4.21, SE = 0.011).

experiment was discontinued. Also, as another experiment, when another sample of *plota* had been left for longer than 30 days, it started to give off stench on the 35th day. As shown in these experiments, though *plota* changes smell and becomes usable for the next fermentation step after keeping for 1 month, the state does not last and starts to rot immediately afterwards. This is why *plota* is not kept and all of it are used as the ingredient for the next step. *Syuka*, on the other hand, can keep for several months after it has reached to the usable state, and will be used little by little for the next few months as the ingredient for the next step of fermentation.

Lactic acid production

Since both samples of *syuka* and *plota* showed pH ranges between the upper side of 3 and the lower side of 4, it is assumed that lactic acid bacteria dominate the flora. Two types of syuka with moringa at 3-weeks and 3-months old and 1-week-old *plota*, with the pHs of which were deemed to be stable, were brought back to Japan and the volumes of lactic acid per 100 g were measured. While 1-week-old purota had only 1.18 g of lactic acid, 3-weeks old *syuka* contained 4.56 g, and 3-months old *syuka* 3.36 g of lactic acid per 100 g (Table 2).

Ethiopian kale(brassica carinata) belongs to the brassica family. Moringa (*moringa stenopetala*) belongs to Moringaceae Martynov, which is closely related to brassicales. Both plants have large populations of lactic acid bacteria living on the leaves. Lactic acid bacteria growing naturally on *syuka* produce a much larger quantity of plant-derived lactic acid compared to those growing on *plota*. Lactic acid bacteria in *syuka* increase quickly to take dominance in the flora, killing other microorganisms that produce harmful substances. However, in *plota*, lactic acid bacteria in the air and on the surface of grains take several days to dominate the flora. While lactic acid bacteria slowly build up, substances produced by other microorganisms take hold, spoiling storage quality and food safety. This is why *nech chaka* does cannot be preserved for a long time.

Table 2. Variation in the Composition of syuka and plota at different fermentation durations per 100 g.

Contained material	Syuka		Plota
	3 weeks old	3 months old	1 week old
Lactic acid (g)	4.54	3.36	1.18
Ethyl acetate (ppm)	60.00	34.00	190.00
Volatile basic nitrogen (mg)	87.00	34.00	–
Ethanol (g)	0.21	0.41	2.00

Selection and proliferation of aciduric yeast under acidic condition

Not only lactic acid bacteria, but also aciduric yeast shows proliferation in *syuka* and *plota*. Two types of *syuka* of 3-weeks and 3-months old and 1-week-old *plota*, which had been prepared in the lower land, were brought back to Japan and the volumes of ethanol per 100 g were measured. In *syuka*, the amount of ethanol was found to increase as it aged, showing 0.21 g in the 3-weeks old sample and 0.41 g in the 3-months old sample. Ethanol in 1-week-old *plota* was as high as 2.00 g (Table 2). In *plota*, lactic acid bacteria come to dominate the flora on the 4th day and with that, almost all other naturally occurring wild yeasts and other microorganisms disappear. However, wild yeasts that come with the sorghum increase, prolifically producing a large amount of ethanol by the 4th day, thus a large amount of ethanol is still being generated even on the 7th day. In *syuka*, on the other hand, lactic acid bacteria come to hold a dominating position in the flora on the second day and kill off wild yeasts, thus the amount of ethanol to be generated is smaller. In both *syuka* and *plota*, once lactic acid bacteria dominate the flora, only aciduric yeasts will grow.

If *syuka* and *plota* made using the lower land method are left for several days, white mold starts to grow on the surface. This is a film yeast, which is aerobic and aciduric and grows especially where growth of other microorganisms is reduced. Thus, the growth of the film yeast on the surface of *syuka* and *plota* indicates the sufficient lowering of pH and dominance of lactic acid bacteria in the flora. The Dirashe people also see the white mold as an indicator of whether the dough becomes good *parshot* or *nech chaka*. At the same time, proliferation of the film yeast is not favored, and as soon as the growth is found, the surface of the dough is washed and folded in and mixed, creating an anaerobic condition to kill off the yeast. One of the smells emitted by the film yeast during the fermentation process is ethyl acetate. A tiny amount of ethyl acetate makes a favorable flavor, but a large amount of that smells like thinner and is not liked. Amount of ethyl acetate per 100 g in the 3-week-old and the 3-month-old syukas with moringa were 60 mg and 34 mg, respectively, showing a decrease as it aged (Table 2).

When *syuka* or *plota* is left alone, anaerobic butyric acid bacteria grow at the bottom of the dough. Though butyric acid bacterium is beneficial for humans, it emits a bad odor that reduces appetite. By mixing and sending air into *syuka* and *plota*, the local people stop the growth of the film yeast as well as the growth of butyric acid bacteria. Also, by adding grain flour when mixing, protein and sugar are provided for the lactic acid bacteria and yeasts, increasing aciduric yeast as well as reducing film yeast and butyric acid bacteria.

Generating of flavor

Fruit flies are attracted to *syuka* within several days into fermentation, but only a small number of them are attracted to *plota* and it takes more than 1 week for them to notice it. The smell of ester compound in fermented vegetable materials attracts fruit flies (Barrows 1907), and, as described earlier, a large amount of ethyl acetate is generated in *syuka* in its early fermentation stage.

Also, around this early stage of fermentation, rancid and ammoniacal odors among other fermentation smells are emitted, very strongly from syuka and faintly from *plota*. According to interviews with 50 local people aged between 30 and 50 by the author, this bad odor "changes to a sweet fermenting smell after 2–3 months in *syuka* and after 1 month in *plota*", and "this is the right time to move on to the next fermentation process". The author could not recognize the change in the smell, but local people could and were using it as the indicator to start the next fermentation process. Two types of *syuka* with particularly strong smell at 3-weeks and 3-months old were selected as samples and brought back to Japan, and the volumes of volatile basic nitrogen per 100 g were measured. The amount of volatile basic nitrogen in the 3-weeks old and 3-months old *syuka* were 87 mg and 34 mg, respectively, showing a decrease as it aged, and it is assumed that rancid and ammoniacal odors that cause the bad smell has decreased accordingly (Table 2).

When grains are fermented, ethanol produced by yeasts, fatty organic acid produced by lactic acid bacteria etc., bind together to form esters that emit aromas (Imai et al. 1983; Higashi 2006). Various kinds of volatile substances are produced and mixed, then create an aroma that promotes appetite (Chavan and Kadam 1989). In *syuka*, it is thought that vegetable-derived lactic acid bacteria break down protein and sugar and during which unpleasant ammoniacal, sulfuric, and ethyl lactate smells are emitted. However, as the resynthesis of organic and fatty acids progresses during the 3 months, the unpleasant odor changes an aroma with a hint of sweetness. It is highly likely that this sweet aroma is a mixture of fruity smells such as ethyl lactate and ethyl acetate, sulfuric compound smells such as methanethiol and dimethylsulfide (Imai et al.1983) and glucosinolate (mustard oil glucoside) that is the hot and bitter taste in the brassica family of plants such as wasabi and cabbage (Imai et al. 1983).

Discussion

Fermentation has been employed as a means to improve storage quality for the last 6000 years (Wood and Holzapfel 1995). Among several types of fermentation, lactic acid fermentation plays an important role with several types of food prepared in that way. The Dirashe people make great use of lactic acid fermentation in making *parshot*, a kind of germinated grain alcoholic drink. *Parshot*, which is the product of the lactic acid fermentation, is low in contamination, maintains high food safety, and keeps well during the fermentation process as well as a finished product. Also, the lactic acid bacteria derived from the vegetable material produce a large amount of lactic acid, break down sugar and protein, resynthesize organic and fatty acids that add taste and flavor to the food. Since alcohol fermentation of *parshot* is a slow process with a small amount of germinated seeds, the alcohol concentration is low at 2.96–3.64%. When the vegetable leaves are not obtainable, *nech chaka* is made. The grain flour, minus vegetable leaves, goes through lactic fermentation, in the same way as *syuka*, for a month to produce *plota*, which is then alcohol fermented to make *nech chaka*. Lactic acid bacteria on the grains increase slowly and cannot maintain dominance in the flora over a long time with lower production of lactic acid. This makes nech chaka inferior in storage quality, food safety, and flavor. Also, it needs speedy alcohol fermentation before the contamination starts. To achieve that, a large quantity of germinated seeds is added, resulting in a higher alcohol concentration than parshot at 4.1%. In this way, the Dirashe people add green leaves that are the source of lactic acid bacteria deliberately to promote lactic acid fermentation, in order to improve storage quality, food safety, and flavor as well as making the alcohol concentration adjustable, thus it enables them to take a large amount of it every day.

But, why do they have to go through all this to take a large amount of alcoholic drink every day? In Ethiopia, there is another ethnic group of people called the Konso people whose staple diet is *chaka*, a germinated grain alcoholic drink made of sorghum. The Konso people, who live in the neighboring Konso region, descend from the Oromo ethnic group as with Dirashe, and there are commonalities in lifestyle, culture, society, and language between them. The Konso people, however, cultivate not only sorghum, maize and moringa, but also grains such as wheat and barley, tubers, legumes, and other vegetables, on stone terraces cut out of the steep hillsides (Shinohara 1998, 2000, 2002). Though they take *chaka* as the staple food twice out of four meals a day, dumplings made of tubers and grains as the staple and legumes and vegetables as secondary are also taken at the other two meals (Shinohara 2000, 2002). By this means, though the Konso people consider *chaka* brewed from sorghum and maize as their staple food, they also take nutrients from other sources.

Contrary to that, the Dirashe people rely almost entirely on the nutrients from parshot and rarely eat any other type of food. Their explanation for this is that it is difficult to grow legumes and other vegetables due to the unstable weather and serious problems of insect damage. However, the main component of sorghum and maize is starch with very little protein (Chavan and Kadam 1989). Since the nutrient values of grains improve when they are fermented (Chavan and Kadam 1989), it is assumed that these people are consuming them in alcohol-fermented forms. Another advantage of alcohol fermentation is to convert the grains into a fluidized form, enabling the people to take a large amount without giving them a feeling of having a full stomach. It is highly possible that the Dirashe people are taking in all the nutrients necessary for everyday life from the grains, by fluidizing and fermenting, thus increasing the consumption to sustain their lives. The weather in the Dirashe region is unstable with good and very poor harvests alternating every few years. The Dirashe people keep their sorghum crop in storage holes in the ground, which are capable of keeping the crop for several years, thus securing a stable food supply every year. Though the crops do not spoil, sorghum kept in the storage holes have an unpleasant smell such as an ammoniacal odor. Explained the flavor changes when the food material is fermented (Chavan and Kadam 1989). The Dirashe people say that by making the crop into *parshot*, the unpleasant smell changes to sweet and nice aroma. They also say that lactic acid fermentation change the grains unpleasant smell, and it is assumed that lactic acid and alcohol fermentations of the grains and resulting improvement of flavor enable them to consume such stored grains. Based on these reasons, they need to have the grains alcohol fermented, as well as consume it in a large quantity. Aside from a risk of damaging the liver if taken excessively, alcohol made in a small-scale home brewing, often in nonsterile conditions, have a possibility of contamination by other microorganisms that may spoil storage quality and food safety. The Dirashe people overcome this shortcoming by lactic acid fermentation prior to the alcohol fermentation.

However, how did they establish this kind of fermentation method? Lactic acid fermentation is a widespread form of food preparation in Ethiopia with many regions taking lactic acid-fermented food as their staple. In the north of the country and the urban areas, a round and thin flatbread which is called injera as big as 1 m in diameter, is eaten as staple food, made of teff flour mixed with water and left for fermentation (Shigeta 2007). The lactic acid fermentation is completed in two stages (Gifawesen and Besrat 1982); lactic acid bacteria, mainly of the *enterobacteriaceae* family are used for the first fermentation for 18 h, then these lactic acid bacteria plus yeasts are used to complete the fermentation that takes 30–33 h (maximum 48 h) (Gifawesen and Besrat 1982). Injera can be preserved for 2–3 days. In the south

of the country, protein stored in the pseudostem of ensete is harvested, wrapped in the ensete leaves, and left in a hole in the ground for several days to be lactic acid fermented, which will then be kneaded into a flat disk and baked into a hard bread called *kocho* (Negash and Niehof 2004; Yewelsew et al. 2006; Shigeta 2007). *Kocho* tastes mildly sour (Negash and Niehof 2004). The way *kocho* is prepared is similar to the *syuka* made in the higher land, in which grain flour with a small amount of dried leaves of Ethiopian kale and water are mixed in and kneaded, then wrapped in an ensete leaf and buried underground for 2 weeks with stones piled on top acting as a lid to make the hole anaerobic. Ensete is also grown in the higher land and *kocho* is eaten sometimes as breakfast there. Ensete is eaten widely in the highland of southern Ethiopia from the old day. Elderly peoples told that food preparation method of *kocho* was transmitted from another area to Dirashe area by flow of people. It can be assumed that the method of *syuka* making in the higher land comes from that of *kocho* making. It can also be assumed that *syuka* making in the lower land is the improved version of the one in the higher land in terms of the more efficient proliferation of lactic acid bacteria. Since the vegetable leaves are easy to obtain in the higher land, a small amount of *syuka* is made every 2 weeks on prioritizing speedier increase in lactic acid bacteria. In the lower land, however, since the vegetable leaves are scarce, the efficient increase in lactic acid bacteria from the smaller amount of leaves takes priority, thus it is made in a large quantity taking 2–3 months. As seen above, the lactic acid fermentation process in *parshot* production has its root in *kocho* making, and is completely separated from the next step, the alcohol fermentation. The alcohol fermentation stage in the production, on the other hand, has its root in a common and widespread method of alcohol making in Africa, in which starch in grains is converted to sugar by adding germinated seeds. As observed the *parshot* making, fermented food in the world is thought to have developed as a result of fermenting technologies of neighboring areas influencing each other.

Acknowledgment

The Author appreciates the help received from Asia African Area Studies in Kyoto University especially Dr. Gen Yamakoshi and Dr. Juichi Itani. This work was supported by JSPS KAKENHI Grant Number 15K16188 and 25300012, The Japan Science Society Sasagawa.

Conflict of Interest

None declared.

References

Aguirre, M., and M. D. Collins. 1993. Lactic acid bacteria and human clinical infection. J. Appl. Bacterial. 75:95–107.

Barrows, W. M. 1907. The reactions of the pomace fly, Drosophila ampelophila Loew, to odorous substances. J. Exp. Zool. 4:515–537.

Bekele, A., A. M. S. McFarland, and A. J. Whisenant. 2006. Impacts of a manure composting program on stream water quality. Trans. ASABE 49:389–400.

Blandino, A., M. E. Al-Aseeri, S. S. Pandiella, D. Cantero, and C. Webb. 2003. Cereal-based fermented foods and beverages. Food Res. Int. 36:527–543.

Caplice, E., and G. F. Fitzgerald. 1999. Food fermentations: role of microorganisms in food production and preservation. Int. J. Food Microbiol. 50:131–149.

Chavan, J. K., and S. S. Kadam. 1989. Nutritional improvement of cereals by fermentation. Food Sci. Nutr. 28:379–400.

Conway, P. L. 1996. Selection criteria for probiotic microorganisms. Asia Pacific J. Clin. Nutr. 5:10–14.

Gifawesen, C., and A. Besrat. 1982. Yeast folra of fermenting tef (Eragrostis tef) dough. SINET: Ethiopia. J. Sci. 5:21–25.

Harlander, S. 1992. Food biotechnology. Pp. 191–207 in J. Lederberg, ed. Encyclopaedia of microbiology. Academic Press, New York.

Higashi, K. 2006. Hakkou to jouzou IV: Jozou ha biseibutu to nougyou no ke ssokutai [Fermentation and brewing IV: Brewing connect microbial activity to agriculture]. Kabushikigaishkourin Press, Tokyo.

Imai, M., S. Hirano, and M. Aiba. 1983. Nukadoko no jukusei ni kansuru kenkyu [Resserch on maturing of nuka-doko (the pickling medium of salted rice bran): changing flavor component in the maturing process]. Nihonnougeikagakukaishi 57:1113.

Mowat, A. M.. 1989. Antibodies to interferon-gamma prevent immunologically mediated damage in murine graft versus host reaction. Immunology 68:18–23.

Negash, A., and A. Niehof. 2004. The significance of enset culture and biodiversity for rural household food and livelihood security in southwestern Ethiopia. Agr. Hum. Values 21:61–71.

Oyewole, O. B. 1997. Lactic fermented foods in Africa and their benefits. Food Control 8:289–297.

Ozaki, M.. 2009. Nyuusankin: kenkou wo mamoru syokuhin no Himitu [Lactic acid bacteria: the secret of food protecting health]. Yasakasyobou Press, Tokyo.

Sharma, A., and A. C. Kapoor. 1996. Level of antinutritional factors in pearl millet as affected by processing treatments and various types of fermentation. Plant Foods Hum. Nutr. 49:241–252.

Shigeta, M. 2007. Syokubunnka: ensete to teff [Food culture: ensete and Teff]. Pp. 36–42 in T. Okakura, ed. Ehiopia wo shirutameno 50 syou [50 chapters to know Ethiopia]. Akashisyoten Press, Tokyo.

Shinohara, T. 1998. Africa de kechi wo kangaeta [Thinking about cheapskate in Africa]. Chikumasyobou Press, Tokyo.

Shinohara, T. 2000. Ethiopia KonsoSyakai no kukou to kachiku [Agriculture and livestock in the Konso society, Ethiopia]. Pp. 69–94 in K. Matui, ed. Shizenkan no jinruigaku [Anthropology of perspective on nature]. Youjyusyorin Press, Okinawa.

Shinohara, T. 2002. Ethiopia Konso Syakai ni okeru noukou no syuyakusei [Specialization of agriculture in the Konso Society, Ethiopia]. Pp. 125–162 in M. Kakeya, ed. Africa noukoumin no sekai: sono zairaisei to henyou [The world of African agricultural people: its indigenous knowledge and transformation]. Kyotodaigakusyuppankai Press, Kyoto.

Simango, C. 1997. Potential use of traditional fermented foods for weaning in Zimbabwe. J. Soc. Sci. Med. 44:1065–1068.

Soni, S. K., and D. K. Sandhu. 1990. Indian fermented foods: microbiological and biochemical aspects. Indian J. Microbiol. 30:135–157.

Uzochukwu, S. V. A., E. Balogh, and P. O. Ngoddy. 1994. the role of microbial gums in the colour and consistency of palm wine. J. Food Quality 17:393–407.

Wood, B. J. B., and W. H. Holzapfel. 1995. Pp. 398–407. The genera of lactic acid bacteria. Blackie Academic and Professional, London.

Yamamoto, M. 1995. Kuchigami-zake no hai wa meguru [The cup of Kuchikami-sake(liquor made by chewing with one's mouth) comes around]. Pp. 39–48 in S. Yoshida, ed. Sakedukuri no minzokushi [People's history of brewing]. Yasakasyoboukabu Press, Tokyo.

Yewelsew, A., B. J. Stoecker, M. J. Hinds, and G. E. Gates. 2006. Nutritive value and sensory acceptability of corn and kocho based foods supplemented with legumes for infant feeding in southern Ethiopia. AJFAND 6:1–19. Available at: http://www.ajfand.net/Volume6/No1/Stoecker1655.pdf (browsed 27 October 2012).

15

Toward better understanding of postharvest deterioration: biochemical changes in stored cassava (Manihot esculenta Crantz) roots

Virgílio Gavicho Uarrota[1], Eduardo da Costa Nunes[2], Luiz Augusto Martins Peruch[2], Enilto de Oliveira Neubert[2], Bianca Coelho[1], Rodolfo Moresco[1], Moralba Garcia Domínguez[3], Teresa Sánchez[3], Jorge Luis Luna Meléndez[3], Dominique Dufour[3,4], Hernan Ceballos[2], Luis Augusto Becerra Lopez-Lavalle[3], Clair Hershey[3], Miguel Rocha[5] & Marcelo Maraschin[1]

[1]Plant Science Center, Plant Morphogenesis and Biochemistry Laboratory, Postgraduate Program in Plant Genetic Resources, Federal University of Santa Catarina, Rodovia Admar Gonzaga 1346, CEP 88.034-001 Florianópolis, SC, Brazil
[2]Santa Catarina State Agricultural Research and Rural Extension Agency (EPAGRI), Experimental Station of Urussanga (EEUR), Rd. SC 446Km 19 S/N, Urussanga, Florianópolis, SC, CEP 88840-000, Brazil
[3]International Center for Tropical Agriculture (CIAT), Apartado Aéreo 6713 Cali, Colombia
[4]Centre de Coopération Internationale en Recherche Agronomique pour le Développement (CIRAD), UMR Qualisud, 73 Rue Jean-Francois Breton, TAB-95/16, 34398 Montpellier Cedex 5, France
[5]Centre of Biological Engineering, University of Minho, Campus de Gualtar, 4710-057 Braga, Portugal

Keywords

Cassava, deterioration, organic acids, polyphenol oxidase, scopoletin, soluble sugars

Correspondence

Virgílio Gavicho Uarrota, Plant Science Center, Plant Morphogenesis and Biochemistry Laboratory, Postgraduate Program in Plant Genetic Resources, Federal University of Santa Catarina, Rodovia Admar Gonzaga 1346, CEP 88.034-001, Florianópolis, SC, Brazil.

E-mail: uaceleste@yahoo.com.br

Funding Information

This work was supported by PEC-PG ("Programa de Estudantes Convênio de Pós-Graduação") coordinated by CAPES ("Coordenação de Aperfeiçoamento de Pessoal de Nível Superior"), CNPq, TWAS-Fellowship for Advanced Research and Training (FR Number 3240268144) and CIAT (International Center for Tropical Agriculture).

Abstract

Food losses can occur during production, postharvest, and processing stages in the supply chain. With the onset of worldwide food shortages, interest in reducing postharvest losses in cassava has been increasing. In this research, the main goal was to evaluate biochemical changes and identify the metabolites involved in the deterioration of cassava roots. We found that high levels of ascorbic acid (AsA), polyphenol oxidase (PPO), dry matter, and proteins are correlated with overall lower rates of deterioration. On the other hand, soluble sugars such as glucose and fructose, as well as organic acids, mainly, succinic acid, seem to be upregulated during storage and may play a role in the deterioration of cassava roots. Cultivar Branco (BRA) was most resilient to postharvest physiological deterioration (PPD), while Oriental (ORI) was the most susceptible. Our findings suggest that PPO, AsA, and proteins may play a distinct role in PPD delay.

Introduction

Cassava (*Manihot esculenta* Crantz.) is a major tropical root crop grown in Africa, Latin America, Oceania, and Asia, feeding more than 800 million people each day. The root, which is the major edible portion of the plant, is an important source of dietary energy and comprises more than 80% starch (Montagnac et al. 2009; Lyer et al. 2010; Harris and Koomson 2011). Historically, cassava has played an important role in food security as a famine reserve crop. In Eastern and Southern Africa where maize is preferred, but drought is recurrent, cassava, which is, to some extent drought tolerant, is harvested when other crops fail (Rosenthal and Ort 2012). Similarly, cassava provides additional food security when armed conflicts lead to the destruction of above-ground crops, as it remains viable below ground for up to 36 months (Rosenthal and Ort 2012). While cassava continues to be a vital subsistence crop for small-scale farmers, it is also an increasingly important crop on both regional and global levels (Rosenthal and Ort 2012). In addition to its role in food security, cassava is being used as a biofuel crop in many countries, including China, Thailand, and Brazil (Dai et al. 2006; Nguyen et al. 2007; Zidenga 2012; Zidenga et al. 2012). Globally, cassava is the fifth most important crop overall in terms of human caloric intake (Rosenthal et al. 2012). However, subsistence and commercial utilization of cassava are affected by its short shelf life that results from a rapid postharvest physiological deterioration (PPD) process, which renders the root unpalatable within 72 h of harvest (Owiti et al. 2011).

Posthaverst physiological deterioration is triggered by mechanical damage, an inevitable result of harvesting operations. PPD then progresses from the site of damage, eventually causing general discoloration of the vascular parenchyma throughout the root. According to previous studies (Sánchez et al. 2013; Uarrota et al. 2014), cassava root deterioration is related to two separate processes: physiological, or primary, deterioration and microbiological, or secondary, deterioration (Acedo and Acedo 2013; Njoku et al. 2014). Physiological deterioration is usually the initial cause of reduced acceptability of roots. It can be observed by the blue-black streaks in the root vascular tissue that later spread and cause a more general brown discoloration finally leading to unsatisfactory cooking quality and adverse taste (Salcedo et al. 2010; Sayre et al. 2011; Naziri et al. 2014). Primary deterioration also involves changes in oxidative enzyme activities which generate phenols, including catechins and leucoanthocyanidins, which polymerize in later stages to form condensed tannins (Zidenga 2012; García et al. 2013; Sánchez et al. 2013). Microbiological deterioration results from pathogenic rot, fermentation, and/or softening of the roots, and generally occurs when the roots have already become unacceptable because of physiological deterioration.

Few reliable estimates can be found that document the extent of postharvest losses. A systematic assessment of physical losses worldwide by the Food and Agriculture Organization (FAO) suggests that losses of root and tuber crops are in the range of 30% to 60%. In the case of cassava in Africa, losses in 2002 were estimated at 19 million tons out of a total production of 101 million tons across the entire continent (NRI 2014; Harris et al. 2015). Yet, the magnitude of losses significantly differs across countries and different value chains within a single country as such losses largely depend on how cassava is produced, processed, and consumed, and on the level of coordination among value chain actors (Naziri et al. 2014). Extending the shelf life of cassava by about 2 to 3 weeks would translate into a reduction in financial losses by about $2.9 billion in Nigeria alone over a 20-year period (Zidenga 2012). The rapid postharvest perishability of freshly harvested cassava roots is a problem not known in any other root and tuber crop. Within 1–3 days of harvest, roots begin to develop an endogenous disorder, typically characterized by blue-black streaking of the vascular tissues of the xylem, which is accompanied by an unpleasant odor and flavor. PPD profoundly impacts processing as well as marketing of the roots (Lyer et al. 2010). Several approaches have been developed to preserve cassava roots, such as underground storage, storage in boxes with moist sawdust, storage in bags combined with the use of fungicides, pruning plants before harvest, cold storage (2–4°C) for up to 2 weeks, freezing or waxing the roots to prevent access to oxygen, and even chemical treatments (Howeler et al. 2013; Sánchez et al. 2013). However, these methods are too expensive or complicated for handling large volumes of roots and have been restricted mostly to high-value product chains, such as the consumption of fresh cassava roots (Sánchez et al. 2013). Thus, a major goal of cassava breeding and biotechnology is to increase its shelf life by delaying the onset of PPD. Such efforts would expand the industrial applications of cassava worldwide (Zidenga 2012).

Molecular and biochemical studies of PPD have pointed to reactive oxygen species (ROS) production as one of the earliest events in the process, and many other compounds have been reported (Buschmann et al. 2000; Reilly et al. 2003). Specific genes involved in PPD have been identified and characterized, and their expression has been evaluated (Reilly et al. 2007; Timothy 2009). Several secondary metabolites, particularly hydroxycoumarins, accumulate in the process (Bayoumi et al. 2008; Bayoumi et al. 2010). However, more research is necessary to better understand the biochemical changes involved in PPD of cassava roots. Accordingly, this study aimed to evaluate the biochemical changes involved in PPD in four cassava cultivars, including fresh roots (hereinafter designated as nonstored samples) and root samples stored up to 11 days. Using metabolomic techniques integrated with chemometric tools, we further assessed biochemical markers of PPD. Supervised and unsupervised methods of data analysis were also used to discriminate among cassava samples during postharvest physiological deterioration.

Material and Methods

Selection of cassava cultivars

Cassava cultivars were provided by the Santa Catarina State Agricultural Research and Rural Extension Agency (EPAGRI), specifically, the experimental station of Urussanga, and were produced over the 2011/2012 growing season. Four cultivars were selected for this study as follows: SCS 253 Sangão (hereinafter designated as SAN), Branco (hereinafter designated as BRA, a landrace), IAC576-70 (hereinafter designated as IAC, a commercial variety), and Oriental (hereinafter designated as ORI, a landrace). The cultivars were selected as they are widely used by small farmers and lacking research efforts.

Plant materials and postharvest physiological deterioration

On-farm trials were carried out at the Ressacada Experimental Farm (Plant Science Center, Federal University of Santa Catarina, Florianópolis, SC, Brazil – 27°35'48" S, 48°32'57"W) in September 2011, using the four EPAGRI cassava cultivars noted above. The experimental design was in complete randomized blocks, with 4 blocks (6.3 × 15 m²/block) spaced at 1 m. Each block consisted of four plots (12 × 1.2 m²/plot), spaced at 0.5 m. Cassava stakes of length 15 cm were planted upright and spaced 1 × 1 m. Each plot was considered an experimental unit to which a treatment was applied, and all land operations were mechanized. Soil fertility had already been determined by chemical analysis, and cultivation was performed manually.

Cassava root samples were harvested from 12-month-old plants for analysis. Immediately after harvest, the roots were washed with sterilized water, and both proximal and distal parts of the root were

removed. Cross sections were made (0.5-1 cm) over the remaining root and stored at room temperature (66-76% humidity, ±25°C). Induction of PPD was performed during 11 days using different roots in the same batch. Monitoring the development of PPD and associated metabolic disturbances was performed daily after induction of PPD. Fresh samples and those at 3, 5, 8, and 11 days postharvest were collected at each point, dried (35-40°C/48 h) in an oven, milled with a coffee grinder (Model DGC-20N series, Cadence, Brazil), and kept for analysis. For enzymatic analysis, nonstored samples were collected and stored (-80°C) until analysis. PPD was also induced using two other methodologies, including storage of the entire root and the method of Wheatley (1982) whereby only the proximal and distal parts of the roots were removed without slicing the remaining part. The experiment was conducted using the same room conditions as to the method described firstly.

Postharvest Physiological Deterioration Scoring

Five independent evaluations of PPD were carried out. For each harvest, a random sample of three sliced roots from each plant variety was scored according to visual observations of sliced cassava roots (from 1 –10% of deterioration to 10–100% of deterioration) at each stage of PPD (i.e., 3, 5, 8, and 11 days postharvest) and imaged with a digital camera (OLYMPUS FE-4020, 14 megapixel, China). The mean PPD score for each root was calculated by averaging the scores for the three transversal sections and five evaluations (see Table S1). Roots showing symptoms of microbial rotting, which would not be reflective of PPD, or those showing inset activity were discarded.

Dry matter content (%)

To obtain the dry matter content of cassava samples, 10–30 g of chopped and grated fresh roots were weighed, and oven dried at 60°C for 48 h. Dry matter was expressed as the percentage of dry weight relative to fresh weight (Morante et al. 2010).

Polyphenol oxidase activity during PPD

For polyphenol oxidase (PPO) analysis, 2 g of fresh tissue were homogenized with 0.6 g of PVPP and 8 mL of 50 mmol/L (pH 7) phosphate buffer, followed by recovery of the supernatant by filtration and centrifugation (3220g, 4°C, 15 min, 18 cm of rotor radius). The product constituted an enzymatic extract. PPO activity was measured using 2.85 mL of 0.2 mmol/L (pH 7) phosphate buffer, 50 μL of catechol (60 mmol/L) as substrate, and 100 μL of enzymatic extract at 25°C. Changes in absorbance (420 nm) were recorded over a 5-min period in a UV-vis spectrophotometer (Spectrumlab D180, BEL Photonics, Brazil; Montgomery and Sgarbieri 1975). Activity was expressed as units of activity (UA), and one unit of PPO was defined as the change in one unit of absorbance per second.

Ascorbic acid determination during PPD

Ascorbic acid (AsA) content was assayed as described previously with slight modifications (Omaye et al. 1979). The extract was prepared by grinding 1 g of sample with 5 mL of 10% TCA, centrifuged (2465g, 18 cm rotor radius, 20 min), and then re-extracted twice. To the supernatant, 1.0 mL of extract and 1 mL of DTC reagent (2,4-dinitrophenylhydrazine–thiourea–$CuSO_4$) were added to a total volume of 10 mL and incubated (37°C, 3 h), followed by the addition of 0.75 mL ice-cold 65% H_2SO_4 (v v^{-1}). The mixture was allowed to stand for 30 min at 30°C. The resulting color was read at 520 nm in the spectrophotometer (Spectrumlab D180, BEL Photonics, Brazil). A standard AsA curve was constructed to determine content ($y = 0.0361x$, $r^2 = 0.99$, 0 to 1000 mg mL^{-1}) and the results were expressed in μg g^{-1} (ppm) of fresh weight.

Protein extraction and quantification from cassava roots during PPD

At each sampling time, root tissues were grated using a food processor (Walita-Master Plus, Brazil) and stored at −80°C before use. The frozen tissue was ground under liquid nitrogen to a fine powder using a prechilled pestle and mortar. Then, 5 g of tissue were added to a prechilled 50 mL tube containing 20 mL of extraction buffer (phosphate buffer 0.1 mol/L, pH 6.4, 0.25 g of PVP), 200 μL of 1 mmol/L DTT, and 200 μL of 1 mmol/L EDTA. Tubes were vortexed vigorously and transferred to a horizontal shaker (300 rpm) for 1 h. Tubes were centrifuged (5000 rpm, 20 min), and the supernatant was recovered by filtration, transferred to a fresh tube, and stored at −20°C (Reilly 2001).

The protein content of each sample was determined using the Bradford protocol (Bradford 1976) with small modifications. A calibration curve was constructed using bovine serum albumin (BSA) (Sigma-Aldrich, St. Louis, MO) as standard. Protein solutions were prepared in 0.15 M NaCl, and a series of dilutions was prepared (0 to 100 mg mL^{-1}, $y = 0.0082x$, $r^2 = 0.96$) to build the standard curve.

Extraction and quantification of soluble sugars and organic acids by HPLC during PPD

Nonstored samples and those at 3, 5, 8, and 11 days postharvest were collected. The outer and inner bark were removed and crushed with a food processor, as previously described, and dried in an oven at 35°C (48 h). After oven drying, samples were again crushed with a coffee grinder to obtain a fine powder, sieved and stored at room temperature for analysis.

Sugars and organic acids were extracted from 0.5 g of cassava root flour samples in 10 mL of mobile phase (H_2SO_4, 5 mmol/L) and determined accordingly (Chinnici et al. 2005). Briefly, the suspension was homogenized using an Ika Works Ultra-Turrax (Ika, China) Digital Homogenizer and mixed slowly using a horizontal shaker (Microplate shaker, 330 rpm) for 30 min. The suspension was centrifuged (12879g, 10 min) and filtered through a 0.22 μm disposable syringe membrane filter, followed by collection of the supernatant. Sugars and organic acids were analyzed by HPLC using a Bio-Rad Aminex HPX-87H HPLC column equipped with a UV detector (MWDG 1365D for organic acids), connected in series with a refractive index detector (RID G 1362A for sugars) and an injection valve fitted with a 15 μL loop. The samples were separated isocratically at 0.6 mL min^{-1} at 30°C.

Retention times and standard curves were prepared for the following sugars and organic acids (see Table 1). Three consecutive injections (10 μL) were performed. Sugars and organic acids were expressed (mg g^{-1}) as mean ± standard deviation.

Scopoletin extraction and quantification during PPD

Cassava root flour samples (1 g) were placed in 50 mL falcon tubes containing 2 mL 98% ethanol and homogenized with an Ika Works Ultra-Turrax T18 (IKA, China) for 30 sec. The suspension was vortexed (1 min), incubated (microplate shaker, 600 rpm, 30 min), and centrifuged (9861, 5 min). The extract was filtered on a Whatman # 1 paper and through a 0.22 μm nylon membrane. Samples were transferred to 1.5 mL vials for HPLC (Agilent 1200 series, Agilent Technologies, Waldbronn, Germany) analysis (Buschmann et al. 2000). To accomplish this, samples (50 μL) were injected into the Agilent 1200 series HPLC equipped with a reverse-phase column (Techsphere BDS C18, 250 mm × 4.6 mm, 5 μm) and a diode array detector. The column was kept at 25°C, and acetonitrile and 0.5% phosphoric acid (v v^{-1}) in aqueous solution were used as mobile phase. The gradient profile was 60 − 1% for 30 min with a 0.5 mL min^{-1} flow and 50 μL

Table 1. The HPLC standard curves prepared for sugars and organic acids studied. Three consecutive injections (10 μL) were performed. Sugars and organic acids were expressed (mg g^{-1}) as mean ± standard deviation.

Group of compound	Name of the compound[1]	Code	Standard curve	r^2
Soluble Sugars	Glucose	G7528	$y = 26748656x-1523663$	0.99
	Fructose	F2543	$y = 26028204x-8253663$	0.99
	Raffinose	R0514	$y = 22680182x + 45255.3$	0.99
	Sucrose	S7903	$y = 22582989x + 727997.7$	0.99
	Citric	CO759	$y = 3281.1x + 46046$	0.99
Organic acids	Malic	240179	$y = 2498.2x + 3816.4$	0.99
	Succinic	S3674	$y = 1737.8x-4255.3$	0.99
	Fumaric	R412205	$y = 4047.85x-5748.3$	0.99

[1]All reagents were acquired from Sigma-Aldrich

injection volume. Scopoletin was detected at 215, 280, and 350 nm according to its retention time with a standard compound sample (Sigma–Aldrich: scopoletin ≥ 99% – No. S2500). Scopoletin quantification was determined through a calibration standard curve (y = 158159.59x, r^2 = 0.9993, 1–75 mg l^{-1}). Three consecutive injections (10 μL) were performed. Quantifications were made on a dry weight basis, and data were represented in nmol g^{-1}, as mean ± standard deviation.

Statistical analysis

Each harvest was considered independent from the others. Roots of the same cultivar in the four field blocks were combined into one bulk volume, and repetitions (n = 3) were made according to the bulk sample. All statistical analyses were carried out using R software (R core team-2015, version 3.1.1), using their respective packages and scripts. All values were presented as mean ± standard deviation of three

repetitions (n = 3). Two-way ANOVA and multivariate analysis were applied when necessary.

Results and Discussion

PPD scoring

Results of PPD scoring of the four genotypes studied showed that ORI was the most susceptible cultivar to PPD, followed by SAN, while BRA and IAC were found to be more tolerant (Table 2). BRA and IAC showed a slower deterioration rate when compared with the faster deterioration rates observed at ORI and SAN. Similar regression coefficients were found among tolerant clones (0.935, 0.863 for ORI and SAN, respectively). Figure 1 shows regression models for the four cultivars during storage. For all PPD methods, an increasing rate of deterioration throughout storage was observed. A clear relationship was found when using the average value of

Table 2. Comparison of methods of PPD induction in four cultivars studied. Values are represented as mean scores of five independent evaluations in percentage (%), from zero to 100% of deterioration during different storage times.

Cultivar	Method of PPD	3 days	5 days	8 days	11 days
SANGÃO	Root slicing	15.70	70.70	89.30	100.00
	Wheatley	9.30	5.70	8.60	17.10
	Entire root	0.00	35.70	15.00	26.00
Average		8.3	37.4	37.6	47.7
BRANCO	Root slicing	34.90	46.40	68.60	81.50
	Wheatley	0.00	0.70	0.00	15.00
	Entire root	0.00	0.00	0.00	1.40
Average		11.6	15.7	22.9	32.6
IAC576-70	Root slicing	31.50	49.30	67.20	91.00
	Wheatley	0.00	2.90	1.40	5.00
	Entire root	0.00	0.00	11.40	3.60
Average		10.5	17.4	26.7	21.0
ORIENTAL	Root slicing	59.90	87.10	100.00	100.00
	Wheatley	7.90	17.20	24.30	33.60
	Entire root	0.00	35.00	35.70	64.30
Average		22.6	46.4	53.3	66.0

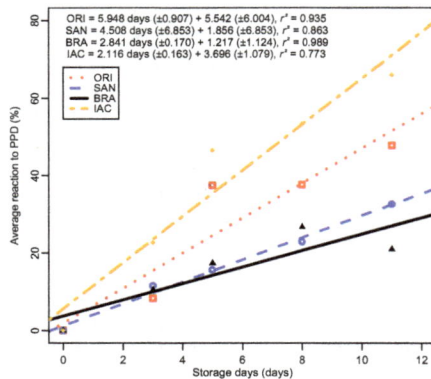

Figure 1. Average reaction to PPD (grouping together the three methods for assessing PPD) through time for the four cultivars involved in this study. The result of the linear regression analysis is also provided. Standard errors of the parameters in the regression analyses are given in parentheses.

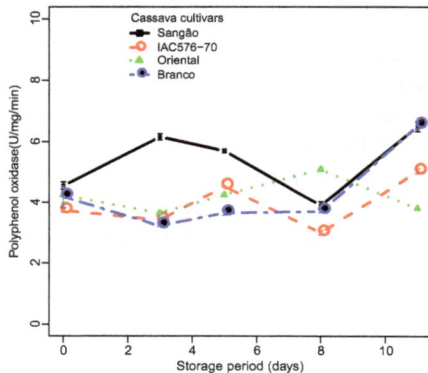

Figure 2. Changes in the activity of polyphenol oxidase in cassava cultivars during PPD. Each data point is presented as mean ± standard deviation ($n = 3$) in units per milligram*minutes (U mg^{-1} min^{-1}).

the three methods as shown in Figure 1. Cassava roots deteriorated faster after they were sliced. The root slicing method was applied to all analyses in this study. The entire root method was found to preserve the best post-harvest quality, and BRA was the most resilient based on all testing methods.

Polyphenol oxidase

Polyphenol oxidase results during storage are summarized in Figure 2. By comparing the mean PPO values from each cultivar, we found that BRA and ORI did not differ significantly ($P < 0.05$). During storage, a small decrease in PPO activity at day 8 was observed for IAC and SAN; however, in general, PPO activity varied similarly in the tolerant (BRA/IAC) and susceptible (SAN/ORI) cultivars.

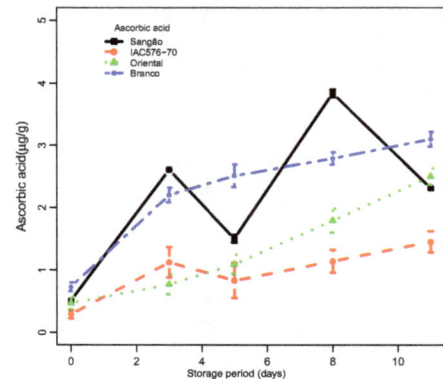

Figure 3. Changes in the concentrations of ascorbic acid in root samples of cassava cultivars during PPD. Each data point is presented as mean ± standard deviation ($n = 3$) in (μg g^{-1}).

PPO activity based on analysis of nonstored samples and those at day 3 of PPD, showed significant difference from samples at days 5, 8, and 11 ($P < 0.05$). SAN showed higher PPO activity than other cultivars, particularly at days 3 and 5, most likely related to the higher PPD scores for this cultivar.

When PPO values were correlated with PPD (see Fig. S1A), we found a high negative correlation and a clear discrimination between tolerant cultivar (BRA/IAC) and susceptible (ORI/SAN) ones. A similar trend in correlation values could be observed (0.790, 0.785, and 0.970, 0.93, respectively), meaning that higher activity of PPO in cassava roots was correlated with the reduced deterioration in that cultivars. Fluctuations in PPO, as shown in Figure 2, could be attributed to differences in genotype, as well as pre- and postharvest handling conditions.

PPO has been identified as a major cause of darkening in raw Asian noodles and other wheat products (Anderson et al. 2006), as well as browning induced in mechanically damaged potatoes (Batistuti and Lourenço 1985). PPO catalyzes the oxidation of phenols into quinones, which subsequently polymerize into brown pigments, a phenomenon that has also been reported in avocado (Gomez-Lopez 2002) and browning in marula fruits (Mdluli 2005).

Ascorbic acid

Our results showed that AsA gradually accumulated over time in all cultivars studied (Fig. 3). Interestingly, the cultivars with tolerance to PPD, IAC, and BRA, demonstrated the most extreme contrast for AsA, while the two susceptible cultivars, SAN and ORI, after several days of evaluation, showed only intermediate values of AsA. ORI showed a sharp increase in AsA after the third day of storage. SAN, on the other hand, showed a strong fluctuation in AsA activity through time, and, as a result, its behavior is difficult to define.

Two-way ANOVA showed differences in AsA between BRA/SAN and ORI/IAC. During storage, nonstored samples were statistically different from samples at days 3, 5, 8, and 11, respectively ($P < 0.05$). Although Figure 3 shows increases in AsA during storage until day 3, followed by fluctuation on other days, a clear correlation was found when assessing the levels of AsA between tolerant and susceptible cultivars in the context of PPD. Specifically, tolerant cultivars (BRA/IAC) behaved similarly and presented low negative correlations (0.785 and 0.793, respectively) when compared with susceptible cultivars (ORI/SAN) (0.969 and 0.932, respectively) (see Fig. S1B). In susceptible cultivars, the levels of AsA appeared to impact the degree of deterioration. Thus, the high levels of AsA presented by the tolerant cultivar (BRA) may be interpreted as causing a delay in PPD.

A wide range of factors, such as genotype, as well as pre- and postharvest conditions, may influence the AsA content. Losses of AsA during storage have been reported previously in many fruits, depending on storage conditions (Kabasakalis et al. 2000). Moreover, AsA has been reported to act as an antioxidant, thus prolonging the shelf life of commercial products (Fung and Luk 1985). In addition, lower storage temperatures have been reported to reduce the loss of AsA and the incidence of storage disorders in peas, broccoli, and spinach (Felicetti and Mattheis 2010). Although many studies have reported on the effects of AsA in many different crops, results of this study indicate that high levels of AsA may have a positive effect on PPD in cassava roots, thus inviting more research to better understand the precise role of AsA.

Changes in total proteins and dry matter content

The results of total protein contents in the roots of the sampled cultivars are summarized in Figure 4. Protein amounts significantly changed among the cultivars over the experimental period, except for days 5 and 11 of storage, revealing a genotype-specific behavior for that variable. Interestingly, while a considerable increase in protein amount was observed on day 3 in BRA, which could possibly be attributed to its tolerance to PPD, IAC showed contrasting behavior. This variable also differed between BRA and ORI, and correlations between PPD and protein levels are all summarized in Figure S1C. Lower negative correlations and similar trends in deterioration can be observed in all cultivars with protein levels, but such results make it difficult to define trends.

Dry matter content was also determined in nonstored samples and was correlated with PPD at days 3, 5, 8, and 11. A positive correlation was established between dry matter and PPD, leading to the implication that cultivars with a high level of dry matter are more prone to suffer from PPD (see Fig. S1D). It was also observed that these correlations decreased as PPD progressed; thus, at days 3, 5, 8, and 11, we found correlation values of 37%, 25%, 17%, and 11%, respectively, and these findings confirm previous studies (Ceballos et al. 2012; Sánchez et al. 2013). Taken together, it can be concluded that cassava roots with lower content of dry matter have longer shelf life.

Scopoletin content during PPD

The HPLC results of scopoletin contents during PPD are summarized in Table 3 and a representative chromatographic profile for the studied cultivars during PPD is provided in Figure S2. It was found that tolerant cultivars presented high levels of scopoletin at the starting point (nonstored samples) and at day 11 of PPD. A different trend was observed for susceptible cultivars, which presented low levels of scopoletin in nonstored samples and at day 11 of PPD. Fluctuations during PPD can be attributed to such factors as genotype and PPD conditions. By correlating scopoletin with PPD (Figure S3), we found that those cultivars with low levels of scopoletin presented a high degree of deterioration (ORI/SAN) when compared with tolerant ones (BRA/IAC). Unlike other variables, the positive correlation between scopoletin and PPD led to a clear separation of tolerant and susceptible cultivars. Two-way ANOVA showed significant differences ($P < 0.05$) in scopoletin levels in all cultivars and during storage. BRA showed a high level of scopoletin when compared to ORI.

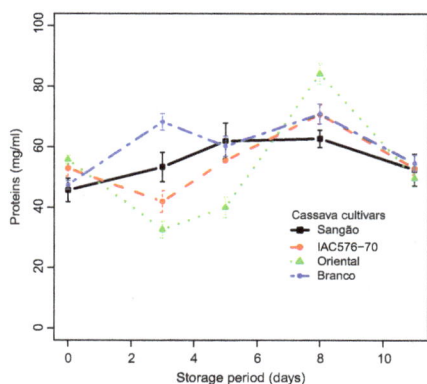

Figure 4. Changes in the concentrations of total proteins in roots of cassava cultivars during storage. Each data point is presented as mean ± standard deviation ($n = 3$) in (mg mL^{-1}).

Table 3. HPLC analysis of scopoletin (mmol g^{-1}) during PPD in cassava root tubers of the four cassava cultivars studied. Data are represented as mean ± standard deviation of two repetitions ($n = 3$). Letters in the column represent significant differences (Tukey HSD test, $P < 0.05$).

PPD days	BRA	ORI	SAN	IAC
0	91.46c	18.59b	25.99d	64.26e
3	92.11c	124.89a	45.40c	81.81d
5	95.00bc	48.17b	120.80a	123.90b
8	193.96ab	81.79ab	125.81a	214.00a
11	223.08a	54.94b	66.65b	98.10c

Values are represented as mean of three repetitions ($n = 3$) in mmol g^{-1} of dry weight. Different letters in the column represent significant statistical differences (Tukey HSD, $P < 0.05$).

PPD has been described as a physiological process that results from altered gene expression (Reilly et al. 2003; Reilly et al. 2004) and the accumulation of secondary metabolites. Among these secondary metabolites are found hydroxycoumarins (e.g., scopoletin) which show antioxidant properties, and by oxidation and polymerization, they confer the typical blue/black phenotype to root cassavas undergoing PPD. Hydroxycoumarins are important in plant defense as phytoalexins by the induction of biosynthesis following various stress events, such as wounding or bacterial and fungal infections. Additionally, they display a wide range of pharmacological activities, including anticoagulation (Mueller 2004), anti-inflammatory (Silvan et al. 1996), antimicrobial (Smyth et al. 2009), and antitumoral (Grazul and Budzisza 2009). However, while their biosynthesis pathway in cassava has not been elucidated (Wheatley 1982; Bayoumi et al. 2010), their accumulation in the biomass of that species during root deterioration has been previously reported (Wheatley 1982; Wheatley and Schwabe 1985; Sánchez et al. 2013). An uptake of scopoletin regulated by interaction among plant hormones, such as salicylic acid, was also reported (Taguchi et al. 2001).

Soluble sugar content during PPD

Soluble sugar contents detected in cassava roots during PPD are summarized in Figure 5A–E and Table S2. Figure S4 shows a typical chromatogram of soluble sugars detected in BRA during PPD, and Figure 5A–E shows changes in soluble sugars, including raffinose, sucrose, glucose, fructose, and total sugars, during storage. Significant differences ($P < 0.05$) were found in soluble sugar amounts during storage for each cultivar. Raffinose content was observed to decrease in all cultivars, except BRA, which showed a small increase up to the third day of storage (Fig. 5A). Decrease in sucrose was also observed, except for ORI where increases were observed until day 3 of storage (Fig. 5B). Glucose, fructose, and total sugar

content showed similar trends. A small decrease (SAN, IAC), followed, in turn, by increases, was observed in all cultivars studied (Fig. 5C–E). Glucose and fructose were the main sugars found in all samples studied. Researchers working with susceptible and tolerant cultivars of cassava stored for 14 days at ambient conditions also reported similar results for soluble sugars (Sánchez et al. 2013).

Organic acids during PPD

Figure 6A–D summarizes the results of the organic acid analysis performed. Changes in the contents of these metabolites for the studied cultivars can also be reviewed in Figure S5 and Table 3. The organic acid profiles of cassava roots sampled significantly differed. The main organic acids predominantly found during PPD were succinic and fumaric acids. In PPD-tolerant BRA, succinic acid and malic acid were the major compounds detected (Fig. 6A). Small decreases, followed by an increase in succinic acid, were observed in those samples. In ORI (Fig. 6B), increases in succinic acid and decreases in fumaric acid during PPD were also observed, while the level of malic acid remained quite constant during PPD. In SAN (Fig. 6C), increases in succinic acid up to day 8 of PPD were found; meanwhile, the levels of fumaric and malic acids decreased. For IAC, no trend was detected for succinic and fumaric acids (Fig. 6D). The chromatographic profile (Fig. S3) of BRA samples (nonstored) at days 3 and 5 of PPD shows other organic acids detected in small amounts, for example, phytic acid (data not shown). In tolerant clones (BRA/IAC), we found succinic acid to be the main acid related to PPD; in susceptible cultivars (ORI/SAN), we found that fumaric acid was the main acid related to PPD.

As primary metabolic products, organic acids play a regulatory role in plant growth and development. Organic acids are metabolically active solutes in cellular osmoregulation and surplus cationic balance, acting as key components in response to nutritional deficiencies, metal ion accumulation, and plant–microorganism interaction. Organic acids can also enhance resistance to diseases and inhibit oxidation during storage at low temperature, resulting in a significant extension of storage life for plant biomasses (Sun et al. 2012). They have also been related to maintenance of membrane integrity in stress conditions (Gunes et al. 2007). Since the postharvest physiological deterioration properties of stored cassava remain largely unknown, studying the metabolic profile of organic acids in postharvest stored cassava roots can lead to a better understanding of PPD.

Multivariate statistical analyses

Chemometric techniques that include multivariate models (e.g., principal component analysis (PCA), hierarchical

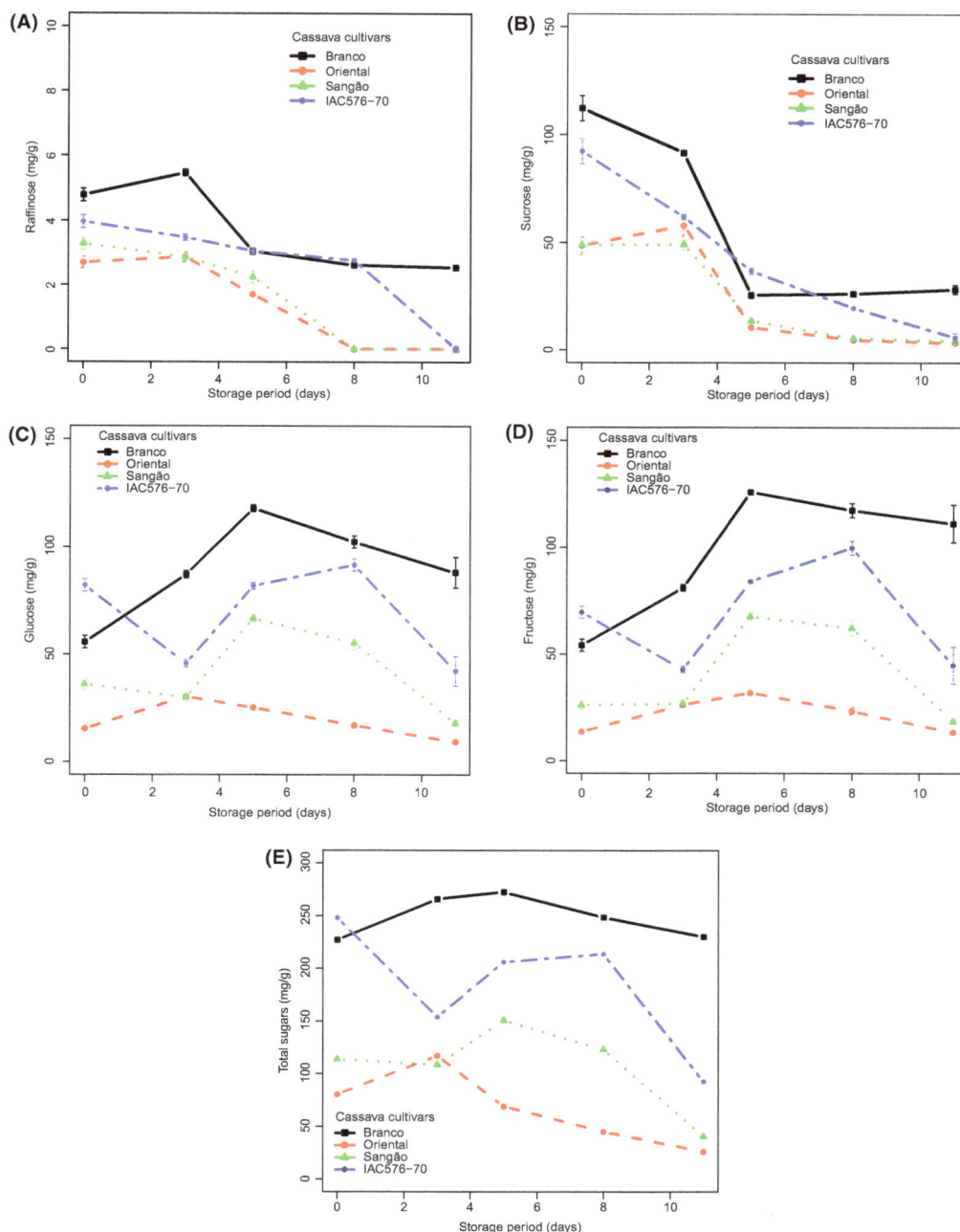

Figure 5. Changes in the concentrations of soluble sugars in cassava cultivars during storage. Each data point is presented as mean ± standard deviation (*n* = 3) in (mg g^{-1}). (A) Raffinose, (B) Sucrose, (C) Glucose, (D) Fructose, and (E) Total sugars.

cluster analysis (HCA), partial least squares discriminant analysis (PLS-DA), linear discriminant analysis (LDA), and support vector machines (SVM)) can be applied to complex and collinear data to extract relevant information. Both nonsupervised (HCA, PCA) and supervised (PLS-DA, LDA, and SVM) methods reduce large datasets by combining collinear variables into a small number of latent

variables (LVs), which are then used in place of the full dataset to build prediction models (Sills and Gossett 2012; Tang et al. 2014).

When PCA was applied to the scaled data in the present work, a clear separation was noted between nonstored samples and those that had undergone 3 days of PPD. A clear separation between tolerant and susceptible

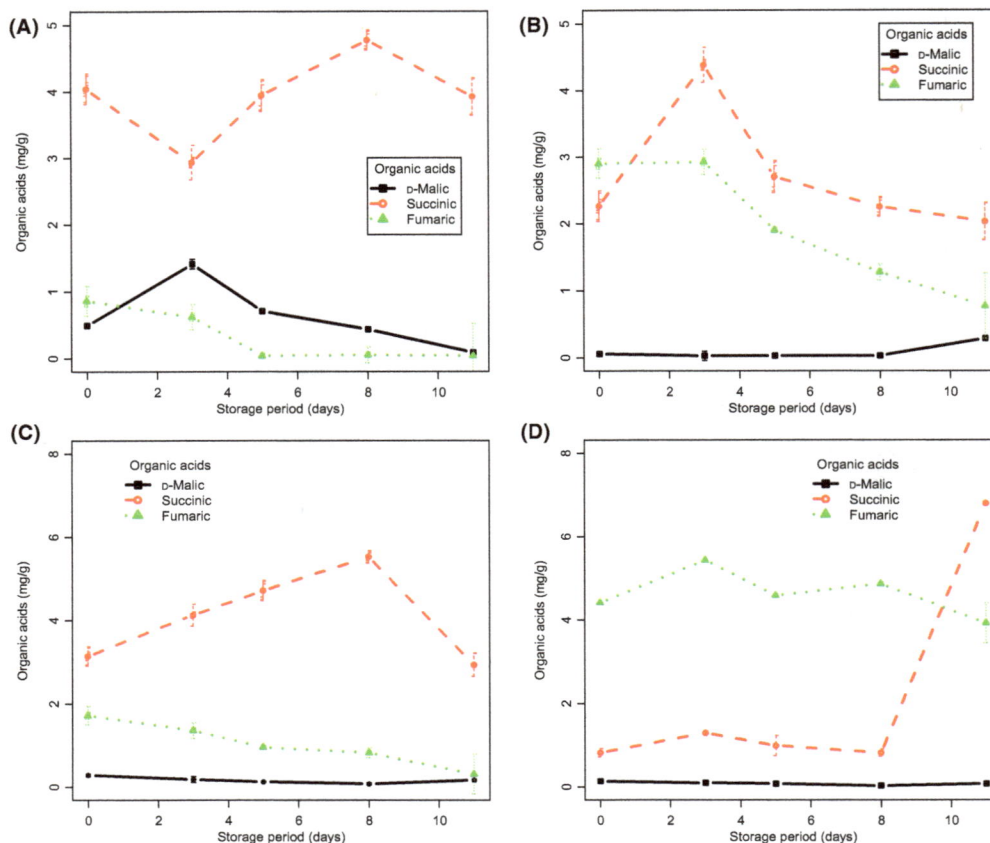

Figure 6. Changes in the concentration of organic acids in roots of cassava cultivars during PPD. Each data point is presented as mean ± standard deviation (n = 3) in (mg g^{-1}). (A) Branco, (B) Oriental, (C) Sangão, and (D) IAC 576-70.

cultivars was found (Fig. 7A). The total variance explained by the two principal components was 63.7%, that is, PC1 (37%) and PC2 (26.7%). The loadings plot showed that samples grouped in PC1+/PC2+ according to their values of polyphenol oxidase; in PC2+/PC1− according to fructose, scopoletin, proteins, and ascorbic acid contents; in PC1-/PC2− according to glucose, raffinose, total sugars, and organic acids (malic and succinic acid). Samples grouped in PC1+/PC2− showed similarity in their fumaric acid concentrations. The clustering method was able to separate nonstored (fresh) samples with those at 3 days of storage. Other samples were clustered together on days 5 and 8 of storage based on similar behavior.

When a serrated cluster heat map was applied to the data (Fig. 7B), results similar to those found in PCA were detected, and four major clusters were detected to occur: group 1 (SAN, ORI, SAN3, and ORI5); group 2 (IAC11,

ORI8, ORI11, and SAN8); group 3 (BRA, IAC, ORI3, and BRA3), and group 4 (SAN8, BRA11, BRA5, SAN5, IAC5, BRA8, and IAC8). A cophenetic correlation coefficient of 78.1% was found. The cluster heat map reveals major metabolic components that influenced the clustering noticed. Proteins and ascorbic acid were the major compounds related to group 1, polyphenol oxidase activity for group 2, glucose, succinic acid, and total sugars contents for group 3, and sucrose, raffinose, and malic acid for the last group.

Using supervised methods, for example, PLS-DA (Fig. 7C), a better separation was found (accuracy of 88.4%), when compared to PCA. BRA grouped with IAC, and ORI grouped with SAN. Most nonstored samples and those at day 3 of storage were found in the same component (x-variate), and those at day 5, 8, and 11 grouped in the same component as their presented similarities. Again, the built model was capable of predicting and separating

Figure 7. (A) Scores plot of a two-component PCA model from the metabolic dataset of cassava roots showing sample clustering according to metabolic fingerprinting and the percentage of variance captured by each PC. (B) A seriated cluster heat map (HCA), with cophenetic correlation coefficient of 78.1%. (C) PLS-DA components score plot of cassava samples during PPD, taking into consideration all the metabolites analyzed. PLS-1 (x-variate 1) = 61.23%; PLS-2 (x-variate2) = 19.43% of variance explained.

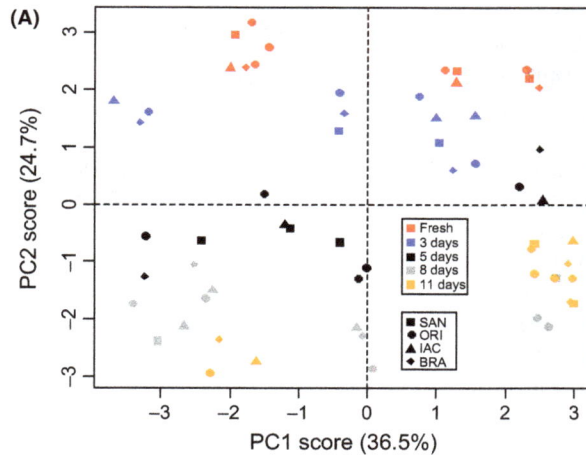

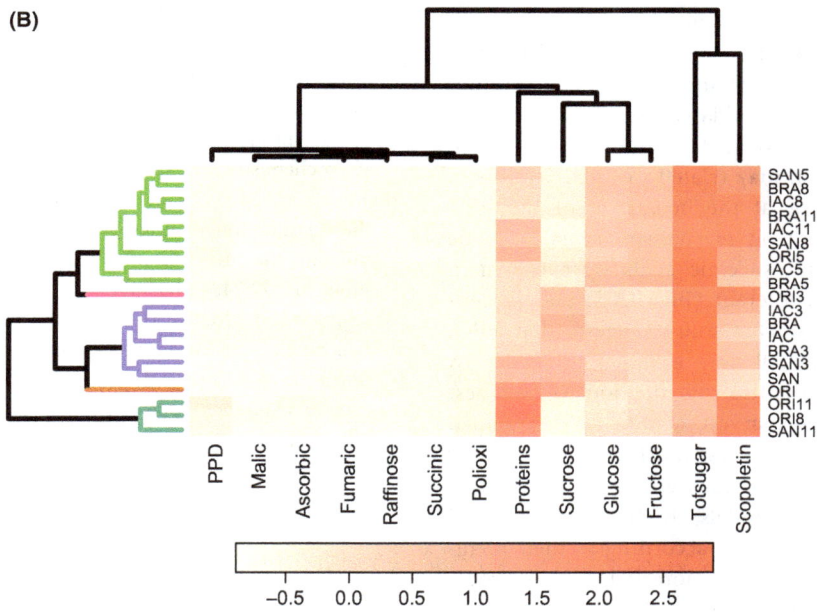

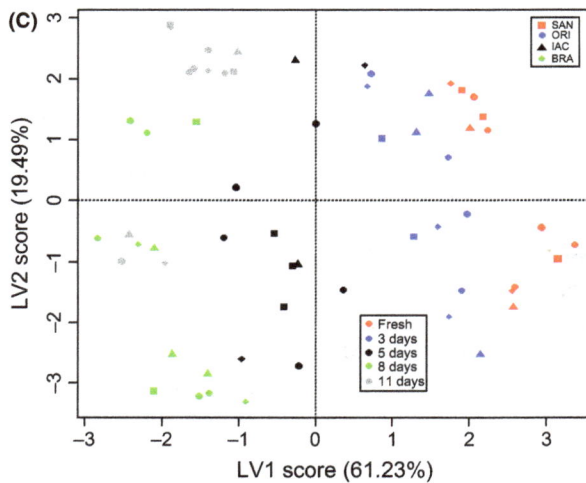

tolerant and susceptible cultivars. The total variances explained from the axis were 80.72%, that is, 61.23% from latent variable 1 (x-variate 1) and 19.49% from latent variable 2 (x-variate 2). The loading values showed that samples grouped in x-variate 1+ according to malic and fumaric acids, raffinose, sucrose, and total sugars, while for the x-variate 1-, samples grouped according to the values of glucose, fructose, scopoletin, ascorbic acid, proteins, polyphenol oxidase, and degree of PPD. Samples were also grouped in x-variate 2+ by similar amounts of malic and fumaric acids, ascorbic acid, and polyphenol oxidase, while in x-variate 2-, samples grouped according to the amounts of sugars, succinic acid, scopoletin, and proteins.

Conclusions

Based on the biochemical data herein presented, metabolic differences in cassava root samples could be correlated with deterioration state and cultivar. Our findings indicate that polyphenol oxidase, ascorbic acid, and proteins are all upregulated in the initial stages of PPD, up to 72 h, and, as such, may related to PPD. Scopoletin biosynthesis is increased at the beginning of the PPD process, but the underlying mechanism remains to be elucidated. Fumaric and succinic acids were the main organic acids found in tolerant (BRA/IAC) and susceptible (ORI/SAN) clones. Our study suggests that PPO, AsA, and proteins may all play a role in PPD delay. We also found that the entire root method worked best for maintaining the postharvest quality as determined by observing samples from the most resilient cultivar to PPD, BRA, and the most susceptible, ORI. Finally, the pattern recognition models, both supervised and unsupervised, classified samples according to their metabolic profiles and degree of deterioration. Cassava roots with lower dry matter have longer shelf life.

Acknowledgments

This work was supported by PEC-PG ("Programa de Estudantes Convênio de Pós-Graduação") coordinated by CAPES ("Coordenação de Aperfeiçoamento de Pessoal de Nível Superior"), CNPq, TWAS-Fellowship for Advanced Research and Training (FR Number 3240268144) and CIAT (International Center for Tropical Agriculture). The excellent technical support from Moralba Garcia Domínguez and Bryan Hanson is acknowledged. The researcher fellowship granted by CNPq on behalf of M. Maraschin is acknowledged.

Conflict of Interest

Authors declare no conflicts of interest.

References

Acedo, J. Z., and A. L. Acedo. 2013. Controlling postharvest physiological deterioration and surface browning in cassava (*Manihot esculenta* Crantz) roots with hot water treatment. Acta Horticulturae 989:357–362.

Anderson, J. V., E. P. Fuerst, W. J. Hurkman, W. H. Vensel, and C. F. Morris. 2006. Biochemical and genetic characterization of wheat (*Triticum spp.*) kernel polyphenol oxidases. J. Cereal Sci. 44:353–367.

Batistuti, J. P., and E. J. Lourenço. 1985. Isolation and purification of polyphenol oxidase from a new variety of potato. Food Chem. 18:251–263.

Bayoumi, S. A. L., M. G. Rowan, I. S. Blagbrough, and J. R. Beeching. 2008. Biosynthesis of scopoletin and scopolin in cassava roots during post-harvest physiological deterioration: the E-Z-isomerisation stage. Phytochemistry 69:2928–2936.

Bayoumi, S., M. G. Rowan, J. R. Beeching, and I. S. Blagbrough. 2010. Constituents and secondary metabolite natural products in fresh and deteriorated cassava roots. Phytochemistry 71:598–604.

Bradford, M. M. 1976. A rapid and sensitive method for the quantitation of microgram quantities of protein utilizing the principle of protein-dye binding. Anal. Biochem. 72:248–254.

Buschmann, H., M. X. Rodriguez, J. Tohmes, and J. R. Beeching. 2000. Accumulation of hydroxycoumarins during post-harvest deterioration of tuberous roots of cassava (*Manihot esculenta*). Ann. Bot. 86:1153–1160.

Ceballos, H., J. Luna, A. F. Escobar, D. Ortiz, J. C. Perez, H. Pachon, et al. 2012. Spatial distribution of dry matter in yellow fleshed cassava roots and its influence on carotenoid retention upon boiling. Food Res. Int. 45:52–59.

Chinnici, F., U. Spinabelli, and C. Riponi. 2005. Optimization of the determination of organic acids and sugars in fruit juices by ion-exclusion chromatography. J. Food Compos. Anal. 18:121–130.

Dai, D., Z. Hu, G. Pu, H. Li, and C. Wang. 2006. Energy efficiency and potentials of cassava fuel ethanol in Guangxi region of China. Energy Convers. Manage. 47:1686–1699.

Felicetti, E., and J. M. Mattheis. 2010. Quantification and histochemical localization of ascorbic acid in Delicious, Golden Delicious, and Fuji apple fruit during on-tree development and cold storage. Postharvest Biol. Technol. 56:56–63.

Fung, Y. S., and S. F. Luk. 1985. Determination of ascorbic acid in soft drinks and fruit juices. Part I. Background correction for direct ultraviolet spectrophotometry. Analyst 110:201–206.

García, J. A., T. Sánchez, H. Ceballos, and L. Alonso. 2013. Non-destructive sampling procedure for biochemical or gene expression studies on post-harvest physiological deterioration of cassava roots. Postharvest Biol. Technol. 86:529–535.

Gomez-Lopez, V. M. 2002. Some biochemical properties of polyphenol oxidase from two varieties of avocado. Food Chem. 77:163–169.

Grazul, M., and E. Budzisza. 2009. Biological activity of metal ions complexes of chromones, coumarins and flavones. Coord. Chem. Rev. 253:2588–2598.

Gunes, A., A. Inal, M. Alpaslan, F. Eraslan, E. G. Bagci, and N. Cicek. 2007. Salicylic acid induced changes on some physiological parameters symptomatic for oxidative stress and mineral nutrition in maize (Zea mays L.) grown under salinity. J. Plant Physiol. 164:728–736.

Harris, M., and C. Koomson. 2011. Moisture-pressure combination treatments for cyanide reduction in grated cassava. J. Food Sci. 76:T20–T24.

Harris, K. P., A. Martin, S. Novak, S. H. Kim, T. Reynolds, and C. L. Anderson. 2015. Cassava bacterial blight and postharvest physiological deterioration. Production Losses and Control Strategies. EPAR Brief No. 298.University of Washington. Evans School Policy Analysis and Research (EPAR), 1–34.

Howeler, R., N. Lutaladio, and G. Thomas. 2013. Save and grow. Food and Agriculture Organization of the United Nations, Rome.

Kabasakalis, V., D. Siopidou, and E. Moshatou. 2000. Ascorbic acid content of commercial fruit juices and its rate of loss upon storage. Food Chem. 70:325–328.

Lyer, S., D. S. Mattinson, and J. K. Fellman. 2010. Study of the early events leading to cassava root postharvest deterioration. Trop. Plant Biol. 3:151–165.

Mdluli, K. M. 2005. Partial purification and characterisation of polyphenol oxidase and peroxidase from marula fruit (Sclerocarya birrea subsp. Caffra). Food Chem. 92:311–323.

Montagnac, J., C. Davis, and S. Tanumihardjo. 2009. Nutritional value of vassava for use as a staple food and recent advances for improvement. Comprehensive Reviews in Food Science and Food Safety 8:181–194.

Montgomery, M. W., and V. C. Sgarbieri. 1975. Isoenzymes of banana polyphenol oxidase. Phytochemistry 14:1245–1249.

Morante, N., T. Sanchez, H. Ceballos, F. Calle, J. C. Pérez, C. Egesi, et al. 2010. Tolerance to postharvest physiological deterioration in cassava roots. Crop Sci. 50:1333–1338.

Mueller, R. L. 2004. First-generation agents: aspirin, heparin and coumarins. Best Practice & Research Clinical Haematology 17:23–53.

Naziri, D., W. Quaye, B. Siwoku, S. Wanlapatit, T. Viet Phu, and B. Bennett. 2014. The diversity of postharvest losses in cassava value chains in selected developing countries. Journal of Agriculture and Rural Development in the Tropics and Subtropics 115:111–123.

Nguyen, T. L. T., S. H. Gheewara, and S. Garivait. 2007. Energy balance and GHG-abatement cost of cassava utilization for fuel ethanol in Thailand. Energy Pol. 35:4585–4596.

Njoku, D. N., C. O. Amadi, J. Mbe, and N. J. Amanze. 2014. Strategies to overcome post-harvest physiological deterioration in cassava (Manihot esculenta) root: a review. Nigeria Agricultural Journal 45:51–62.

NRI. Postharvest Loss Reduction Centre. Roots and Tubers. Postharvest.nri.org. Available at: http://postharvest.nri.org/scenarios/roots-and-tubers, Accessed on December 4, 2014.

Omaye, S. T., J. D. Tumbull, and H. E. Sauberilich. 1979. Selected methods for the determination of ascorbic acid in animal cells, tissues and fluids. Methods Enzymol. 62:3–11.

Owiti, J., J. Grossmann, P. Gehrig, C. Dessimoz, C. Laloi, M. Hansen, et al. 2011. iTRAQ-based analysis of changes in the cassava root proteome reveals pathways associated with post-harvest physiological deterioration. Plant J. 67:145–156.

R Core Team (2015). R: A language and environment for statistical computing. R Foundation for Statistical Computing, Vienna, Austria. URL http://www.R-project.org/.

Reilly, K. 2001. Oxidative stress related genes in cassava post-harvest physiological deterioration. PhD thesis. University of Bath. p. 291.

Reilly, K., R. Gomez-Vasquez, H. Buschmann, J. Tohme, and J. R. Beeching. 2003. Oxidative stress responses during cassava post-harvest physiological deterioration. Plant Mol. Biol. 53:669–685.

Reilly, K., R. Gomez-Vasquez, H. Buschmann, J. Tohme, and J. R. Beeching. 2004. Oxidative stress responses during cassava post-harvest physiological deterioration. Plant Mol. Biol. 56:625–641.

Reilly, K., D. Bernal, D. F. Cortés, R. Gómez-Vásquez, J. Tohme, and J. R. Beeching. 2007. Towards identifying the full set of genes expressed during cassava post-harvest physiological deterioration. Plant Mol. Biol. 64:187–203.

Rosenthal, D. M., and D. R. Ort. 2012. Examining cassava's potential to enhance food security under climate change. Trop. Plant Biol. 5:30–38.

Rosenthal, D. M., R. A. Slattery, R. E. Miller, A. K. Grennan, T. R. Cavagnaro, C. M. Fauquetk, et al. 2012. Cassava about-FACE: greater than expected yield stimulation of cassava (Manihot esculenta) by future CO_2 levels. Glob. Change Biol. 18:2661–2675.

Salcedo, A., A. Del Valle, B. Sanchez, V. Ocasio, A. Ortiz, P. Marquez, et al. 2010. Comparative evaluation of physiological post-harvest root deterioration of 25 cassava (Manihot esculenta) accessions: visual vs. hydroxycoumarins fluorescent accumulation analysis. African Journal of Agriculture Research 5:3138–3144.

Sánchez, T., D. Dufour, J. L. Moreno, M. Pizarro, I. J. Aragón, M. Domínguez, et al. 2013. Changes in extended shelf life of cassava roots during storage in ambient conditions. Postharvest Biol. Technol. 86:520–528.

Sayre, R., J. R. Beeching, E. B. Cahoon, C. Egesi, C. Fauquet, J. Fellman, et al. 2011. The BioCassava PlusProgram: biofortification of Cassava for Sub-Saharan Africa. Annual Review of Plant Biology. 62:251–272.

Sills, D. L., and J. M. Gossett. 2012. Using FTIR to predict saccharification from enzymatic hydrolysis of alkali-pretreated biomasses. Biotechnol. Bioeng. 109:353–362.

Silvan, A. M., M. J. Abad, P. Bermejo, M. Sollhuber, and A. Villar. 1996. Antiinflammatory activity of coumarins from Santolina oblongifolia. J. Nat. Prod. 59:1183–1185.

Smyth, T., V. N. Ramachandran, and W. F. Smyth. 2009. A study of the antimicrobial activity of selected naturally occurring and synthetic coumarins. Int. J. Antimicrob. Agents 33:421–426.

Sun, X. H., J. J. Xiong, A. D. Zhu, L. Zhang, Q. L. Ma, J. Xu, et al. 2012. Sugars and organic acids changes in pericarp and endocarp tissues of pumelo fruit during postharvest storage. Sci. Hortic. 142:112–117.

Taguchi, G., K. Yoshizawa, R. Kodaira, N. Hayashida, and M. Okazaki. 2001. Plant hormone regulation on scopoletin metabolism from culture medium into tobacco cells. Plant Sci. 160:905–911.

Tang, K., L. Ma, Y. Han, Y. Nie, J. Li, and Y. Xu. 2014. Comparison and chemometric analysis of the phenolic compounds and organic acids composition of Chinese wines. J. Food Sci. 80:C20–C28.

Timothy, M. 2009. Modulation of root antioxidant status to delay cassava post-harvest physiological deterioration. PhD Thesis. University of Bath.

Uarrota, V. G., R. Moresco, B. Coelho, E. C. Nunes, L. A. M. Peruch, E. O. Neubert, et al. 2014. Metabolomics combined with chemometric tools (PCA, HCA, PLS-DA and SVM) for screening cassava (Manihot esculenta Crantz) roots during postharvest physiological deterioration. Food Chem. 161:67–78.

Wheatley, C. C. 1982. Studies on cassava (Manihot esculenta Crantz) root post-harvest physiological deterioration. University of London. PhD thesis.

Wheatley, C. C., and W. W. Schwabe. 1985. Scopoletin involvement in post-harvest physiological deterioration of cassava root (Manihot esculenta Crantz). J. Exp. Bot. 36:783–791.

Zidenga, T. 2012. Delaying postharvest physiological deterioration in cassava. ISB News Report, 1–4.

Zidenga, T., E. Leyva-Guerrero, H. Moon, D. Siritunga, and R. T. Sayre. 2012. Extending cassava root shelf life via reduction of reactive oxygen species production. Plant Physiol. 159:1396–1407.

16

Antibiotic resistance and multidrug-resistant efflux pumps expression in lactic acid bacteria isolated from pozol, a nonalcoholic Mayan maize fermented beverage

Maria del Carmen Wacher-Rodarte[1], Tanya Paulina Trejo-Muñúzuri[1], Jesús Fernando Montiel-Aguirre[2], Maria Elisa Drago-Serrano[3], Raúl L. Gutiérrez-Lucas[3], Jorge Ismael Castañeda-Sánchez[3] & Teresita Sainz-Espuñes[3]

[1]Depto de Alimentos y Biotecnología, Facultad de Química, UNAM, Ciudad Universitaria Coyoacán, 04510 México, Distrito Federal, México
[2]Depto de Bioquímica, Facultad de Química, UNAM, Ciudad Universitaria Coyoacán, 04510 México, Distrito Federal, México
[3]Depto. Sistemas Biológicos, UAM-XochimilcoCalzada del Hueso No.1100, Coyoacan, 04960 Mexico, Distrito Federal, México

Keywords
Antibiotic resistance, efflux pumps, ethidium-bromide, lactic-acid-bacteria, Mayan-pozol

Correspondence
Sainz-Espuñes Teresita, Calz. del Hueso 1100, 04960, Mexico City, Mexico.

E-mail: trsainz@correo.xoc.uam.mx

Funding Information
No funding information provided.

Abstract

Pozol is a handcrafted nonalcoholic Mayan beverage produced by the spontaneous fermentation of maize dough by lactic acid bacteria. Lactic acid bacteria (LAB) are carriers of chromosomal encoded multidrug-resistant efflux pumps genes that can be transferred to pathogens and/or confer resistance to compounds released during the fermentation process causing food spoiling. The aim of this study was to evaluate the antibiotic sensibility and the transcriptional expression of ABC-type efflux pumps in LAB isolated from pozol that contributes to multidrug resistance. Analysis of LAB and *Staphylococcus (S.) aureus* ATCC 29213 and ATCC 6538 control strains to antibiotic susceptibility, minimal inhibitory concentration (MIC), and minimal bactericidal concentration (MBC) to ethidium bromide were based in "standard methods" whereas the ethidium bromide efflux assay was done by fluorometric assay. Transcriptional expression of efflux pumps was analyzed by RT-PCR. LAB showed antibiotic multiresistance profiles, moreover, *Lactococcus (L.) lactis* and *Lactobacillus (L.) plantarum* displayed higher ethidium bromide efflux phenotype than *S. aureus* control strains. Ethidium bromide resistance and ethidium bromide efflux phenotypes were unrelated with the overexpression of *lmrD* in *L. lactics*, or the underexpression of *lmrA* in *L. plantarum* and *norA* in *S. aureus*. These findings suggest that, moreover, the analyzed efflux pumps genes, other unknown redundant mechanisms may underlie the antibiotic resistance and the ethidium bromide efflux phenotype in *L. lactis* and *L. plantarum*. Phenotypic and molecular drug multiresistance assessment in LAB may improve a better selection of the fermentation starter cultures used in pozol, and to control the antibiotic resistance widespread and food spoiling for health safety.

Introduction

Pozol is a handcraft traditional nonalcoholic beverage produced from fermented maize dough and consumed mainly in southeastern Mexico, in the Maya region, as an important component of their diet (Ulloa et al. 1996). Pozol production entails a complex fermentation process of more than 40 different species of lactic acid bacteria, yeasts, and fungi (Wacher et al. 1993).

Analysis of *Escherichia (E.) coli* pathotype strains and other intestinal bacteria isolated from pozol indicated that a 60% were resistant to one antibiotic whereas 4% were antibiotic multiresistant (Sainz et al. 2001). In lactic acid bacteria (LAB), antibiotic resistance has been ascribed to multidrug resistant (MDR) efflux pumps involved in the expulsion of structurally unrelated compounds antibiotics, biocides, toxic agents like ethidium bromide (Mazurkiewicz et al. 2005). Efflux pumps including the chromosomally encoded ABC-type transporters LmrA and LmrCD described in *Lactococcus* (*L.*) *lactis* (Poelarends et al. 2002; Lubelski et al. 2006). LmrA transporter is associated with the resistance to wide variety of clinically relevant antibiotics (Poelarends et al. 2002), whereas LmrCD confers resistance to toxic compounds like daunomycin, cholate, and ethidium bromide (Lubelski et al. 2006). In *Lactobacillus* (*L.*) *brevis* and *L. plantarum*, the ABC transporter HorA confer resistance to toxic compounds generated during beer fermentation (Ulmer et al. 2000, 2002; Sakamoto et al. 2001). Lactic acid bacteria especially *Lactobacillus spp.* with resistance to toxic compounds are regarded as spoil strains for the production of fermented beverages like beer (Sakamoto and Konings 2003). Moreover, LAB are carriers of antibiotic-resistant genes that can be transferred to other bacteria including human pathogens (Toomey et al. 2009, 2010). Thus, analysis of MDR genes in LAB is relevant not just for preventing the spoiling of foods and beverages, but also to control the potential dissemination of resistance-associated genes to harmless bacteria from the intestinal microbiota and even foodborne pathogens. ABC efflux pumps are extensively described in *Lactococcus lactis*, so this may be a first approach to study this system in pozol LAB, however, other efflux transporters (Piddock 2006) will be investigated in future work.

The aim of this study was to evaluate the antibiotic resistance and the expression of the ABC-type pump genes that contributes to the multidrug resistance in LAB isolated from pozol in order to achieve a better selection of the fermentation starter cultures.

Material and Methods

Selection of strains

Strains belonging to *Weissella, Lactococcus, Lactobacillus, Leuconostoc, Streptococcus,* and *Enterococcus* genus were obtained from Dr. Wacher's pozol strains collection. *Staphylococcus aureus* ATCC 29213 was used as control strain.

Microbiologic procedures

Reagents and media were obtained from Sigma Chemical Co., St. Louis, Mo., and BD Bioscience, Sparks, MD. APT broth was used for *Weisella, Lactococcus, Lactobacillus,* and *Leuconostoc* cultures, and was incubated at 30°C for 24 h. MRS broth was used for *Enterococcus* and *Streptococcus* cultures, and incubated at 37°C for 24 h.

Susceptibility testing

Antimicrobial Disk Susceptibility Tests in accordance with the procedures outlined by Clinical and Laboratory Standards Institute (CLSI 2006) were performed for the pozol strains and controls. APT agar and MRS agar were used for this method. Inocula were adjusted to 1.5×10^8 CFU/mL (0.5 McFarland). BD BBLTM Sensi-Discs (Becton Dickinson and Company, Sparks, MD) were used. The antibiotics tested were: Ampicilin (AM), Penicilin (PE), Dicloxacilin (DC), Cloxacilin (CX), Cefotaxime (CFX), Cefalotine (CF), Ciprofloxacin (CPF), Clindamicin (CLM), Eritromicin (E), Tetraciclin (TE), Gentamicin (GE), Netilmicin (NET), Kanamicin (K), Neomicin (N), Trimetoprim/Sulfametoxazol (SXT), Vancomicin (VA), and Chloramphenicol (C). Plates were incubated as mention before. Determinations were performed by triplicate. The definition for R/S character for antibiotic susceptibility followed CLSI criteria (CLSI 2006).

Minimal inhibitory or bactericide concentration to Ethidium bromide

Ethidium bromide (EB) efflux assay

The ability of some microorganisms to efflux ethidium bromide, which inside the cell intercalates with double-stranded nucleic acids, thus determining fluorescence increase when properly excited. Ethidium bromide is a substrate for a variety of efflux pumps so, once inside the cell, it is extruded with the result of decreasing the overall measurable fluorescence.

For EB MICs, 50 mL of suitable broth were inoculated with a 24 h overnight culture adjusted to 0.5 McFarland, and variable concentrations of ethidium bromide (5, 10, 20, 40, and 80 μg/mL) were added to each test tube. Culture media and EB tubes were used as negative controls and inoculated tubes with each strain without EB were used as positive controls. Tubes were incubated at 30°C or 37°C as described and growth (turbidity) was measured during 24 h, 48 h, and 1 week later. MIC was defined as the lowest concentration of EB in which no growth was present after 48 h of incubation time.

Ethidium bromide uptake

For this test, a modification of the procedure described by Kaatz et al. (2000) and Patel et al. (2010) was used. Cells were grown overnight in suitable broth and adjusted to 0.4 McFarland. Ten μg/mL of EB (final concentration) was added and incubated for 25 min at room temperature. Cells were harvested by centrifugation (7000 ×g) and washed with fresh culture media, and then 4 mL of suitable broth was added. The suspension was maintained at

30°C or 37°C, and the fluorescence of aliquots was determined at frequent intervals of time (5–30 min) in a Hitachi F-4500 fluorometer, Tokyo (excitation wavelength, 540 nm; emission wavelength 545 nm).

Expression of efflux with significant differences was in accordance with that described by Patel et al. (2010) where at least a 20% difference of the fluorescence intensity has to be achieved between the tested strains and the control.

Gene expression

Total RNA was isolated using the SV Total RNA Isolation System (Promega, Madison, WI) following the manufactures instructions with additional modifications, lysozyme (10 mg/mL) and lysostaphin (0.5 mg/mL) were added to a final volume of 100 μL in TE buffer to assure all the bacterial walls were disrupted. RNA concentration was determined using a 2000 Nanodrop spectrophotometer (Thermo Scientific, Wilmington, DE) and was stored at −70°C until used for RT-PCR assay.

RT-PCR analysis

Lactococcus lactis secY and *lmrD*, *Lactobacillus plantarum secY* and *lmrA*, *Staphylococcus aureus secY* and *norA* mRNAs were examined by one-step reverse transcription (Qiagen OneStep RT-PCR Kit, Valencia, CA.) following the manufacturer's instructions.

A total of 0.5 μg was reverse transcribed with Superscript II reverse transcriptase at 50°C for 30 min; followed by amplification with specific primers listed in Table 1. An initial step of 15 min at 95°C followed by 30 amplification cycles consisting of 30 sec of denaturation at 94°C, 30 sec of annealing temperature (Table 2), and 1 min of extension at 72°C were used. After amplification, RT-PCR

Table 1. Primers used in this study.

Gen	Oligonucleotide sequences	Amplicon (bp)	Ref
	Staphylococcus aureus		
secY	F: ATCCCCAAGGTTCTCAAGGT	174	This study
	R: CACCTTGTTTTGCCCATTCT		
norA	F: TTATATCGCCGTTTGGTGGT	246	
	R: TCGCTGACATGTAGCCAAAG		
	Lactococcus lactis		
secY	F: GTGGTCAAAACAAGGGGAAA	217	This study
	R: TTGTTCACCCATCCAAGTGA		
lmrD	F: GGCAACTTCACATGCTGCTA	232	
	R: AGAGGTGAAACGAGCAAGGA		
	Lactobacillus plantarum		
secY	F: GCCGGGGTTATTCCTGTTAT	180	This study
	R: GAACGTGAAGAGCACGATCA		
lmrA	F: CTAACGCTTTTCCGCAAGTC	184	
	R: GCTAAAGCATCTTGGCGTTC		

Table 2. Annealing temperatures.

Microorganism	Gen	°C
Staphylococcus aureus	secY	63.5
	norA	63.5
Lactococcus lactis	secY	55.0
	lmrD	63.5
Lactobacillus plantarum	secY	60.0
	lmrA	52.0

products were analyzed on 1.5% agarose gels, and bands were visualized after ethidium bromide staining.

Densitometry analysis

Digital gel photographs of the stained RT-PCR products were taken under UV exposure by using a Kodak EDAS (Eastman Kodak Company, Molecular Imaging Systems, NY, USA) 290 System. Amplicons (cDNA bands) were determined as the integrated area (pixels) of the band intensities by densitometric analysis with Kodak Digital Science1D 3.6 software (Eastman Kodak Company, Molecular Imaging Systems, NY, USA). The numerical values for cDNA band intensities were corrected with the values for the sec Y bands, as the secY gene is expressed at a relatively constant level in cells and is commonly used in semiquantitative RT-PCR systems to assess the relative efficiency of each individual PCR. Tenfold logarithmic dilutions of the cDNA mixture were used to verify the linear correlation between the intensity (pixels) of the bands and the initial amount of cDNA.

Results

Antibiogram tests were done as first step to evidence the sensibility or resistance of acid lactic bacteria to conventional antibiotics. According to CLSI susceptible antimicrobial susceptibility test, interpretive category (S) implies that isolates are inhibited by the usually achievable concentrations of the antimicrobial agent when the recommended dosage is used. The intermediate category (I) includes isolates for which response rates may be lower than susceptible isolates. The resistant category (R) implies that isolates are not inhibited by the usually achievable concentrations of the antimicrobial agent present in the commercial disks used. As shown in Table 3, *S. aureus* control strains had sensibility to penicillins, cephalosporins, erythromycin, tetracycline, kanamycin (ATCC 25923 and ATCC 6538), vancomycin, and chloramphenicol. Moreover, *S. aureus* showed resistance (R) to ciprofloxacin (*S. aureus* 29213); clindamycin, gentamicin, and neomycin or intermediate resistance (IR) to ciprofloxacin (*S. aureus* 25923 and ATCC 6538), netilmicin, kanamycin (*S. aureus* ATCC 29213), and sulfamethoxazole-trimethoprim.

Table 3. Antibiogram tests of *Staphylococcus aureus* control strains and lactic acid bacteria strains.

Strain	1				2			3	4	5	6	7			8	9	10
	AM	PE	DC	CX	CFX	CF	CPF	CLM	E	TE	GE	NET	K	N	SXT	VA	C
S. aureus ATCC25923	S	S	S	S	S	S	I	R	S	S	R	I	S	R	I	S	S
S. aureus ATCC6538	S	S	S	S	S	S	I	R	S	S	R	I	S	R	I	S	S
S. aureus ATCC29213	S	S	S	S	S	S	R	R	S	S	R	I	I	R	I	S	S
E. italicus A	S	S	S	S	S	S	S	S	S	S	R	S	S	R	R	S	S
E. italicus B	S	S	S	S	S	S	S	S	S	S	R	S	S	R	R	S	S
S. bovis	R	R	S	I	S	R	R	I	S	S	R	R	R	R	R	S	S
W. confusa A	R	S	R	R	R	R	R	I	S	S	R	S	R	R	R	R	S
W. confusa B	S	R	R	R	S	R	R	R	S	I	R	S	R	S	I	R	S
L. plantarum A	S	S	R	R	S	R	R	R	S	S	R	S	R	R	I	R	S
L. plantarum B	S	S	R	I	I	R	R	R	S	I	R	S	R	S	R	R	S
L. lactis	S	S	R	R	S	R	R	I	S	S	I	S	S	R	R	I	S
L. mesenteroides A	S	S	R	R	S	S	R	S	S	S	I	S	R	S	R	R	S
L. mesenteroides B	S	S	R	R	S	I	R	I	S	I	R	I	R	R	R	R	S

(S) Sensible, (R) Resistant, (I) Intermediate resistance.

Enterococcus italicus strains were sensible to all antibiotics except gentamicin, neomycin, and sulfamethoxazole-trimethoprim.

Streptococcus bovis showed (1) sensibility to dicloxacillin, cefotaxime, erythromycin, tetracycline, vancomycin, and chloramphenicol; (2) resistance to ampicillin, penicillin, cefalotin, ciprofloxacin, all aminoglycosides tested and sulfamethoxazole-trimethoprim; and (3) intermediated resistance to cloxacillin and clindamycin.

Weisella confusa showed (1) sensibility to ampicillin (strain B), penicillin (strain A); cefotaxime (strain B), erythromycin, tetracycline (strain A), netilmicin, neomycin (strain B), chloramphenicol; (2) resistance to ampicillin (strain A), penicillin (strain B), dicloxacillin, cloxacillin, cefotaxime (strain A), cefalotin, ciprofloxacin, clindamycin (strain B), gentamicin, kanamycin, neomycin (strain A), sulfamethoxazole-trimethoprim (strain A), vancomycin; and (3) intermediate resistance to clindamycin (strain A), tetracycline (strain B), Sulfamethoxazole-trimethoprim (strain B) (Table 3).

Lactobacillus plantarum showed (1) sensibility to ampicillin, penicillin, cefotaxime (strain A), erythromycin, tetracycline (strain A), netilmicin, neomycin (strain B), chloramphenicol; (2) resistance to dicloxacillin, cloxacillin, cefalotin, ciprofloxacin, clindamycin, gentamicin, kanamycin, neomycin (strain A), sulfamethoxazole-trimethoprim (strain B), vancomycin; and (3) intermediate resistance to cefotaxime (strain B), tetracycline (strain B) and sulfamethoxazole-trimethoprim (strain A).

Lactococcus lactis had (1) sensibility to ampicillin, penicillin, cefotaxime, erythromycin, tetracycline, netilmicin, kanamycin, chloramphenicol; (2) resistance to dicloxacillin, cloxacillin, cefalotin, ciprofloxacin, neomycin, sulfamethoxazole-trimethoprim; and (3) intermediate resistance to clindamycin, gentamicin, and vancomycin.

Leuconostoc mesenteroides showed (1) sensibility to ampicillin, penicillin, cefotaxime, cefalotin (strain A), clindamycin (strain A), erythromycin, tetracycline (strain A), netilmicin (strain A), neomycin (strain A), chloramphenicol; (2) resistance to dicloxacillin, cloxacillin, ciprofloxacin, gentamicin (strain B), kanamycin, neomycin (strain B), sulfamethoxazole-trimethoprim, vancomycin; and (3) intermediate resistance to cefalotin (strain B), clindamycin (strain B), tetracycline (strain B), gentamicin (strain A) and netilmicin (strain B).

Sensibility test to ethidium bromide

Once the antimicrobial resistance to conventional antibiotics is estimated, the test for efflux phenotype was evidenced by assessing the sensibility or resistance to ethidium

Table 4. Assays of MIC and MBC to ethidium bromide.

Strain	MIC (µg/mL)	MBC (µg/mL)
S. aureus ATCC25923	5	40
S. aureus ATCC6538	10	20
S. aureus ATCC29213	5	20
E. italicus A	<5	<5
E. italicus B	<5	10
S. bovis	<5	10
W. confusa A	40	>80
W. confusa B	40	>80
L. plantarum A	>80	>80
L. plantarum B	>80	>80
L. lactis	40	>80
L. mesenteroides A	>80	>80
L. mesenteroides B	>80	>80

MIC, minimal inhibitory concentration; MBC, minimal bactericidal concentration.

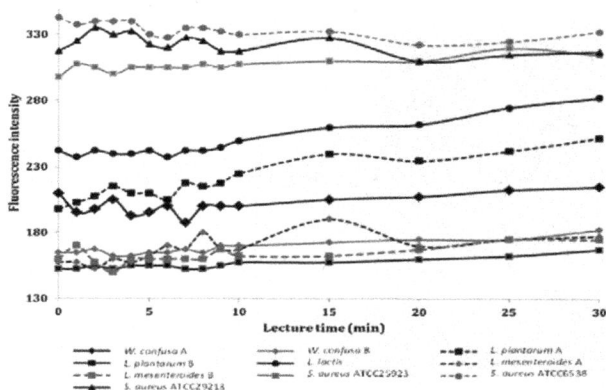

Figure 1. Accumulative test of ethidium bromide in bacterial strains with significant differences in fluorescence intensity in regard with the *Staphylococcus aureus* control strains assessed for five minutes each during thirty minutes.

bromide, a DNA intercalating agent. Minimal inhibitory concentration (MIC) and minimal bactericidal concentration (MBC) to ethidium bromide from strains are summarized in the Table 4. As shown, *E. italicus* and *S. bovis* strains, showed MIC and MBC values twice or even lower in comparison with those from *S. aureus* control strains. Instead, *W. confusa*, *L. plantarum*, *L. lactis,* and *L. mesenteroides* showed MIC and MBC values twice or even greater than those from the *S. aureus* control strains.

Accumulative test of ethidium bromide was done given that this DNA intercalating toxic agent is used as substrate to detect phenotypically the expression of efflux pumps (Patel et al. 2010). In comparison with *S. aureus* control strains, accumulation of ethidium bromide was lower in *L. lactis*, *L. plantarum* (strain A), and *W. confusa* (strain A) or even lower in *W. confusa* (strain B), *L. plantarum* (strain B), and *L. mesenteroides* (strain B) (Fig. 1).

Agarose gel analysis of mRNA encoding efflux pumps amplified by RT-PCR.

Accumulation of ethidium bromide led us to the molecular characterization of efflux pumps in acid lactic bacteria. Efflux pumps have been more studied in *L. lactis* and *L. plantarum* strains, therefore they were selected to analyze the mRNA amplification by RT-PCR encoding efflux pumps associated with antimicrobial multiresistance. Each strain showed two bands with divergent intensity indicating differential mRNA expression without or with ethidium bromide (Fig. 2).

Quantitative analysis by densitometry was done to estimate significant differences on the relative expression from the particular mRNA product of acid lactic bacteria regarding the constitutive expression of *secY* mRNA included as control. As shown in the Figure 3, no differences on the relative mRNA expression without or

with ethidium were found in all strains. Thus, ethidium bromide is not a determining factor for their expression. In comparison with the relative *norA* mRNA expression of *S. aureus* ATCC 6538 control strain, no differences were found on the relative *lmrA* mRNA expression of *L. plantarum*. Moreover, relative mRNA expression found in *S. aureus* and *L. plantarum* was less than 1.0 in regard with the constitutive expression of *secY* control. Instead, significant differences were found concerning the relative *lmrD* mRNA expression of *L. lactis* in comparison with that observed in *S. aureus* and *L. plantarum* strains without and with ethidium bromide.

Discussion

According to the results, most acid lactic bacteria strains (except *E. italicus*) displayed a profile of multiresistance as they showed (1) resistance (or intermediate resistance) to a greater antibiotic number; and (2) the antibiotic resistance was distributed in more than three antibiotic families by comparison with the *S. aureus* control strains.

Moreover, to exhibit resistance to antibiotics, *Lactococcus*, *Lactobacillus*, *Leuconostoc,* and *Weissella genus*, displayed higher resistance to ethidium bromide as well lower ethidium bromide accumulation in comparison with *S. aureus* control strains. In *L. lactis*, the resistance to antibiotics and to ethidium bromide as well as the high ethidium bromide efflux were independent from the overexpression of *lmrD* mRNA (Figs. 2, 3). In *L. lactis*, *lmrD* gene encodes D protein subunit of the heterodimeric ABC transporter LmrCD responsible for the efflux of a wide spectrum of toxic agents (ethidium bromide, daunomycin, cholate, acid bile), but not to some antibiotics (tetracycline kanamicin, chloramphenicol) (Lubelski et al. 2006; Zaidi et al. 2008). Transcriptional *lmrC* and *lmrD* gene expression is under the control of *lmrR* that encodes a LmrR repressor protein that in the absence of toxic compounds it binds to the promoter regions of the *lmrCD* genes to repress their transcription (Agustiandari et al. 2008). In the presence of toxic compounds, LmrR is released from the promoter regions to induce the transcription of the *lmrCD* genes (Agustiandari et al. 2008). In this study, molecular mechanism underlie the constitutive *lmrD* mRNA overexpression without or with ethidium bromide may have resulted from different modes of binding of LmrR to *lmrR* and *lmrCD* control regions resulting in the generation of different transcripts that encode different structural genes either with or without the *lmrR* transcriptional regulator gene (Agustiandari et al. 2011; Takeuchi et al. 2014). Another presumable mechanism may include defective LmrR that is unable to bind the promoter/operator region of the lmrCD to accomplish their repression (Lubelski et al. 2006).

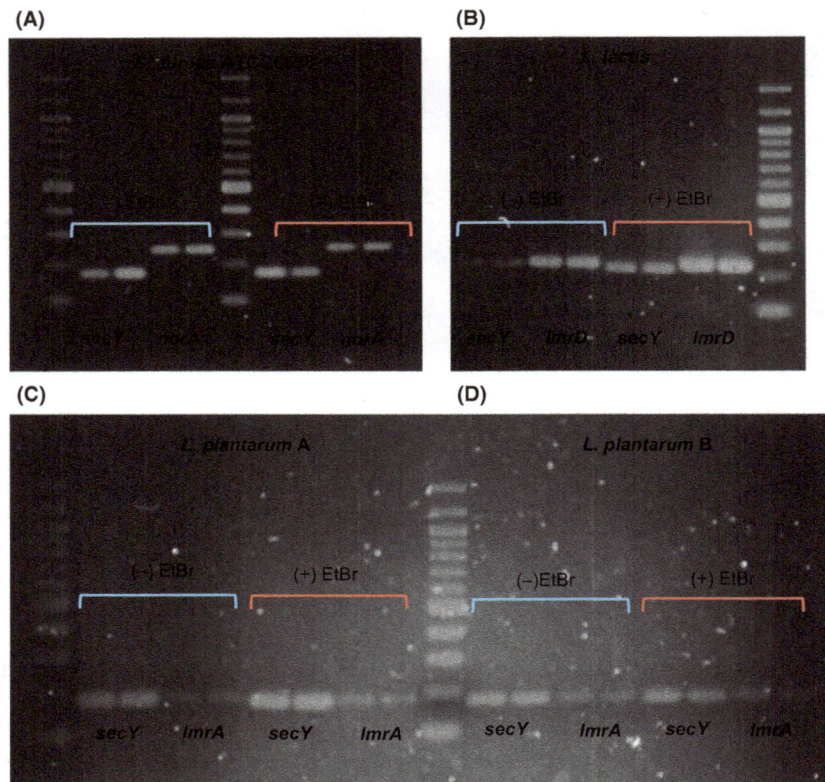

Figure 2. Agarose gel analysis of RT-PCR products: (A) *Staphylococcus aureus* ATCC6538 control strain; (B) *Lactococcus lactis*; *Lactobacillus plantarum* strain A (C) and strain B (D).

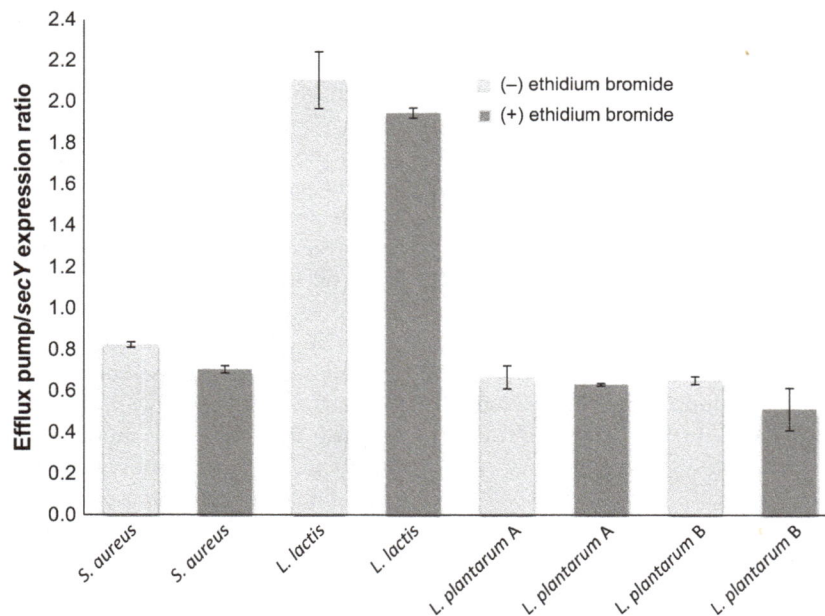

Figure 3. mRNA expression ratios of efflux pumps from each bacterial strain relative to *sec Y* mRNA expression with *Staphylococcus aureus* ATCC 6538 as control strain. Data are depicted as mean values plus standard deviation (SD).

In this study, the mRNA expression of *lmrA* was analyzed, an ATP-dependent multidrug resistance transporter of the ABC family conferring resistance on beer spoilage *Lactobacillus* strains to toxic hop compounds (Ulmer et al. 2000; Sakamoto et al. 2001). In *L. plantarum* strains, resistance to antibiotics and to ethidium bromide as well the high ethidium bromide efflux phenotype were accompanied by a lower expression of *lmrA* mRNA. Moreover, *lmrA* mRNA expression was independent from the presence of ethidium bromide in spite of this is a substrate for LmrA protein transporter as described in *L. plantarum* for a highly homolog gene *horA* (Ulmer et al. 2000, 2002; Sakamoto et al. 2001). These findings indicate that in the analyzed *L. plantarum* strains, LmrA-independent ethidium bromide efflux phenotype may have resulted by alternative mechanisms including an altered fluidity and composition of the cytoplasmic membrane as well-modified cell wall composition of lipoteichoic acids as described in *L. brevis* (Behr et al. 2006).

In comparison with some lactic acid strains, *S. aureus* control strains displayed higher sensibility to ethidium bromide and weaker ethidium bromide efflux phenotype. Analysis of sensibility to ethidium bromide in *S. aureus* ATCC 25293 has been also described previously (Couto et al. 2008). In this assay, phenotypical features of resistance to ethidium bromide found in *S. aureus* strains were unrelated with the *norA* mRNA expression, which was unaltered without or with ethidium bromide. As it is known, *norA* provides resistance to fluoroquinolones and even, *norA* overexpression is associated with multidrug resistance of mutants obtained by the exposition of parental strains with increasing concentrations of biocides and toxic dyes (Bhateja et al. 2006; Huet et al. 2008). The findings of this assay may reflect the wide array of *norA*-independent mechanisms that in the case of *S. aureus* have been described more than 10 efflux pumps systems (Andersen et al. 2015). Importantly, although these pumps show different substrate specificity, most of them are capable of extruding compounds of different chemical classes, thus providing the cell with the means to develop a MDR phenotype or to survive in a hostile environment (Poole 2007).

In conclusion, the findings showed that the antibiotic resistance and the efflux phenotype were independent from the overexpression of *lmrD* mRNA in *L. lactis* or the lower expression of *lmrA* or *norA* in *L. plantarum* and *S. aureus*, respectively. Substantive role on antibiotic resistance by the efflux-associated genes were not confirmed, however, this study provides the experimental findings that other unknown drug resistance mechanisms may underlie the antibiotic resistance and the ethidium bromide efflux phenotype in *L. lactis* and *L. plantarum* isolates from pozol.

In vitro assays have evidenced the substantive transfer of drug-resistant genes of lactic acid bacteria to other lactic acid bacteria and pathogenic strains like *Listeria* spp (Toomey 2009, Toomey et al. 2010). Thus, further studies focused on the phenotypic and molecular characterization of mechanism of antimicrobial resistance of *L. lactic* and *L. plantarum* may impact the safety consumption of pozol in most cases of handcraft production and in the control of the potential dissemination of antimicrobial resistance factors in foodborne pathogens.

Conflict of Interest

None declared.

References

Agustiandari, H., J. Lubelski, B. van den Berg, H. van Saparoea, O. P. Kuipers, and A. J. Driessen. 2008. LmrR is a transcriptional repressor of expression of the multidrug ABC transporter LmrCD in Lactococcus lactis. J. Bacteriol. 190:759–763.

Agustiandari, H., E. Peeters, J. G. de Wit, D. Charlier, and A. J. Driessen. 2011. LmrR-mediated gene regulation of multidrug resistance in *Lactococcus lactis*. Microbiology 157(Pt 5):1519–1530. doi:10.1099/mic.0.048025-0.

Andersen, J. L., G. X. He, P. Kakarla, R. KC, S. Kumar, W. S. Lakra, et al. 2015. Multidrug efflux pumps from Enterobacteriaceae, *Vibrio cholerae* and *Staphylococcus aureus* bacterial food pathogens. Int. J. Environ. Res. Public Health 12:1487–1547. doi: 10.3390/ijerph120201487.

Behr, J., M. G. Gänzle, and R. F. Vogel. 2006. Characterization of a highly hop-resistant *Lactobacillus brevis* strain lacking hop transport. Appl. Environ. Microbiol. 72:6483–6492.

Bhateja, P., K. Purnapatre, S. Dube, T. Fatma, and A. Rattan. 2006. Characterisation of laboratory-generated vancomycin intermediate resistant *Staphylococcus aureus* strains. Int. J. Antimicrob. Agents 27:201–211.

CLSI. 2006. Methods for dilution antimicrobial susceptibility tests for bacteria that grow aerobically approved standard M7-A7, 7th ed. Clinical and Laboratory Standards Institute, Wayne, PA.

Couto, I., S. S. Costa, M. Viveiros, M. Martins, and L. Amaral. 2008. Efflux-mediated response of *Staphylococcus aureus* exposed to ethidium bromide. J. Antimicrob. Chemother. 62:504–513. doi:10.1093/jac/dkn217.

Huet, A. A., J. L. Raygada, K. Mendiratta, S. M. Seo, and G. W. Kaatz. 2008. Multidrug efflux pump overexpression in *Staphylococcus aureus* after single and multiple in vitro exposures to biocides and dyes. Microbiology 154(Pt 10):3144–3153. doi:10.1099/mic.0.2008/021188-0.

Kaatz, G., S. M. Seo, L. O'brien, M. Wahiduzzaman, and T. J. Foster. 2000. Evidence for the Existence of a Multidrug

Efflux Transporter Distinct from NorA in *Staphylococcus aureus*. Antimicrob. Agents Chemother. 44:1404–1406.

Lubelski, J., A. de Jong, R. van Merkerk, H. Agustiandari, O. P. Kuipers, J. Kok, et al. 2006. LmrCD is a major multidrug resistance transporter in *Lactococcus lactis*. Mol. Microbiol. 61:771–781.

Mazurkiewicz, P., K. Sakamoto, G. J. Poelarends, and W. N. Konings. 2005. Multidrug transporters in lactic acid bacteria. Mini Rev. Med. Chem. 5:173–181.

Patel, D., C. Kosmidis, S. M. Seo, and G. W. Kaatz. 2010. Ethidium bromide MIC screening for enhanced efflux pump gene expression or efflux activity in *Staphylococcus aureus*. Antimicrob. Agents Chemother. 54:5070–5073. doi:10.1128/AAC.01058-10.

Piddock, L. J. V. 2006. Clinically relevant chromosomally encoded multidrug resistance efflux pumps in bacteria. Clin. Microbiol. Rev. 19:382–402.

Poelarends, G. J., P. Mazurkiewicz, and W. N. Konings. 2002. Multidrug transporters and antibiotic resistance in *Lactococcus lactis*. Biochim. Biophys. Acta 10:1555.

Poole, K. 2007. Efflux pumps as antimicrobial resistance mechanisms. Ann. Med. 39:162–176.

Sainz, T., C. Wacher, J. Espinoza, D. Centurión, A. Navarro, J. Molina, et al. 2001. Survival and characterization of *Escherichia coli* strains in a typical Mexican acid-fermented food. Int. J. Food Microbiol. 71:169–176.

Sakamoto, K., and W. N. Konings. 2003. Beer spoilage bacteria and hop resistance. Int. J. Food Microbiol. 89:105–124.

Sakamoto, K., A. Margolles, H. W. van Veen, and W. N. Konings. 2001. Hop resistance in the beer spoilage bacterium *Lactobacillus brevis* is mediated by the ATP-binding cassette multidrug transporter HorA. J. Bacteriol. 183:5371–5375.

Takeuchi, K., Y. Tokunaga, M. Imai, H. Takahashi, and I. Shimada. 2014. Dynamic multidrug recognition by multidrug transcriptional repressor LmrR. Sci. Rep. 4:6922. doi:10.1038/srep06922.

Toomey, N., A. Monaghan, S. Fanning, and D. J. Bolton. 2009. Assessment of antimicrobial resistance transfer between lactic acid bacteria and potential foodborne pathogens using in vitro methods and mating in a food matrix. Foodborne Path. Dis. 6:925–933. doi:10.1089/fpd.2009.0278.

Toomey, N., D. Bolton, and S. Fanning. 2010. Characterisation and transferability of antibiotic resistance genes from lactic acid bacteria isolated from Irish pork and beef abattoirs. Res. Microbiol. 161:127–135. doi:10.1016/j.resmic.2009.12.010.

Ulloa, M., F. Ramírez, and K. H. Steinkraus. 1996. Mexican pozol. Pp. 252–259 *in* K. H. Steinkraus, ed. Handbook of indigenous fermented foods, 2nd ed. Marcel Dekker, New York.

Ulmer, H. M., M. G. Gänzle, and R. F. Vogel. 2000. Effects of high pressure on survival and metabolic activity of *Lactobacillus plantarum* TMW1.460. Appl. Environ. Microbiol. 66:3966–3973.

Ulmer, H. M., H. Herberhold, S. Fahsel, M. G. Gänzle, R. Winter, and R. F. Vogel. 2002. Effects of pressure-induced membrane phase transitions on inactivation of HorA, an ATP-dependent multidrug resistance transporter *Lactobacillus plantarum*. Appl. Environ. Microbiol. 68:1088–1095.

Wacher, C., A. Cañas, P. E. Cook, E. Barzana, and J. D. Owens. 1993. Sources of microorganisms in pozol, a traditional Mexican fermented maize dough. World J. Microbiol. Biotechnol. 9:226–274.

Zaidi, A. H., P. J. Bakkes, J. Lubelski, H. Agustiandari, O. P. Kuipers, and A. J. Driessen. 2008. The ABC-type multidrug resistance transporter LmrCD is responsible for an extrusion-based mechanism of bile acid resistance in *Lactococcus lactis*. J. Bacteriol. 190:7357–7366. doi:10.1128/JB.00485-08.

17

Screening of microorganisms from Antarctic surface water and cytotoxicity metabolites from Antarctic microorganisms

Lanhong Zheng[1,a], Kangli Yang[1,a], Jia Liu[2,a], Mi Sun[1], Jiancheng Zhu[1], Mei Lv[1], Daole Kang[1], Wei Wang[1], Mengxin Xing[1] & Zhao Li[1]

[1]Key Laboratory for Sustainable Utilization of Marine Fisheries Resources, Ministry of Agriculture, Yellow Sea Fisheries Research Institute, Chinese Academy of Fishery Sciences, Qingdao 266071, China
[2]Medical College, Qingdao University, Qingdao 266021, China

Keywords
Antarctic surface water, cytotoxicity, fermented active products, microorganism screening, MTT method

Correspondence
Lanhong Zheng, Key Laboratory for Sustainable Utilization of Marine Fisheries Resources, Ministry of Agriculture, Yellow Sea Fisheries Research Institute, Chinese Academy of Fishery Sciences, Qingdao 266071, China.

E-mail: zhenglh@ysfri.ac.cn

Funding Information
This work was supported by the Special Scientific Research Funds for Central Non-profit Institutes, Chinese Academy of Fishery Sciences (2014B01YQ01) and the Science and Technology Development Plans of Shandong Province (2014GSF121016), the Special Scientific Research Funds for Central Non-profit Institutes, Yellow Sea Fisheries Research Institutes (No.20603022013017 and 20603022012013). The authors are grateful to all members of the laboratory for their continuous technical advice and helpful discussion.

[a]These authors contribute equally to this work.

Abstract

The Antarctic is a potentially important library of microbial resources and new bioactive substances. In this study, microorganisms were isolated from surface water samples collected from different sites of the Antarctic. 3-(4,5-dimethyl-2-thiazolyl)-2,5-diphenyl-2H-tetrazolium bromide (MTT) assay-based cytotoxicity-tracking method was used to identify Antarctic marine microorganism resources for antitumor lead compounds. The results showed that a total of 129 Antarctic microorganism strains were isolated. Twelve strains showed potent cytotoxic activities, among which a Gram-negative, rod-shaped bacterium, designated as N11-8 was further studied. Phylogenetic analysis based on 16S rRNA gene sequence showed that N11-8 belongs to the genus *Bacillus*. Fermented active products of N11-8 with molecular weights of 1–30 kDa had higher inhibitory effects on different cancaer cells, such as BEL-7402 human hepatocellular carcinoma cells, U251 human glioma cells, RKO human colon carcinoma cells, A549 human lung carcinoma cells, and MCF-7 human breast carcinoma cells. However, they displayed lower cytotoxicity against HFL1 human normal fibroblast lung cells. However, they displayed lower cytotoxicity against HFL1 human normal fibroblast lung cells. Microscopic observations showed that the fermented active products have inhibitory activity on BEL-7402 cells similar to that of mitomycin C. Further studies indicated that the fermented active products have high pH and high thermal stability. In conclusion, most strains isolated in this study may be developed as promising sources for the discovery of antitumor bioactive substances. The fermented active products of Antarctic marine *Bacillus* sp. N11- 8 are expected to be applied in the prevention and treatment of cancer.

Introduction

Because of the unique geographic location and harsh natural environment — including low temperature, drought, and strong radiation — the polar region has not been polluted by humans and retains the original state. The Arctic and Antarctic have become not only an important place for recording the historical evolution process of earth system, but also a place of new microbial species resource (Anderson 1995; Gosink et al. 1998). The special habitat leads to unique molecular mechanisms and physiological and biochemical characteristics in genetic composition, properties, and metabolic regulation of polar microorganism. Microbiologists have conducted a lot of research work in the polar region, which has proved that the polar microorganisms in these special environments has a broad prospect in the development of novel drugs, new enzyme models, health food and other basic research and applications (Humphry et al. 2001; Shivaji et al. 2005). Polar region, therefore, becomes a potential, important microbial resources database, and a new bioactive substances and microbial potential provenance (Alan et al. 2000; Van Trappen et al. 2004; Wu et al. 2013).

In recent years, a number of polar microorganisms and the related metabolites were found to possess antitumor activity. Because of the increasing concern on marine and extreme environmental microbial resources, more and more compounds with antitumor activity are being identified from polar microorganisms (Zeng et al. 2008). Antibacterial activity of the Antarctic bacterium *Janthinobacterium* sp. SMN 33.6 was studied against multiresistant Gram-negative bacteria (Asencioa et al. 2013). 259 strains were isolated from the samples collected from the Antarctic soil and the South Ocean water, 11% of which showed strong antitumor activity. The results showed that the Antarctic microorganisms have a good potential for bioactive metabolite research (Zhu et al. 2006; Ding et al. 2014). Even so, the number of marine antineoplastic drugs is still not comparative to antitumor drugs of terrestrial sources. In this study, a large number of microorganisms were isolated from surface water samples obtained from different sites in the Antarctic. A MTT assay-based antitumor activity-tracking method was used to identify new Antarctic marine microorganism resources for antitumor lead compounds.

Materials and Methods

Sample collection and isolation of Antarctic marine microorganisms

No specific permissions were required for the collection of sample from the location. The field studies did not involve endangered or protected species. The Antarctic surface seawater samples were collected from several different stations in Antarctic in March 2012 (Table 1). The surface water was filtered using 0.22 μm membranes, and then the membranes were preserved in sterile vials containing 20% glycerol with Nansen bottles at $-80°C$ before use.

The microorganism strains were isolated by plating 200 μL of different dilutions of the samples preserved in the sterile vials containing 20% glycerol (from 10^{-1} to 10^{-7}) on 2216E seawater-based medium and beef extract-peptone medium agar plates in triplicates, respectively. The inoculated plates were incubated at 4, 15, and 25°C, respectively, for 3–6 days. Colonies arising on all solid plates were selected based on their physiological features and morphological characteristic — including rate of growth, shape, size pigmentation, and margin (Button et al. 1993).

Inoculum, fermentation and screening of bioactive product preparation

The basal medium used for the production of antitumor products consisted of 1.0% tryptone, 0.3% beef extract, and 0.5% NaCl (pH 6.5–7.0). Incubation was carried out in a rotary shaker at 25°C and 180–200 rpm. A loopful of cells from a slant was transferred to 3 mL of the above mentioned sterile medium in a 12 mL test tube and incubated at 25°C and 200 rpm for 24 h. This was used as the inoculum. Fermentation was carried out in 250-mL Erlenmeyer flasks, each containing 50 mL of sterile production medium. The medium was inoculated with 4–10% (v/v) of the old culture. The inoculated flasks were kept on a rotary shaker (Thermo, Waltham, Massachusetts, America) at 25°C and 200 rpm for 48 h (Button et al. 1993). The fermented broth was collected and centrifuged at 11,180 g for 15 min, using a centrifuge (Hitachi, Eastportcity, Honshu, Japan). After centrifugation, the supernatant and the microorganism precipitation were separately collected. Microorganism

Table 1. Antarctic surface seawater samples collected from several different stations in Antarctic.

Station	Latitude	Longitude	Quantity of sea surface water was filtered (mL)
5#	6002	6002	200
6#	6002	5902	200
11#	6001	5402	200
16#	6100	6458	200
18#	6100	6257	200
26#	6100	5456	200
35#	6201	6301	400
39#	6303	6659	400
49#	6234	6158	200
51#	6100	6657	400

Provide the source of bacteria.

precipitation was resuspended in 20 mL of 20 mmol/L PBS solution, and broken by ultrasonication for 10 min. The supernatants were collected after centrifugation. Extracellular and intracellular fermentation products of each strain were filtered using 0.22 μm membrane, packed in tubules, and frozen at −20°C, prepared as the sample to be tested for its antitumor activity.

Cell lines and culture conditions

BEL-7402 human hepatocellular carcinoma cells, RKO human colon carcinoma cells, A549 human lung carcinoma cells, U251 human glioma cells, MCF-7 human breast carcinoma cells, and HFL1 human normal fibroblast lung cells were provided by the Cell Bank of Chinese Academy of Sciences (Shanghai, China). BEL-7402, RKO, A549, U251, and MCF-7 cells were cultured in DMEM (Hyclone, South Logan, Utah, America) supplemented with 10% heat-inactivated fetal bovine serum (Gibco, California, America), 100 U/mL of penicillin, and 100 mg/mL streptomycin (Sigma, St. Louis, Missouri, America). HFL1 cells were cultured in F12K (Hyclone) supplemented with 10% heat-inactivated fetal bovine serum, 100 U/mL of penicillin, and 100 mg/mL streptomycin. Cells were grown in a carbon dioxide incubator (Thermo Forma, Waltham, Massachusetts, America) under a humidified atmosphere containing 5% CO_2 at 37°C (Fogh et al. 1977).

The screening of antitumor fermentation products from isolated strains

The inhibitory effects of fermentation products from the isolated strains on the viability of BEL-7402, RKO, A549, U251, MCF-7 cells, and normal HFL1 cells were evaluated using the MTT assay (Mosmann 1983). Some improvement was made in our study. In brief, the above cancer cells (2.2×10^4 CFU/mL) in 180 μL of DMEM culture media were seeded into each well on 96-well microplates and cultured for 24 h. Then, 20 μL SBP with certain concentrations of the fermentation products were added to the media. After incubation for 48 h, MTT solution (20 μL, 0.5 mg/mL) was added to each well and the cells were cultured for another 4 h at 37°C. The media were removed and 150 μL DMSO was added to dissolve the formazan crystals. The OD_{570} was measured with an Infinite M200 PRO microplate reader (TECAN Group Ltd, Mannerdorf, Switzerland) with subtraction of the background absorbance. All experiments were performed in triplicate. The cytotoxicity of the fermentation products from the isolated strains was expressed as inhibition ratio of cell viability (IR). IR = [(A_{570} value of the control-A_{570} value of the experimental samples)/A_{570} value of the control] × 100%.

Morphological, physiological and biochemical characteristics of strain N11-8

Cell morphology was determined by SEM (scanning electron microscopy). The cells were grown onto poly-L-lysine-coated coverslips in plate for 24 h to allow firm attachment. The cells were fixed with glutaraldehyde. After overnight fixation at 4°C, the coverslips were dehydrated using ethanol and dried in a critical point dryer. Cells on coverslips were coated with gold and analyzed, using S-3400N SEM (Hitachi).

Standard protocols (Tindall et al. 2007) were used to assess oxidase activities. Activities of constitutive enzymes, substrate oxidation, carbon source utilization, and other physiological properties were determined using the API 20E and API 20NE, the API ZYM strips (bioMe'rieux, Lyon, France), and the Gram-negative MicroPlates (Biolog, Hayward, California, America), according to the manufacturer's instructions.

Phylogenetic analysis

Chromosomal DNA of the strain N11-8 was extracted (Ausubel et al. 1995) and the 16S rRNA gene was amplified by PCR, (Biometra, Göttingen, Germany), using the universal primers 27f 5′-AGAGTTTGATCCTGGTCAG-3′ and 1492r 5′-CGGCTACCTTGTTACGAC-3′ (Weisburg et al. 1991). Purified PCR products were ligated into pMD 18-T (TaKaRa, Minami kusatsu, Japan), according to the manufacturer's instructions. Sequencing reactions were carried out using ABI BigDye 3.1 Sequencing kits (Applied BioSystems, San Francisco, California, America) and an automated DNA sequencer (model ABI3730; Applied BioSystems). The near-complete 16S rRNA gene sequence of the strain N11-8 was submitted to GenBank/EMBL to search for the similar sequences using the BLAST algorithm. The identification of phylogenetic neighbors and the calculation of pairwise 16S rRNA gene sequence similarities were achieved using the EzTaxon server (http://www.ez-taxon.org/) (Chun et al. 2007). The sequences were aligned using CLUSTAL X1.8 (Thompson et al. 1997). Phylogenetic trees were constructed using the neighbour-joining (Fig. 2) methods implemented in the program MEGA version 6 (Tamura et al. 2013). In each case, bootstrap values were calculated based on 1000 replicates.

Preparation of the active substance

For the production of secondary metabolites, the Antarctic marine *Bacillus* sp. N11-8 was cultured as above. Fermentation products were centrifuged at 10,000 g for 15 min at 4°C, and the supernatant was then collected. The supernatant was ultrafiltered first using a 50 kDa molecular weight cutoff membrane, then with a 30 kDa molecular weight cutoff membrane, and finally with a 1 kDa molecular weight cutoff membrane. The filtrate with different molecular weight was determined for their antitumor activity. The fraction with highest cytotoxicity was collected for further experiments.

Analysis of inhibitory effect of the fermentation products of Antarctic microorganism N11-8 on the proliferation of various tumor cells

The inhibitory effects of the fermentation products of the isolated strains on the viability of BEL-7402, RKO, A549, U251, MCF-7 cells, and normal HFL1 cells were evaluated by MTT assay as described above. The IC_{50} value was defined as the concentration causing a 50% reduction in cell viability.

Morphological changes in BEL-7402 cells treated with the fermentation products of Antarctic microorganism N11-8

The BEL-7402 cells (2.2×10^4 CFU/mL) in 180 μL of DMEM culture media were seeded into each well on 96-well microplates and cultured in a constant temperature incubator at 37°C and 5% CO_2 for 24 h. After incubation with the fermentation products for 48 h, the morphology and number of tumor cells were observed and counted, respectively, under an inverted phase contrast microscope (Nikon, Tokyo, Japan).

Acid stability analysis of the fermentation products of Antarctic microorganism N11-8

The pH of 150 μg/mL fermentation products of Antarctic microorganism N11-8 (molecular weight 1–30 kDa) was adjusted to 2, 4, 6, 8, 10, or 12, placed at room temperature for 24 h, and then adjusted backed to the original pH value. The cytotoxicity of the fermentation products on BEL-7402 tumor cells was determined by MTT method. Each experiment was repeated for three times.

Thermal stability analysis of the fermentation products of Antarctic microorganism N11-8

One hundred and fifty μg/mL fermentation product of Antarctic microorganism N11-8 (molecular weight: 1–30 kDa) were placed at 40, 60, 80, 100, or 121°C for 30 min. After rapid cooling, the cytotoxicity of the fermentation products on BEL-7402 tumor cells was determined by MTT method. Each experiment was repeated for three times.

Statistical analysis

All of the tests were conducted in triplicate, and the experimental data were expressed as the mean ± SD. The statistical significance of the values between the control and the treated groups was determined by a paired t-test; $P < 0.05$ was considered to be statistically significant.

Results and Discussion

Isolation of Antarctic microorganisms

Isolation and cultivation of a new marine microorganism may be a shortcut to discover novel natural products. A variety of pretreatment methods, for example physical and chemical enrichment, are employed to facilitate the isolation of specific marine microorganisms, especially less abundant bacteria [13, 14]. Using 2216E seawater-based medium and beef extract-peptone culture medium, a total of 129 microorganism strains were isolated from the samples collected from 10 different stations in Antarctic (Table 2). In the process of acquisition of Antarctic surface seawater samples, the microorganisms were retained in the membrane using a seawater filter. The enrichment process of microorganisms facilitated the storage, carrying, and is convenient for microorganism screening and follow-up study.

The screening of metabolites of Antarctic microorganisms with antitumor activity by MTT

The fermentation products of 129 strains of Antarctic microorganisms were screened for antitumor activity using MTT method. The inhibition rates of two Antarctic microorganism strains were more than 50% on BEL-7402 human hepatocellular carcinoma cells; the inhibition rates

Table 2. Separation of microorganisms from the surface seawater samples from different stations in Antarctic.

Station	Latitude	Longitude	Microbial quantity (Strain)
5#	6002	6002	11
6#	6002	5902	13
11#	6001	5402	15
16#	6100	6458	13
18#	6100	6257	12
26#	6100	5456	12
35#	6201	6301	10
39#	6303	6659	14
49#	6234	6158	12
51#	6100	6657	17

Provide the source of bacteria.

Table 3. Antitumor activity (means ± SD) of the fermented products of Antarctic microorganisms.

Microbial strain number	Inhibition rate on BEL-7402 cell (%)
N5-6	31.6 ± 5.12[a]
N6-2	43.2 ± 5.96[b]
N11-8	68.1 ± 7.42[c]
N16	65.5 ± 7.12[c]
N18-2	32.3 ± 4.85[a]
N26-7	34.8 ± 4.13[a]
N49-1	35.7 ± 5.02[a]

Data within the same column with different superscripts are significantly different ($P < 0.05$).

of five strains were more than 30% (Table 3); The fermentation products of strain N11-8 (No. 8) that was isolated from station 11# in the Antarctic showed 68.1% of IR on BEL-7402 cells. According to the results, there are microbial germplasm resources for producing antitumor bioactive substances from the surface waters in Antarctic.

Phenotypic and phylogenetic characterization of Antarctic microorganism N11-8

We isolated a new strain, N11-8, from the surface water of Antarctic. The near-complete 16S rRNA gene sequence (1434 nt) of strain N11-8 was uploaded to GeneBank (GenBank accession number JX974351). Phylogenetic analysis based on the 16S rRNA sequence indicated that strain N11-8 belonged to the genus *Bacillus*, with the highest sequence similarities to *Bacillus licheniformis* ATCC14580T (98.67%), *Bacillus aerius* 24kT (98.6%), *Bacillus siamensis* KCTC13613T (97.7%), *Bacillus amyloliquefaciens* subsp. *amylol* DSM7T (97.63%), and *Bacillus*

amyloliquefaciens subsp. *palnta* FZB42T (97.49%). The neighbour-joining phylogenetic tree further confirmed that the strain N11-8 was phylogenetically related to the genus *Bacillus* (Fig. 1).

In addition to the features that define the strain, the following characteristics are observed. The morphological characters of the cells were clearly visualized by SEM. Cells are ovoid or irregular short rods, approximately 0.5–2.5 mm long and 0.5–0.8 mm wide (Fig. 2). The colonies are uniformly round, 0.8–2.1 mm in diameter, and have smooth, glossy, and mucoid appearance, with texture like cream, white, and translucence. Oxidase is negative.

The temperature range for growth was 20–30°C. The pH range for growth was 7.0–8.0. According to API 20E and API 20NE: Urease and Voges–Proskauer reactions were positive; They were negative for β-galactosidase, arginine dihydrolase, lysine decarboxylase, ornithine decarboxylase, tryptophan deaminase, and gelatinase; citrate was not utilized; neither H_2S or indole was produced; acid was produced from D-glucose, melibiose, L-arabinose, and sucrose, but not from D-mannitol, inositol, D-sorbitol, L-rhamnose, and amygdalin. The fermentation media contained D-glucose, sucrose, melibiose and L-arabinose; it is negative for assimilation experiments: L-arabinose, N-acetylglucosamine, potassium gluconate, capric acid, adipic acid, malic acid, trisodium citrate, D-mannitol, D-mannose, maltose, and phenylacetic acid. According to API ZYM, the isolated strain was positive for alkaline phosphatase, lipase (C14), leucine arylamidase, valine arylamidase, cystine arylamidase, trypsin, chymotrypsin, acid phosphatase, naphthol-AS-BI-phosphohydrolase, α-galactosidase, β-glucuronidase, α-glucosidase, β-fucosidase, and N-acetylglucosaminidase; and negative for esterase (C4), esterase lipase (C8), β-glucosidase and α-mannosidase.

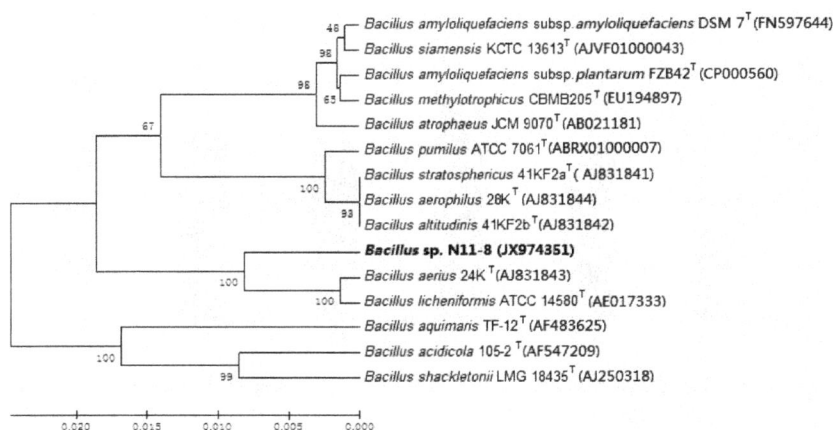

Figure 1. Phylogenetic dendrogram of *Bacillus* sp. N11-8 and its related species based on 16S rRNA gene sequence similarities. The tree was constructed using the neighbour-joining method implemented in the program MEGA version 6. Bar, 0.01 nt substitutions per site.

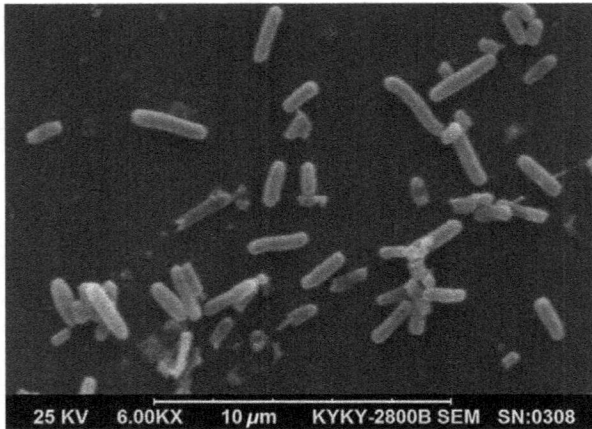

Figure 2. Scanning electron micrographs of Antarctic marine *Bacillus* N11-8. Scale Bar = 10 *μm*.

Fermentation and preparation of the fermentation products of Antarctic marine *Bacillus* sp. N11-8

About 30 L of production media was fermented. The products of different molecular weight <1 kDa, 1–30 kDa, 30–50 kDa and >50 kDa were separated using ultrafiltration. The fermentation products of molecular weight of 1–30 kDa showed

highest antitumor activity on BEL-7402 cells as determined by MTT assay and was chosen for further studies.

The cytotoxicity of the fermentation products of Antarctic marine *Bacillus* sp. N11-8 against tumor cells

Marine microorganisms are a major source of natural products (Waters et al. 2010; Xiong et al. 2013). Anticancer ε-poly-l-lysine (ε-PL) was produced by a marine Bacillus subtilis sp. isolated from sea water in Alexandria (El-Sersy et al. 2012). Sungsanpin, a new 15-amino-acid peptide, was discovered from a Streptomyces species isolated from deep-sea sediment collected off Jeju Island, Korea. Sungsanpin displayed inhibitory activity in a cell invasion assay with the human lung cancer cell line A549 (Um et al. 2013). The cytotoxicity of the fermentation products of N11-8 (molecular weight 1–30 kDa) was determined by MTT assay. As shown in Figure 4, the fermentation products of N11-8 reduced the viability of five selected cancer cell lines in a dose-dependent manner. The IC_{50} values of BEL 7402, RKO, A549, U251 and MCF-7 cells were detected. For instances, the IC_{50} value is 30.15 ± 1.51 μg/mL on MCF-7 cells. However, relatively low cytotoxicity was detected against HFL1 cells (Fig. 3). The MTT results suggested that the fermentation products of Antarctic marine Bacillus sp. N11-8 are expected to be used

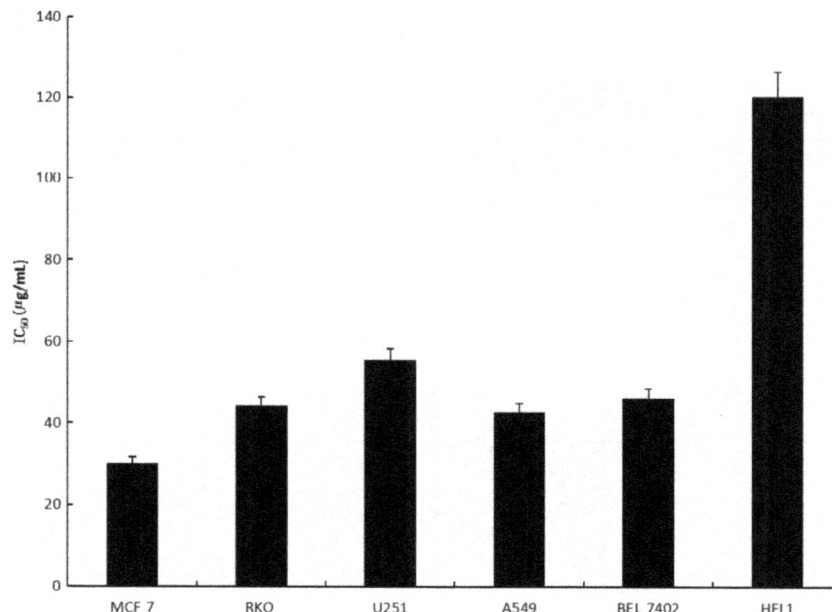

Figure 3. Cytotoxicity of the fermentation products of Antarctic marine *Bacillus* sp. N11-8 on tumor cells. BEL-7402 Human hepatocellular carcinoma cells, RKO human colon carcinoma cells, A549 human lung carcinoma cells, U251 human glioma cells, MCF-7 human breast carcinoma cells, and HFL1 human normal fibroblast lung cells were treated with certain concentrations of the fermentation products of N11-8 for 48 h. The cell inhibitory rate was determined using the MTT assay as described in Materials and methods. Data were presented as means ± SD of three independent experiments; *P < 0.05, compared with HFL1 human normal fibroblast lung cells.

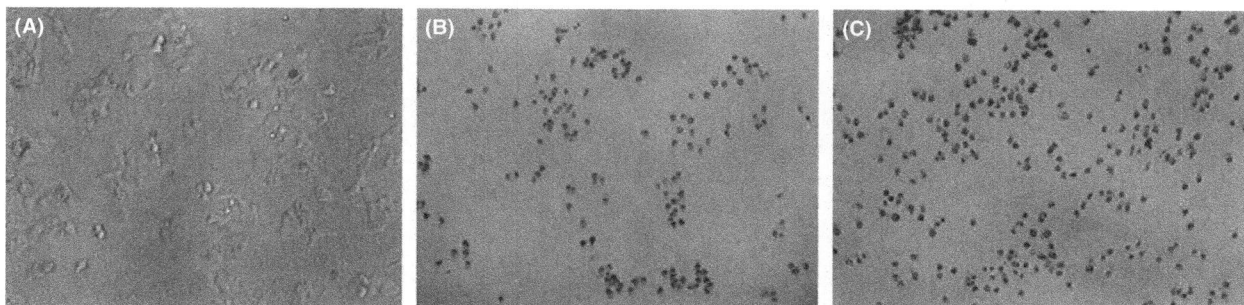

Figure 4. Morphological changes in BEL-7402 cells induced by fermented active products of the Antarctic marine *Bacillus* sp. N11-8 (A) Negative control; (B) Fermented active products of the Antarctic marine *Bacillus* sp. N11-8; (C) Positive control.

for the treatment of liver cancer, glioma, lung cancer, and breast cancer. Additionally, the inhibition effect of the fermentation products of N11-8 on cell viability exhibited some specificity to tumor cells.

Phase contrast microscopy analysis of the effect of the fermentation products of Antarctic marine *Bacillus* sp. N11-8 on the morphology of hepatoma cells

MTT results showed that the fermentation products of Antarctic marine Bacillus sp. N11-8 have an inhibitory activity on BEL-7402 cells in vitro. Therefore, the morphological changes of tumor cells caused by the fermentation products were observed under phase contrast microscope. In comparison with control groups (Fig. 4A),

after the treatment with the fermentation products, cells became round, and then shrank, and finally ruptured (Fig. 4B); in the positive control group (50 μg/mL mitomycin C), cells displayed similar changes (Fig. 4C). These results showed that the fermentation products of Antarctic marine Bacillus sp. N11-8 can change the morphology of BEL-7402 cells, similar to mitomycin C.

The pH stability of the fermentation products of Antarctic marine *Bacillus* sp. N11-8

The pH stability of the fermentation products of Antarctic marine *Bacillus* sp. N11-8 is shown in Figure 5. The antitumor activity of the fermentation products of Antarctic marine *Bacillus* sp. N11-8 changed little between pH 6 and

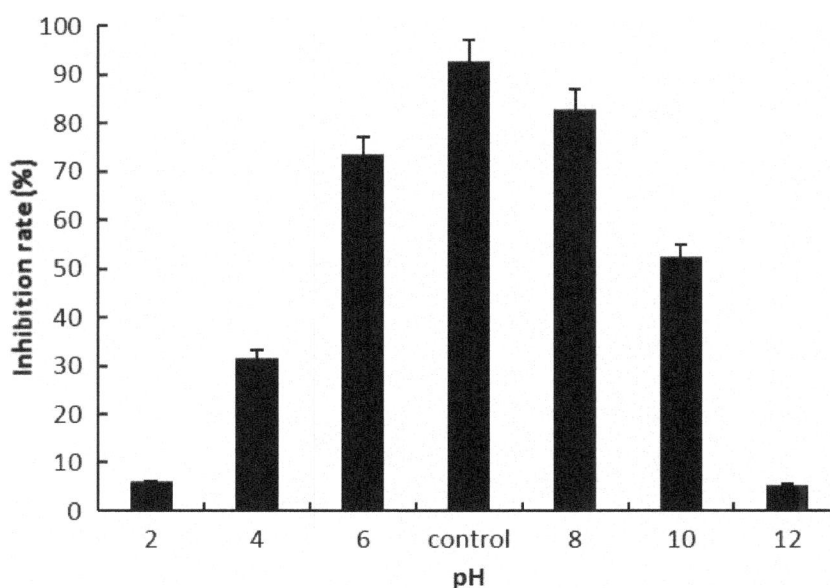

Figure 5. PH stability of the fermented active products of Antarctic marine *Bacillus* sp. N11-8. Data were presented as means ± SD of three independent experiments; *$P < 0.05$, compared with the controls.

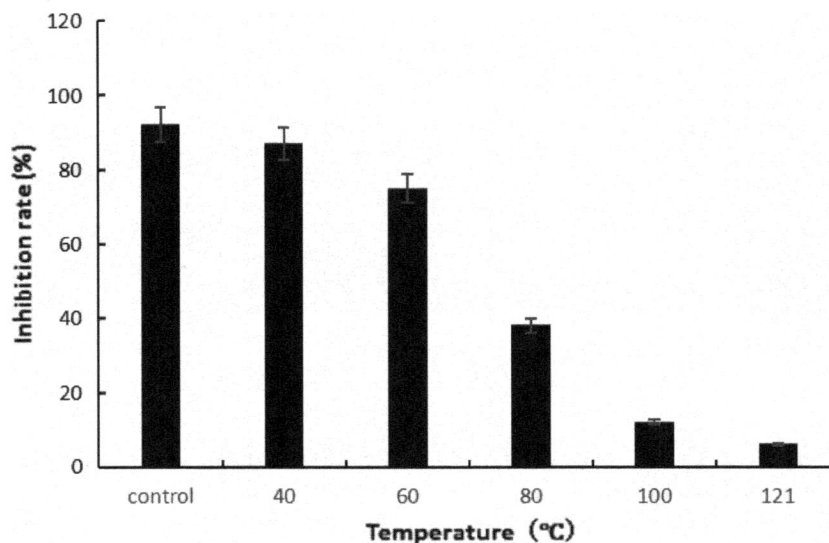

Figure 6. Thermal stability of the fermented active products of the Antarctic marine *Bacillus* sp. N11-8. Data were presented as means ± SD of three independent experiments; *P < 0.05, compared with the controls.

pH 9 (no significant difference). When pH is lower than 6 or higher than 9, the cytotoxicity was Gradually decreased. The active component was stable in weak acidic, neutral and alkaline conditions, but was sensitive to strong acidic and alkaline conditions. Therefore, purification should be kept in a neutral, weak acidic, or weak basic condition.

The thermal stability of the fermentation products of Antarctic marine *Bacillus* sp. N11-8

The fermentation products of Antarctic marine *Bacillus* sp. N11-8 has good thermal stability, as shown in Figure 6. Compared with the controls, 60°C had no effect on the antitumor activity of the fermentation products. But along with the increase in temperature, its antitumor activity decreased gradually, which may be due to the change in the structure of active substances in the fermentation products.

Conclusions

Most Antarctic marine strains obtained in this study may be developed as promising sources for discovery of antitumor bioactive substances. Fermented active products from the *Bacillus* sp. N11-8 is expected to be used in tumor prevention and treatment, which has important research and application value.

Acknowledgments

This work was supported by the Special Scientific Research Funds for Central Non-profit Institutes, Chinese Academy of Fishery Sciences (2014B01YQ01) and the Science and Technology Development Plans of Shandong Province (2014GSF121016), the Special Scientific Research Funds for Central Non-profit Institutes, Yellow Sea Fisheries Research Institutes (No. 20603022013017 and 20603022012013). The authors are grateful to all members of the laboratory for their continuous technical advice and helpful discussion.

Conflict of Interest

The authors declare no conflict of interest.

References

Alan, B., W. Alan, and G. Michael. 2000. Search and discovery strategies for biotechnology: the paradigm Shift. Microbiol. Mol. Biol. Rev. 64:573–606.

Anderson, I. 1995. New frontier in drug research. World Press Rev. 42:37.

Asencioa, G., P. Lavina, K. Alegríab, M. Domínguezb, H. Bellob, G. González-Rochab, et al. 2013. Antibacterial activity of the Antarctic bacterium Janthinobacterium sp. SMN 33.6 against multi-resistant Gram-negative bacteria. Electron. J. Biotechnol. 17:1–5.

Ausubel, F. M., R. Brent, R. E. Kingston, D. D. Moore, J. G. Seidman, J. A. Smith, et al. 1995. Short protocols in molecular biology. Jone Wiley & Sons, New York.

Button, D. K., F. Schut, P. Quang, R. Martin, and B. R. Robertson. 1993. Viability and isolation of marine bacteria by dilution culture: theory, procedures, and initial results. Appl. Environ. Microbiol. 59:881–891.

Chun, J., J. H. Lee, Y. Jung, M. Kim, S. Kim, B. K. Kim, et al. 2007. EzTaxon: a web-based tool for the identification of prokaryotes based on 16S ribosomal RNA gene sequences. Int. J. Syst. Evol. Microbiol. 57(Pt 10):2259–2261.

Ding, Z., D. Li, Q. Gu, and T. Zhu. 2014. Research progress of polar microorganisms secondary metabolites and their bioactivities. Chin. J. Antibiot. 9:6–13.

El-Sersy, N. A., A. E. Abdelwahab, S. S. Abouelkhiir, D.-M. Abou-Zeid, and S. A. Sabry. 2012. Antibacterial and Anticancer activity of ε -poly-L-lysine (ε -PL) produced by a marine Bacillus subtilis sp. J. Basic Microbiol. 52:513–522.

Fogh, J., J. M. Fogh, and T. Orfeo. 1977. One hundred and twenty-seven cultured human tumor cell lines producing tumors in nude mice. J. Natl Cancer Inst. 59:221–226.

Gosink, J., C. Woese, and J. Staley. 1998. Polaribacter gen. nov., with three new species, *P. irgensii* sp. nov., *P. franzmannii* sp. nov. and *P. filamentus* sp. nov., gas vacuolate polar marine bacteria of the Cytophaga-Flavobacterium-Bacteroides group and reclassification of 'Flectobacillus glomeratus' as Polaribacter glomeratus comb. nov. Int. J. Syst. Evol. Microbiol. 48:223–235.

Humphry, D., A. George, G. Black, and S. Cummings. 2001. Flavobacterium frigidarium sp nov., an aerobic, psychrophilic, xylanolyticand laminarinolytic bacterium from Antarctica. Int. J. Syst. Evol. Microbiol. 51:1235–1243.

Mosmann, T. 1983. Rapid colorimetric assay for cellular growth and survival: application to proliferation and cytotoxicity assays. J. Immunol. Methods 65:55–63.

Shivaji, S., G. Reddy, K. Suresh, P. Gupta, S. Chintalapati, P. Schumann, et al. 2005. Psychrobacter vallis sp. nov. and Psychrobacter aquaticus sp. nov., from Antarctica. Int. J. Syst. Evol. Microbiol. 55:757–762.

Tamura, K., G. Stecher, D. Peterson, A. Filipski, and S. Kumar. 2013. MEGA6: molecular evolutionary genetics analysis version 6.0. Mol. Biol. Evol. 30:2725–2729.

Thompson, J. D., T. J. Gibson, F. Plewniak, F. Jeanmougin, and D. G. Higgins. 1997. The CLUSTAL_X windows interface: flexible strategies for multiple sequence alignment aided by quality analysis tools. Nucleic Acids Res. 25:4876–4882.

Tindall, B. J., J. Sikorski, R. A. Smibert, and N. R. Krieg. 2007. Phenotypic characterization and the principles of comparative systematics. Methods Gen. Mol. Microbiol. 3:330–393.

Um, S., Y.-J. Kim, H. Kwon, H. Wen, S.-H. Kim, H. C. Kwon, et al. 2013. Marine microorganisms are a major source for natural products. J. Nat. Prod. 76:873–879.

Van Trappen, S., T. Tan, J. Yang, J. Mergaert, and J. Swings. 2004. Alteromonas stellipolaris sp. nov., a novel, budding, prosthecate bacterium from Antarctic seas, and emended description of the genus Alteromonas. Int. J. Syst. Evol. Microbiol. 54:1157–1163.

Waters, A. L., R. T. Hill, A. R. Place, and M. T. Hamann. 2010. The expanding role of marine microbes in pharmaceutical development. Curr. Opin. Biotechnol. 21:780–786.

Weisburg, W. G., S. M. Barns, D. A. Pelletier, and D. J. Lane. 1991. 16S ribosomal DNA amplification for phylogenetic study. J. Bacteriol. 173:697–703.

Wu, Y., P. Yu, Y. Zhou, L. Xu, C. Wang, M. Wu, et al. 2013. Muricauda antarctica sp. nov., a marine member of the Flavobacteriaceae isolated from Antarctic seawater. Int. J. Syst. Evol. Microbiol. 63:3451–3456.

Xiong, Z. Q., J. F. Wang, Y. Y. Hao, and Y. Wang. 2013. Recent advances in the discovery and development of marine microbial natural products. Marine Drugs 11:700–717.

Zeng, Y., B. Chen, Y. Zou, and T. Zheng. 2008. Polar microorganisms, a potential source for new natural medicines–a review. J. Neurol. Psychopathol. 48:695–700.

Zhu, T., Q. Gu, W. Zhu, Y. Fang, and P. Liu. 2006. Isolation of Antarctic microorganisms and screening of antitumor activity. Chin. J. Mar. Drugs 25:25–28.

18

Explicit numerical solutions of a microbial survival model under nonisothermal conditions

Si Zhu & Guibing Chen*

Center for Excellence in Post-Harvest Technologies, North Carolina A&T State University, The North Carolina Research Campus, 500 Laureate Way, Kannapolis, North Carolina 28081

Keywords

Explicit, Geeraerd model, microbial survival, nonisothermal, numerical solutions, parameter identification

Correspondence

Guibing Chen, Center for Excellence in Post-Harvest Technologies, North Carolina A&T State University, The North Carolina Research Campus, 500 Laureate Way, Kannapolis, NC 28081.

E-mail: gchen@ncat.edu

Funding Information

No funding information provided.

Abstract

Differential equations used to describe the original and modified Geeraerd models were, respectively, simplified into an explicit equation in which the integration of the specific inactivation rate with respect to time was numerically approximated using the Simpson's rule. The explicit numerical solutions were then used to simulate microbial survival curves and fit nonisothermal survival data for identifying model parameters in Microsoft Excel. The results showed that the explicit numerical solutions provided an easy way to accurately simulate microbial survival and estimate model parameters from nonisothermal survival data using the Geeraerd models.

Introduction

Kinetic models for microbial survival during food pasteurization and sterilization processes are essential for design, assessment, optimization, and control of the processes. Microbial survival mostly exhibits nonlinear behavior (Van Boekel 2002; Heldman and Newsome 2003) which has been described by different mathematical models including the Weibull model (Peleg and Cole 1998), the biphasic model (Lee et al. 2001), the log–logistic model (Cole et al. 1993), the modified Gompertz (Linton et al. 1995), and the Geeraerd model (Geeraerd et al. 2000), among others. The selection of a suitable survival model is usually based on how well a model fits experimental survival data.

Parameters involved in a survival model depend on environmental conditions including presence of salt or acid, growth phase of the cells, the products or laboratory media used, and others because the heat resistance of a pathogen is influenced by these factors (Doyle et al. 2001). These parameters must be accurately estimated in order to use the models to evaluate the efficacy of a thermal process. Traditionally, they were estimated from a series of static survival curves. Because a true static condition is impossible to create and there may be more than one combination of parameters that give identical results (Dolan 2003), great efforts have been made to identify them by simultaneously fitting a survival model to each set of dynamic survival data using either software packages or self-written computer programs (Peleg and Normand 2004; Valdramidis et al. 2008; Chen and Campanella 2012).

The Geeraerd model (Geeraerd et al. 2000) is frequently used to describe a type of non log-linear microbial survival curves that show a shoulder and/or a tailing and it encompasses the first-order inactivation when specific parameter values are selected. Under nonisothermal conditions, the model was formulated as a set of two coupled differential equations which could be solved using the Runge–Kutta method. The original Geeraerd model was also modified by incorporating a parameter expressed as a function of the heating rate to depict physiological adaptation induced by mild heat stress (Valdramidis et al. 2007).

The objectives of this study were to derive explicit numerical solutions of the original and modified Geeraerd models and to identify model parameters from nonisothermal microbial survival data using the numerical solutions in Microsoft Excel (Microsoft Corporation, Redmond, WA).

Materials and Methods

Microbial survival models

Under nonisothermal conditions, the Geeraerd model for microbial survival is expressed as the following equations (Geeraerd et al. 2000):

$$\frac{dN(t)}{dt} = -k_{max}(T(t))\frac{1}{1+C_c(t)}(N(t) - N_{res}) \quad (1)$$

$$\frac{dC_c(t)}{dt} = -k_{max}(T(t))C_c(t) \quad (2)$$

where $N(t)$ (CFU mL^{-1}) represents the microbial cell density at time t, C_c (−) is related to the physiological state of cells, N_{res} (CFU mL^{-1}) denotes the residual

population density, and k_{max} (min^{-1}) the specific inactivation rate which is temperature dependent. From equations (1) and (2), equation (3) was obtained (see Appendix S1).

$$\log_{10}\frac{N(t)}{N(0)} = \log_{10}\frac{1 + C_c(0) + \left(e^{\int_0^t k_{max}(T(t))dt} - 1\right)\frac{N_{res}}{N(0)}}{C_c(0) + e^{\int_0^t k_{max}(T(t))dt}} \quad (3)$$

where $N(0)$ (CFU mL^{-1}) is the initial cell density and $N(t)/N(0)$ survival ratio usually denoted by $S(t)$, and $N_{res}/N(0)$ can be expressed as $10^{\log_{10}\frac{N_{res}}{N(0)}}$ when $N_{res} \neq 0$.

It was reported that exposure of *Escherichia coli* K12 to a mild thermal stress induced an increase in heat resistance and therefore a factor k was incorporated into equation (1) to account for this physiological adaption

(Valdramidis et al. 2007).

$$\frac{dN(t)}{dt} = -k_{max}(T(t))\frac{1}{1+C_c(t)}(N(t) - N_{res})k \quad (4)$$

where

$$k = k_1\frac{dT/dt}{k_2 + dT/dt} \quad (5)$$

where k_1 and k_2 are constants and dT/dt is the applied constant heating rate to raise temperature to a target value. From equations (2) and (4), equation (6) was obtained (see Appendix S2). When $k = 1$, there is no detectable adaption of microbial cells to thermal stress and in this case equation (6) becomes equation (3).

$$\log_{10}\frac{N(t)}{N(0)} = \log_{10}\left\{\left(\frac{1 + C_c(0)}{C_c(0) + e^{\int_0^t k_{max}(T(t))dt}}\right)^k \right.$$
$$\left. + \left[1 - \left(\frac{1 + C_c(0)}{C_c(0) + e^{\int_0^t k_{max}(T(t))dt}}\right)^k\right]\frac{N_{res}}{N(0)}\right\} \quad (6)$$

Under nonisothermal conditions, a temperature profile can be represented by a series of discrete temperature points separated by sufficiently small time intervals Δt. In this case, based on the Simpson's rule which approximates the value of a definite integral using quadratic polynomials, the integral term in equation (6) can be calculated by the following equation starting from $\int_0^{t_1=0}k_{max}(T(t))dt = 0$.

$$\int_0^{t_n}k_{max}(T(t))dt = \int_0^{t_{n-1}}k_{max}(T(t))dt + \frac{\Delta t}{6}\left[k_{max}(T_{n-1}) + 4k_{max}\left(\frac{T_{n-1}+T_n}{2}\right) + k_{max}(T_n)\right] \quad (7)$$

where n is the number of temperature points ($n \geq 2$) and T_n the temperature value at time t_n. Incorporating the values of the integral corresponding to a given discrete temperature profile into equation (3) results in the growth curve.

The specific inactivation rate k_{max} can be described by the Bigelow model (Bigelow 1921):

$$k_{max} = \frac{ln10}{AsymD_{ref}}\exp\left[\frac{ln10}{z}(T - T_{ref})\right] \quad (8)$$

where $AsymD_{ref}$ (min^{-1}) denotes the asymptotic decimal reduction time at a reference temperature T_{ref} (°C) and z (°C) the temperature required for a 10-fold change in $AsymD_{ref}$ value. Parameters $AsymD_{ref}$, z, $C_c(0)$, and $\log_{10}N_{res}$ need to be determined. The value of $\log_{10}N(0)$ for each survival test can be experimentally measured at time zero or determined by curve fitting.

Solving the differential equations using a MATLAB solver

The Geeraerd model (Geeraerd et al. 2000) under nonisothermal conditions was solved using the function ode45, a MATLAB's (MathWorks, Natick, MA) standard solver which uses a variable step Runge–Kutta method to solve differential equations numerically. The results obtained were compared with those calculated using equation (3) in Microsoft Excel under the same conditions.

Microbial survival data

Equations (3), (7), and (8) and equations (5)–(8) were fitted to nonisothermal survival data for *E. coli* K12 reported by Valdramidis et al. (2006, 2008), respectively, using the Microsoft Excel Solver. The data which were originally presented in plots were digitized by using the Digitizer Tool of Origin software (OriginLab Corporation, Northampton, MA) following the user guide.

Parameter estimation using the Microsoft Excel solver

The procedure consists of the following four steps. A demonstration of the similar procedure was reported by Zhu and Chen (2015).

(a) Enter guesses for the model parameters to be identified in consecutive cells in Excel.
(b) Generate each survival table: For a survival test, enter discrete temperature profile in two adjacent columns of an Excel spreadsheet, time points being separated by intervals Δt (min) (1/60 min for the present study). Then, calculate $\log_{10}N(t)$ or $\log_{10}S(t)$ at each time point by incorporating the guessed parameter values into equations (3), (7), and (8) or equations (5)–(8).
(c) Look up calculated values of $\log_{10}N(t)$ or $\log_{10}S(t)$ for each survival test in its survival table: For a survival test, enter survival data in two adjacent columns of an Excel spreadsheet. For any data point $\log_{10}N(t)$ or $\log_{10}S(t)$, its calculated value is located in the $\log_{10}N$ column (assumed as X starting from row Y) of its growth table and row "=ROUND(t/Δt, 0)+Y". The formula: =ROUND (t/Δt, 0) in Excel returns the nearest integer of t/Δt. If the row number is contained in an assumed cell D#, then the formula: =INDIRECT ("X"&D#) returns the calculated $\log_{10}N(t)$ or $\log_{10}S(t)$.

(d) Minimize the overall sum of squared errors (SSE): Adding the SSE for each survival test yields the overall SSE, which is then minimized using the Excel Solver by changing the model parameter values. This optimization process results in the best-fit of the model to the entire data sets.

The root mean squared error (RMSE) was used to evaluate the goodness of fitting of the model to microbial growth data using a reported formula (Neter et al. 1996).

Results and Discussion

Validation of the explicit numerical solutions

Microbial survival curves under two nonisothermal conditions were calculated by solving equations (1), (2), and (8) using the ode45 solver in MATLAB for given model parameters $AsymD_{60} = 8$ min, $z = 5°C$, $\log_{10}(N_{res}/N(0)) = -7$, and $C_c(0) = 1$. Under the same conditions, survival curves were also calculated using equations (3) and (7) ($\Delta t = 1/60$ min), and (8) in Excel. Survival curves obtained in these two methods are illustrated in Figure 1. Equation (3) was mathematically derived from equations (1) and (2). So they should be equivalent to each other. Such an equivalence was also visualized by Figure 1 which showed overlapped

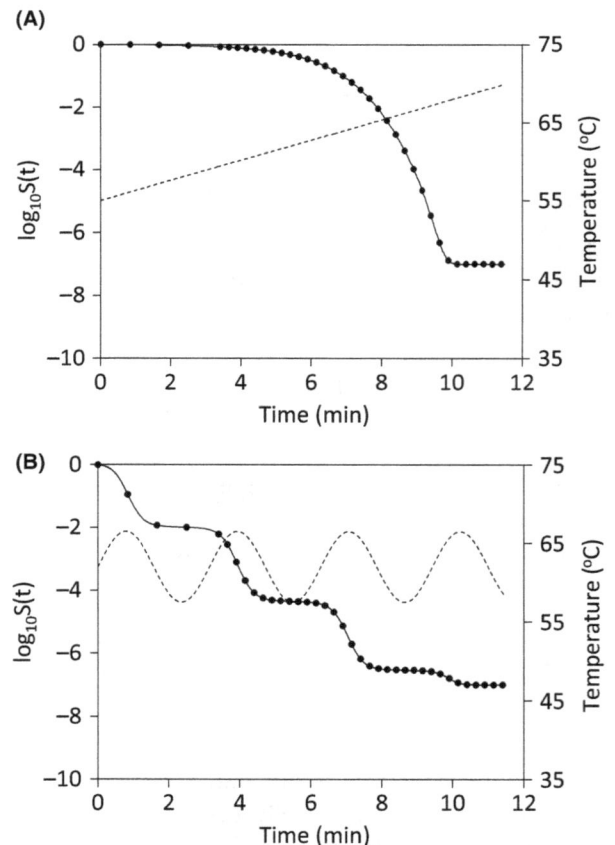

Figure 1. Survival curves calculated using equations (1), (2), and (8) (solid lines) and those using equations (3), (7), and (8) (filled symbols) under the same temperature conditions (dashed lines) (A) linear temperature profile, (B) sine temperature profile.

survival curves obtained in the two methods under the same conditions. Obviously, the explicit equation (3) provided a simpler method for the calculation. The accuracy of equation (3) depends on the accuracy of equation (7) which was used to numerically estimate $\int_0^t k_{max}(T(t))dt$. Theoretically, the error of the numerical approximation becomes negligible when time intervals are sufficiently small. Therefore, it is advised that temperature profiles are measured at small time intervals as possible. However, to reduce computation time, time intervals could be determined by gradually increasing it from a small value until the root mean squared difference between $\log_{10}S(t)$ in two consecutive calculations is smaller than a specified error tolerance.

Identification of model parameters using the Microsoft Excel solver

Figure 2 shows the fitting of equations (3) ($N_{res} = 0$), (7), and (8) to survival data for *E. coli* K12 (Valdramidis et al. 2008) under heat treatments with varying heating rates using the Microsoft Excel Solver. In the calculation, each temperature profile was converted to a series of discrete temperature points separated by constant time intervals of 1/60 min. The obtained model parameters and RMSE for curve fitting are illustrated in Table 1. As shown in the table, the value of RMSE (0.214 $\log_{10}$ CFU mL^{-1}) obtained was relatively low, indicating the model fits the data well and this was also shown by the good agreement between the data points and the fitted curves in Figure 2. When an optimization procedure is used for curve fitting, model parameters must be assigned initial values which should be sufficiently close to their "true" values in order to make the optimization process converge to the "true" values (Chen and Campanella 2012). So, it is necessary to try different sets of guesses for the model parameters to find one that results in a desirably small RMSE for curve fitting. Results obtained from the same data by Valdramidis et al. (2008) were also included in Table 1. The table showed that the RMSE (1.16 $\log_{10}$ CFU mL^{-1}) was four times greater than that obtained in the present study. So, the present study provided a more accurate curve fitting. The reason might be because the value of $C_c(0)$ was not accurately identified in that report.

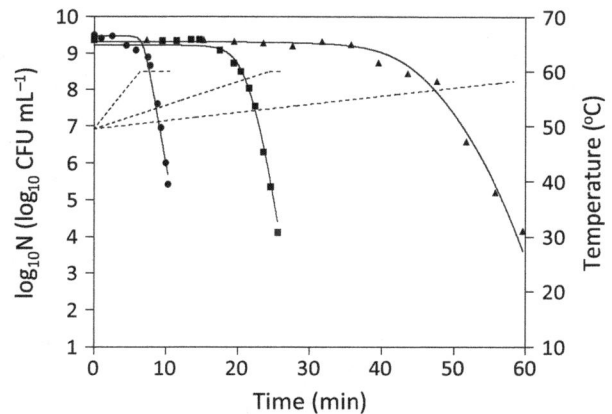

Figure 2. Published survival data for *Escherichia coli* K12 (filled symbols) (Valdramidis et al. 2008) under heat treatments (dashed lines) with a heating rate of 1.64°C min^{-1} (●), 0.43°C min^{-1} (■), and 0.15°C min^{-1} (▲), respectively. Solid lines denote the fitted survival curves using equations (3), (7), and (8).

Thermal stress may increase heat resistance of microorganisms (Valdramidis et al. 2007; Corradini and Peleg 2009). To account for such effect, Valdramidis et al. (2007) proposed equations (4) and (5) to describe survival of *E. coli* K12 during heat treatment. Figure 3A–F illustrate nonisothermal survival curves of *E. coli* K12 measured at varying heating rates which delivered different extents of thermal stress to the microorganism (Valdramidis et al. 2006). Physiological parameters k_1 and k_2 in equation (5) were estimated by fitting equations (5)–(8) ($N_{res} = 0$) to the survival data. In the curve fitting, other model parameters including $AsymD_{ref}$, z, $C_c(0)$, and $N(0)$ were adapted from a previous report (Valdramidis et al. 2006) and fixed. The obtained values of k_1, k_2, and RMSE are listed in Table 2. The small RMSE indicated a good curve fitting. Because the microbial cells' adaptation takes time, when the heating rate is sufficiently high, that is, $dT/dt \gg k_2$, k should be equal to 1 which requires $k_1 = 1$. This meant that k_1 is constantly equal to 1 and thus is redundant in equation (5). As shown in Table 2, k_1 obtained in this study was equal to 0.967 which agreed with the theoretical analysis.

Table 1. Values of model parameters and RMSE obtained by fitting equations (3), (7), and (8) ($N_{res} = 0$) to published survival data of *Escherichia coli* K12 (Valdramidis et al. 2008) and those reported by Valdramidis et al. (2008).

Method	$AsymD_{54.75}$	z	$C_c(0)$	$\log_{10}N(0)_1$	$\log_{10}N(0)_2$	$\log_{10}N(0)_3$	RMSE
	(min)	(°C)		($\log_{10}$ CFU mL^{-1})	($\log_{10}$ CFU mL^{-1})	($\log_{10}$ CFU mL^{-1})	($\log_{10}$ CFU mL^{-1})
This study	10.35	4.97	70.13	9.48	9.23	9.32	0.214
Reported	10.05	5.02	1.92	9.41	9.23	9.30	1.16*

*The RMSE was calculated using the given model parameters.

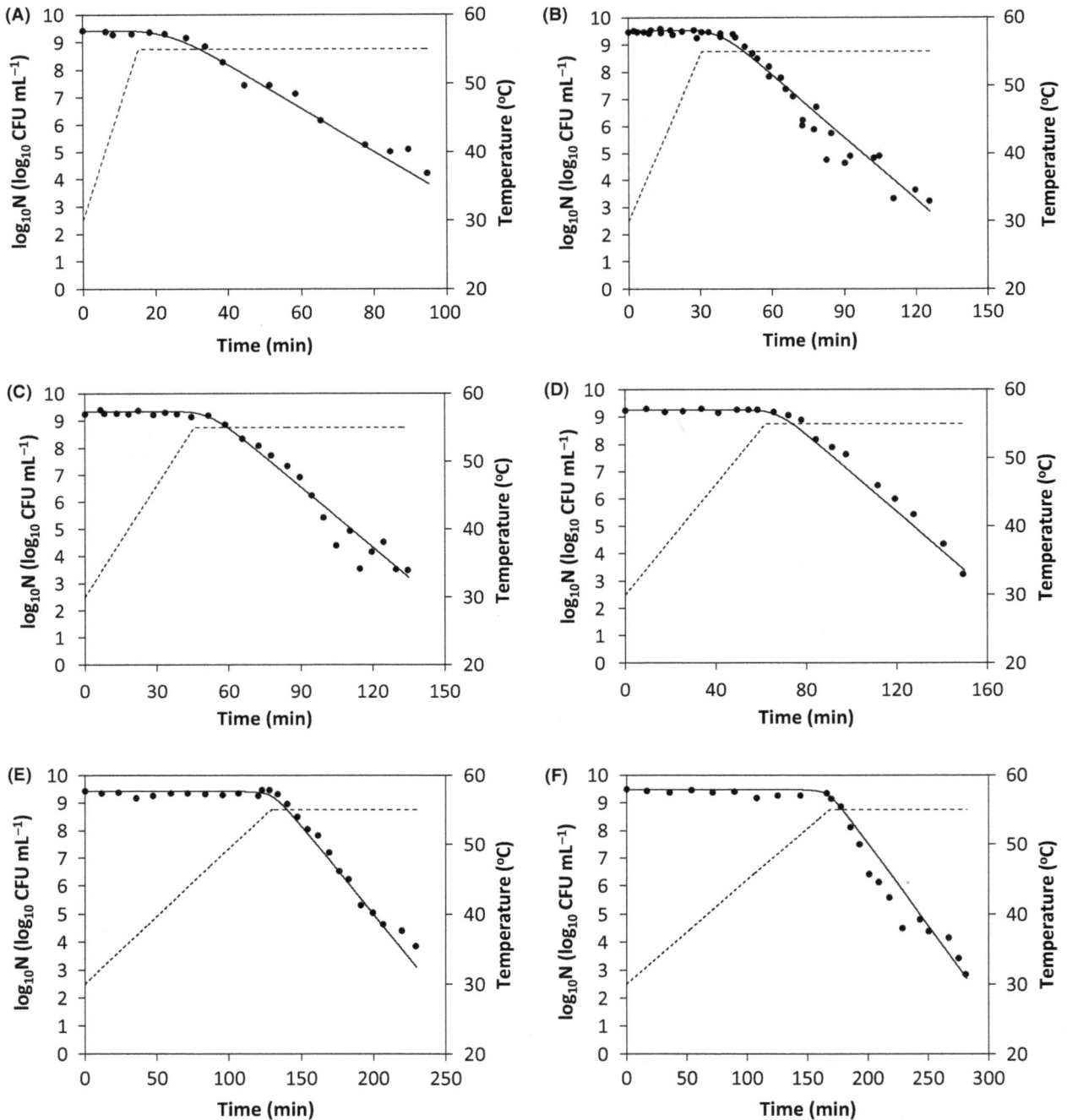

Figure 3. Published survival data for *Escherichia coli* K12 (filled symbols) (Valdramidis et al. 2006) and the fitted curves (solid lines) using equations (5)–(8). Dashed lines denote temperature profiles with a heating rate of 1.64 (A), 0.82 (B), 0.55 (C), 0.40 (D), 0.20 (E), and 0.15 (F) °C min^{-1}, respectively.

Parameters k_1 and k_2 were also identified from the same survival data by Valdramidis et al. (2007) using a two-step method. The reported results are also listed in Table 2. Because this RMSE for curve fitting was one time greater than that in the present study, simultaneously

fitting the survival model to all data sets resulted in more accurate parameter estimation.

The Geeraerd model (Geeraerd et al. 2000) essentially describes microbial survival curves that follow the traditional log-linear model but have also a tailing and a

Table 2. Values of k_1, k_2, and RMSE obtained by fitting equations (5)–(8) to published survival data of *Escherichia coli* K12 (Valdramidis et al. 2006) and those reported by Valdramidis et al. (2007).

Method	k_1 (–)	k_2 (°C min^{-1})	RMSE (log$_{10}$ CFU mL^{-1})
This study	0.969	0.060	0.368
Reported	0.696	0.042	0.787*

*The RMSE was calculated using the reported values of k_1 and k_2.

shoulder. Explicit numerical solutions of both the original and the modified Geeraerd models provide a convenient and accurate way to identify model parameters and predict survival curves that follow the model under practical nonisothermal conditions.

Conclusion

This study demonstrated that the two coupled differential equations used to describe the original or modified Geeraerd models could be simplified into an explicit equation. By numerically integrating the specific inactivation rate with respect to time involved in the equations, the obtained explicit numerical solutions could be conveniently used to accurately simulate microbial survival and estimate model parameters from nonisothermal survival data using only built-in functions in Microsoft Excel. Because there is no need to solve differential equations, the explicit equations simplify the calculation and thus should facilitate practical applications of the Geeraerd models.

Acknowledgments

This work was a contribution of the Center for Excellence in Post-Harvest Technologies of North Carolina A&T State University.

Conflict of Interest

None declared.

References

Bigelow, W. D. 1921. The logarithmic nature of thermal death time curves. J. Infect. Dis. 29:528–536.

Chen, G., and O. H. Campanella. 2012. An optimization algorithm for estimation of microbial survival parameters during thermal processing. Int. J. Food Microbiol. 154:52–58.

Cole, M. B., K. W. Davies, G. Munro, C. D. Holyoak, and D. C. Kilsby. 1993. A vitalistic model to describe thermal inactivation of Listeria monocytogenes. J. Ind. Microbiol. 12:232–237.

Corradini, M. G., and M. Peleg. 2009. Dynamic model of heat inactivation kinetics for bacterial adaptation. Appl. Environ. Microbiol. 75:2590–2597.

Dolan, K. 2003. Estimation of kinetic parameters for nonisothermal food processes. J. Food Sci. 68:728–741.

Doyle, M. E., A. S. Mazzotta, T. Wang, D. W. Wiseman, and V. N. Scott. 2001. Heat resistance of Listeria monocytogenes. J. Food Prot. 64:410–429.

Geeraerd, A. H., C. H. Herremans, and J. F. Van Impe. 2000. Structural model requirements to describe microbial inactivation during a mild heat treatment. Int. J. Food Microbiol. 59:185–209.

Heldman, D. R., and R. L. Newsome. 2003. Kinetic models for microbial survival during processing. Food Technol. 57:40–46.

Lee, D. U., V. Heinz, and D. Knorr. 2001. Biphasic inactivation kinetics of Escherichia coli in liquid whole egg by high hydrostatic pressure treatments. Biotechnol. Prog. 17:1020–1025.

Linton, R. H., W. H. Carter, M. D. Pierson, and C. R. Hackney. 1995. Use of a modified Compertz equation to model nonlinear survival curves for Listeria monocytogenes Scott, A. J. Food Prot. 58:946–954.

Neter, J., M. H. Kutner, C. J. Nachtsheim, and W. Wasserman. 1996. Applied linear regression models. McGraw-Hill Co., Inc, Chicago, IL.

Peleg, M., and M. B. Cole. 1998. Reinterpretation of microbial survival curves. Crit. Rev. Food Sci. Nutr. 38:353–380.

Peleg, M., and M. D. Normand. 2004. Calculating microbial survival parameters and predicting survival curves from non-isothermal inactivation data. Crit. Rev. Food Sci. Nutr. 44:409–418.

Valdramidis, V. P., A. H. Geeraerd, K. Bernaerts, and J. F. Van Impe. 2006. Microbial dynamics versus mathematical model dynamics: the case of microbial heat resistance induction. Innov. Food Sci. Emerg. Technol. 7:118–125.

Valdramidis, V. P., A. H. Geeraerd, and J. F. Van Impe. 2007. Stress-adaptive responses by heat under the microscope of predictive microbiology. J. Appl. Microbiol. 103:1922–1930.

Valdramidis, V. P., A. H. Geeraerd, K. Bernaerts, and J. F. Van Impe. 2008. Identification of non-linear microbial inactivation kinetics under dynamic conditions. Int. J. Food Microbiol. 128:146–152.

Van Boekel, M. A. J. S. 2002. On the use of the Weibull model to describe thermal inactivation of microbial vegetative cells. Int. J. Food Microbiol. 74:139–159.

Zhu, S., and G. Chen. 2015. Numerical solution of a microbial growth model applied to dynamic environments. J. Microbiol. Methods 112:76–82.

19

Effect of hydrocolloids and emulsifiers on the shelf-life of composite cassava-maize-wheat bread after storage

Maria Eduardo[1,2], Ulf Svanberg[2] & Lilia Ahrné[2,3]

[1]Departamento de Engenharia Química, Faculdade de Engenharia, Universidade Eduardo Mondlane, Maputo, Moçambique
[2]Department of Biology and Biological Engineering/Food and Nutrition Science, Chalmers University of Technology, Gothenburg, Sweden
[3]Process and Technology development, SP Technical Research Institute of Sweden, Food and Bioscience, Gothenburg, Sweden

Keywords
Baking improvers, bread quality, Cassava flour, starch retrogradation, storage

Correspondence
Lilia Ahrné, Process and Technology development, SP Technical Research Institute of Sweden, Food and Bioscience, Gothenburg, Sweden.

E-mail: lilia.ahrne@sp.se

Funding Information
Swedish International Development Agency (SIDA).

Abstract

The objective of this study was to evaluate the effect of hydrocolloids and/or emulsifiers on the shelf-life of composite cassava-maize-wheat (ratio 40:10:50) reference bread during storage. Added hydrocolloids were carboxymethylcellulose (CMC) and high methoxyl pectin (HM pectin) at a 3% level (w/w) and/or the emulsifiers diacetyl tartaric acid esters of monoglycerides (DATEM), lecithin (LC), and monoglycerides (MG) at a 0.3% level (w/w). After 4 days of storage, composite breads with MG had comparatively lower crumb moisture while crumb density was similar in all breads. The reference bread crumb firmness was 33.4 N, which was reduced with an addition of DATEM (23.0 N), MG (29.8 N), CMC (24.6 N) or HM pectin (22.4 N). However, the CMC/DATEM, CMC/LC, and HM pectin/DATEM combinations further reduced crumb firmness to <20.0 N. The melting peak temperature was increased from 52 C to between 53.0 C and 57.0 C with added hydrocolloids and/or emulsifiers. The melting enthalpy of the retrograded amylopectin was lower in composite bread with hydrocolloids and emulsifiers, 6.7–11.0 J/g compared to 20.0 J/g for the reference bread. These results show that emulsifiers in combination with hydrocolloids can improve the quality and extend the shelf-life of composite cassava-maize-wheat breads.

Introduction

Partial replacement of wheat flour by flours produced from locally grown crops such as cassava and maize has a major economic interest to reduce the dependence on expensive wheat imports. The challenge in substituting wheat flour lies in the fact that the bread quality is mainly governed by the gluten content of wheat, which becomes gradually lower with increasing amounts of alternative flours (>20%), leading to poor ability of the gluten proteins to form a cohesive and viscoelastic dough during baking and retain the gas formed during the fermentation. Consequently, the bread produced has a lower volume and a compact crumb structure.

To compensate for the poor gluten network Eduardo et al. (2014a) studied the effect of hydrocolloids and emulsifiers on improvement of the quality of composite bread evaluated as specific loaf volume, crumb moisture

and firmness, and crust color. Addition of hydrocolloids (3% w/w), carboxymethylcellulose (CMC), or high methoxyl pectin (HM pectin), combined with different types of emulsifiers (0.1–0.5% w/w), diacetyl tartaric acid ester of monoglycerides (DATEM), sodium stearoyl lactylate (SSL) or lecithin (LC) showed that the specific loaf volume and bread firmness of the composite bread were significantly improved by the combination of hydrocolloids and emulsifiers. Based on the results of this study, two optimized bread formulations consisting of either CMC/DATEM or HM pectin/LC, both at ratio of 3:0.3%, were selected for sensorial and consumer studies in Mozambique. The results showed a high acceptability and willingness to purchase composite bread based on cassava flour among Mozambican consumers (Eduardo et al. 2014b).

Replacement of wheat by starch rich flours like cassava is also expected to affect the shelf-life of bread due to the increased amount of starch which can undergo retrogradation during storage and cause bread firmness to increase and consequently results in a loss of quality. Retrogradation of starch includes the short-term development of a gel network structure of amylose (crystallization) and a long-term reordering of amylopectin, which is a much slower process involving recrystallization of the outer branches of this polymer (Miles et al. 1985; Ring et al. 1987). Retrogradation of starch is also affected by the redistribution of water between starch and gluten and, as a result, the crumb will become increasingly firm with time (Eliasson and Larsson 1993; Davidou et al. 1996; Purhagen et al. 2012). According to Gray and Bemiller (2003), amylopectin is the major factor in the retrogradation process but is not solely responsible for the observed change in texture. However, the mechanisms for these processes are still not completely understood.

The addition of hydrocolloids and emulsifiers to cassava composite bread are expected to influence the retrogradation of starch and consequently the shelf-life of bread. Hydrocolloids can increase water retention capacity influencing the water redistribution and consequently the retrogradation. Davidou et al. (1996) reported a decrease firmness and starch retrogradation during storage in wheat bread by addition of locust bean gum, alginate, and xanthan.

Emulsifiers are commonly used in bakery products to improve softness of the crumb (Demirkesen et al. 2010). They are composed of both hydrophobic and hydrophilic residues, which allow the interaction and formation of complexes with starch, protein, shortening, and water. The improving effect of emulsifiers seems to be related to their effect in reducing the repulsing charges between gluten proteins by causing them to aggregate in composite dough flour as the wheat gluten has been diluted. For instance, interaction of an emulsifier with the protein can improve dough strength and allow better retention of carbon dioxide (Demirkesen et al. 2010).

The combination of hydrocolloids and emulsifiers might have synergistic effects leading to longer shelf-life, but no studies on composite bread have been found in literature. The objective of this study was therefore to investigate the effect of hydrocolloids (HM pectin and CMC) and emulsifiers (DATEM, LC, and MG) and their combined effect on extension of shelf-life of optimized composite cassava-maize-wheat bread up to 4 days of storage.

Materials and Methods

Materials

Wheat flour of 10.5% protein (Bagerivetemjöl, Frebaco Kvarn, Sweden), yellow maize flour of 7.14% ± 0.05 protein)(AB Risenta, Sweden), roasted cassava flour of 1.35% ± 0.07 protein (kjeldahl method N × 6.25) (AOAC, 1990), instant dry yeast, salt, sugar, vegetable oil, and ascorbic acid (Merck Chemicals, Germany) were used for bread making.

Hydrocolloids were HM pectin (degree of esterification 68–75%) (GENU pectin type BIG, CP Kelco, Denmark), CMC (degree of substitution 0.75–0.85) (CEKOL 50000 W, CP Kelco, Finland), and emulsifiers were DATE M (MULTEC HP 20, Puratos, Belgium), soy lecithin, LC (phosphatidylcholine min. 18%, phosphatidylinositol min. 13%, and phosphatidylethanolamine min. 15%) (LECICO P 900 IPM, Lecico GmbH, Hamburg, Germany), and monoglyceride, MG (total monoglyceride min. 90%, free glycerol max. 1% and acid value max. 3 mg KOH/g) (Dimodan® PH200, DANISCO, Denmark). The emulsifiers were selected according to their difference in hydrophilic lipophilic balance values (HLB). DATEM (HLB value of 9.2) and lecithin (HLB value between 3 and 4) are both anionic oil-in-water emulsifiers, which might influence protein denaturation (Pisesookbunterng and D'Appolonia 1983). Monoglyceride (HLB value between 2.8 and 3.8) is commonly used in bakeries to delay staling. This is a nonionic water-in-oil emulsifier. Fresh cassava roots were obtained from local producers in Mozambique and processed into roasted cassava flour as previously described (Eduardo et al. 2013) and was as follows. The roots (~100 kg) was peeled, washed in potable water and manual chipped, which was moist fermented for about 2 days. The fermented cassava was pressed, screened with a mechanical machine, and toasted in a frying pan until cooked and crisp (~10 min). The toasted material was milled in a laboratory mill with a sieve DIN 4188 (0.125 mm aperture sieve). The flour (~20 kg) was then packed in polyethylene bags until use.

Test baking

The test baking experiments were randomized. Composite breads were produced from cassava-maize-wheat flours that contained either HM pectin or CMC at a level of 3% on a flour weight basis. Subsequent tests involved the combination of either 0.3% of DATEM, LC or MG with each of the above levels of HM pectin and CMC. A model system prepared with each different emulsifier was also examined. A control loaf was produced that contained no improvers. The recipe is given in Table 1.

The dough was mixed in a KSM9 mixer (KitchenAid, USA) for two minutes at low speed followed by eight minutes of mixing at medium speed. The dough (1500 g) was covered with a kitchen cloth and allowed to rest at room temperature for 45 min. At the end of the resting period, the dough was divided (into pieces of 50 g), molded by hand and placed in aluminum pans. The loaves were

Table 1. Bread formulation.

Ingredients	%
Flour (50% wheat, 40% cassava and 10% maize)	100.0
Dry yeast	1.6
Salt	1.5
Sugar	2.0
Oil	3.0
Ascorbic acid	0.1
Hydrocolloids (CMC or HM pectin)	3.0
Emulsifiers (DATEM, LC, and MG)	0.3
Water (at 15.5°C)	88.3

proofed for 45 min at 30°C and 80% relative humidity (RH) in a fermentation cabinet (Labrum Klimat Ab, Stockholm, Sweden) and baked at 220°C for 7 min in a rotating convention oven (Dahlen S400, Sveba Dahlen AB, Sweden) with air circulation. Before measurements, the breads were cooled for 1 h at room temperature. Afterward, the unpacked bread loaves were stored in a room with controlled relative humidity (50%) and temperature (23°C) for 4 days until further characterization.

The composite breads were analyzed for weight, crumb moisture, density, and melting enthalpy of the retrograded amylopectin on the baking (day 0) and 4 days later. Firmness was measured after 3 h and 1, 2, 3, and 4 days of storage.

Analysis of composite bread

Weight and volume

The weight of the loaves ($n = 6$) was measured after cooling on day 0 and day four. The volume ($n = 6$) was measured using the rapeseed displacement method, where alfalfa seed was used instead of millet. Each loaf was weighed, and the specific loaf volume (cm^3/g) was calculated as loaf volume (cm^3)/loaf weight (g) taken after 1 h of baking.

Crumb density

The density (ρ) (g/cm^3) corresponding to the density of the material of the cell walls ($n = 2$) was determined with a gas pycnometer (AccuPyc II 1340, CIAB, Sweden) using nitrogen as the displacement fluid.

Crust color

The instrumental measurement of the bread crust color ($n = 4$) was carried out with Digital Colour Imaging System (DigiEye) (Cromocol Scandinavia AB, Borås, Sweden). The controlled illumination cabinet on the DigiEye equipment was utilized to capture high-resolution

images of the fresh bread surface. The DigiEye 2.53b software (Cromocol Scandinavia AB, Borås, Sweden) allows for storage of specific color standards with given L* (lightness), a* (redness-greenness), and b* (yellowness-blueness) values according to the CIELab system definition. The results were reported as the browning index (BI) of the bread crust as calculated by Maskan (2001).

Crumb moisture

The crumb moisture (g of water/100 g, wet sample) ($n = 3$) was determined in triplicate by drying the samples overnight under a vacuum oven at 70°C under 29 in. of Hg (AACC method 44–40, 1995).

Crumb texture

The texture properties of the crumb were measured using an Instron 5542 universal testing machine (Canton, MA, USA). A modified AACC standard method 74–09 was used with a cylindrical probe (diameter 15 mm). The crumb samples ($n = 4$) of composite bread (2.5 cm) were compressed to 40% at a crosshead speed of 1.7 mm/sec. Firmness is the parameter that describes the resistance to compression of the bread crumbs.

Thermal properties

Analyses were performed in a DSC-1 (Mettler Toledo AB, Sweden) using a medium pressure pan. The equipment was calibrated with Indium (enthalpy of fusion 28.41 J/g, melting point 156.4°C), and an empty pan was used as a reference. Approximately 90 mg of crumb:water (ratio 1:2 w/w) was weighed in the pan, which was hermetically sealed in order to avoid moisture loss. Samples ($n = 3$) were heated from 20 to 130°C with a 5°C/min scanning rate (Sahlström et al. 2003). The onset temperature (T_o), the peak temperature (T_p), and the transition enthalpy (J/g, dry sample) of amylopectin crystals (retrogradation) (ΔH_{retro}) were evaluated from the thermograms using the program Mettler Star^e (Mettler-Toledo GMbH, Schwerzenbach, Switzerland).

Statistical analysis

SPSS software version 16.0 (SPSS Inc., Chicago, IL, USA) was used to analyze the data obtained. One-way ANOVA was used in data from the composite bread analysis. Tukey HSD (honesty significant difference) post hoc mean comparison tests were used to detect significant differences at a confidence level of 95% ($P < 0.05$). The mean values tested were calculated based on at least two to six individual measurements of one batch of bread.

Results and Discussion

Effect of hydrocolloids and emulsifiers on the characteristics of composite breads

Adding either hydrocolloids or emulsifiers or combinations of hydrocolloid/emulsifier to composite dough formulation significantly increased the specific volume compared with the reference bread (with no improver) (Table 2). The largest increase in specific volume was obtained with a combination of CMC/MG (28%), followed by CMC/ DATEM (21%) and CMC/LC (19%). However, the specific volume of the loaves with either emulsifiers or hydrocolloids was not significantly different. These results are in agreement with Rosell et al. (2001), Guarda et al. (2004), Bárcenas and Rosell (2005) and Correa et al. (2012), who found an increased loaf volume of wheat bread with an addition of the hydrocolloids HPMC (hydroxypropyl methylcellulose), HM pectin and κ-carrageenan, and similar findings have been reported in gluten-free breads with additions of DATEM (Nunes et al. 2009; Demirkesen et al. 2010), distilled monoglyceride (MG), and lecithin (Nunes et al. 2009). However, no effect was observed with an addition of MG and DATEM in gluten-free bread formulations with mixtures of rice flour and cassava starch (Sciarini et al. 2012). The positive effect of hydrocolloids and/or emulsifiers on the volume of the bread is explained by an increased stability of the dough system during proofing (hydrocolloids) (Guarda et al. 2004) and by the

formation of a stabilized liquid film lamellae/gas cell interface (emulsifiers) (Selmair and Koehler 2010). As a result, additional strength was conferred to the gas cells of the dough, thereby increasing gas retention and/or oven spring. This led as a consequence to higher bread volume (Gómez et al. 2004; Guarda et al. 2004).

Hydrocolloids and emulsifiers or combinations of them also affected the crust color, that is, the brownness index (BI) of the composite cassava-maize-wheat breads. The BI was higher (≥58) in the bread crust of CMC combined with DATEM or MG and HM pectin with LC; BI values above of 58 can be considered to be preferred by consumers (Eduardo et al. 2014b). The increased BI value could be attributable to a more favorable water distribution due to the hydrocolloids, which affects Maillard browning reactions and caramelization (Sciarini et al. 2010).

Effect of the hydrocolloids and emulsifiers on the characteristics of stored composite breads

Weight

The weight of fresh loaves of bread (with or without improvers) ranged between 44.5 g and 45.7 g (Table 3), whereas the weight of the loaf of the reference bread was 45.4 g. The weight of the bread was in general slightly lower in composite breads with improvers, which means that the weight loss (approx. 10% w/w) during the baking process was in general 1% w/w higher in breads with improvers.

Over 4 days of storage, a significant reduction in weight was observed for all bread samples as a result of moisture loss. However, as compared with the reference bread, a significantly higher bread weight was obtained in loaves containing CMC and LC (about 4%), and significantly lower weights for all loaves with MG (about 3%). All the other bread loaves had a weight similar to that of the reference bread, which might indicate that the addition of improvers had little effect on water loss during 4 days of storage. This may be explained by the very high water binding capacity of pregelatinized starch (Seyhun et al. 2005), since the cassava flour used in this study was partially pregelatinized; hydrocolloids have otherwise been shown to reduce the crumb dehydration rate during storage (Guarda et al. 2004).

Crumb density

The density of the fresh composite bread crumb was 1.30 g/cm^3. In general, no significant differences were found between the reference bread and breads baked with hydrocolloids, emulsifiers, or their combinations. The

Table 2. Specific volume and brownness index of fresh composite bread samples as affected by hydrocolloids, emulsifiers, and combinations of both improvers.

Bread formulations	Specific volume (cm^3/g)	Brownness index (color units)
No emulsifier or hydrocolloid	1.93 ± 0.06[a]	37.9 ± 2.1[a]
Emulsifiers (0.3%):		
DATEM	2.08 ± 0.10[b]	47.6 ± 1.2[b]
LC	2.07 ± 0.06[b]	44.1 ± 1.9[ab]
MG	2.07 ± 0.05[b]	43.2 ± 1.8[ab]
Hydrocolloids (3%):		
CMC	2.10 ± 0.07[b]	43.6 ± 3.8[ab]
HM pectin	2.12 ± 0.04[bc]	56.1 ± 3.3[cd]
Hydrocolloids (3%) + Emulsifiers (0.3%):		
CMC/DATEM	2.34 ± 0.06[d]	63.4 ± 2.4[e]
HM pectin/DATEM	2.11 ± 0.01[b]	49.9 ± 1.9[bc]
CMC/LC	2.30 ± 0.05[d]	55.1 ± 3.4[cd]
HM pectin/LC	2.23 ± 0.02[cd]	58.4 ± 1.0[de]
CMC/MG	2.46 ± 0.11[e]	60.7 ± 5.6[de]
HM pectin/MG	2.09 ± 0.06[b]	44.3 ± 3.6[ab]

Values in the same column followed by different letters are significantly different ($P < 0.05$). CMC, carboxymethyl cellulose; HM pectin, high methoxyl pectin; DATEM, diacetyl tartaric acid esters of monoglycerides; LC, lecithin; MG, monoglycerides.

Table 3. Weight, crumb density, crumb moisture, and crumb firmness of fresh and stored composite bread samples as affected by hydrocolloids, emulsifiers, and combinations of both improvers.

Bread formulations	Weight (g)		Crumb density (g/cm³)		Crumb moisture (% wet basis)		Crumb firmness (N)	
	Fresh bread	4-d storage	Fresh bread	4-d storage	Fresh bread	4-d storage	Fresh bread	4-d storage
No emulsifier or hydrocolloid	45.4 ± 0.2[cd]	32.2 ± 0.3[b]	1.30 ± 0.01[ab]	1.38 ± 0.00[a]	48.6 ± 0.2[abc]	27.7 ± 0.2[c]	6.9 ± 0.2[f]	33.4 ± 0.6[g]
Emulsifiers (0.3%):								
DATEM	44.9 ± 0.2[ab]	32.4 ± 0.3[b]	1.31 ± 0.02[ab]	1.36 ± 0.00[a]	48.8 ± 0.2[bc]	29.3 ± 0.1[cd]	5.0 ± 0.3[d]	23.0 ± 0.3[d]
LC	45.4 ± 0.3[bcd]	33.5 ± 0.6[cd]	1.29 ± 0.01[ab]	1.35 ± 0.00[a]	47.8 ± 0.1[ab]	28.8 ± 1.1[cd]	6.0 ± 0.5[e]	36.5 ± 0.6[h]
MG	44.8 ± 0.2[a]	30.8 ± 0.2[a]	1.28 ± 0.02[ab]	1.39 ± 0.03[a]	48.6 ± 0.6[abc]	24.8 ± 0.9[a]	5.7 ± 0.2[e]	29.8 ± 0.4[f]
Hydrocolloids (3%):								
CMC	45.7 ± 0.3[d]	33.9 ± 0.5[d]	1.29 ± 0.01[ab]	1.36 ± 0.01[a]	47.5 ± 0.8[a]	29.5 ± 0.3[cd]	4.2 ± 0.2[bc]	24.6 ± 0.4[e]
HM pectin	44.8 ± 0.1[a]	32.2 ± 0.3[b]	1.31 ± 0.00[ab]	1.38 ± 0.01[a]	48.0 ± 0.3[ab]	27.5 ± 1.5[bc]	3.8 ± 0.1[abc]	22.3 ± 0.5[d]
Hydrocolloids (3%) + Emulsifiers (0.3%):								
CMC/DATEM	44.9 ± 0.3[abc]	32.5 ± 0.5[b]	1.30 ± 0.00[ab]	1.34 ± 0.00[a]	49.7 ± 0.2[c]	31.2 ± 1.3[d]	3.6 ± 0.1[a]	12.4 ± 0.4[a]
HM pectin/DATEM	45.0 ± 0.3[abc]	32.8 ± 0.3[bc]	1.33 ± 0.00[b]	1.38 ± 0.01[a]	48.2 ± 0.5[ab]	29.3 ± 1.0[cd]	4.0 ± 0.1[abc]	17.0 ± 0.5[b]
CMC/LC	45.0 ± 0.1[abc]	32.3 ± 0.2[b]	1.28 ± 0.02[ab]	1.37 ± 0.00[a]	47.8 ± 0.4[ab]	27.3 ± 0.5[bc]	3.8 ± 0.2[abc]	16.6 ± 0.5[b]
HM pectin/LC	44.9 ± 0.3[ab]	32.8 ± 0.3[bc]	1.30 ± 0.02[ab]	1.38 ± 0.02[a]	48.1 ± 0.6[ab]	29.0 ± 0.5[cd]	3.7 ± 0.1[ab]	20.1 ± 0.6[c]
CMC/MG	44.5 ± 0.3[a]	31.3 ± 0.4[a]	1.27 ± 0.01[a]	1.36 ± 0.00[a]	47.9 ± 0.2[ab]	25.3 ± 0.9[ab]	4.3 ± 0.1[c]	21.8 ± 0.4[d]
HM pectin/MG	44.9 ± 0.2[abc]	31.3 ± 0.2[a]	1.27 ± 0.01[a]	1.38 ± 0.05[a]	47.7 ± 0.4[ab]	24.7 ± 0.7[a]	5.7 ± 0.1[e]	29.2 ± 0.6[f]

Values in the same column followed by different letters are significantly different ($P < 0.05$). CMC, carboxymethyl cellulose; HM pectin, high methoxyl pectin; DATEM, diacetyl tartaric acid esters of monoglycerides; LC, lecithin; MG, monoglycerides.

crumb density of all the composite breads with baking improvers increased after 4 days in relation to the initial value (the baking day) and was in the range of 1.34 g/cm³ and 1.39 g/cm³, similar to the reference bread (1.38 g/cm³). Lagrain et al. (2012) reported that bread density is a major determining factor of bread crumb structure and texture during storage. In breads with a similar density and crumb structure, the evolution of crumb stiffness during storage was determined by changes in the starch component.

Crumb moisture

The crumb moisture content of the fresh breads varied between 47.5% and 49.7%. None of the tested hydrocolloids, emulsifiers and their combinations affected the bread moisture content significantly due to the fact that all bread samples were produced using approximately the same amount of water ($\approx$ 88% of flour weight).

However, the moisture content decreased in all bread samples stored for 4 days in comparison with the initial value (Table 3), and a significantly lower moisture content was found for all breads with MG as compared to the reference bread. In contrast, bread with DATEM/CMC had a higher moisture content. A higher moisture content in the crumb is preferred for better quality during storage, as it will give softer crumbs with less crumb hardening (Guarda et al. 2004). The decrease in crumb moisture is known to affect the crumb firming rates (He and Hoseney 1990) by the formation of cross links between partially solubilized starch and gluten proteins (Martin et al. 1991).

Firmness

The firmness of the fresh reference composite bread crumb was 6.9 N, whereas all breads with baking improvers had a significantly lower crumb firmness, between 3.7 and 6.0 N (Table 3). All composite bread samples increased in hardness during storage in relation to the initial value (the first day). However, composite breads baked with baking improvers had a crumb firmness after 2 days of storage similar to fresh reference bread with no improvers (Fig. 1). After 4 days of storage, the crumb firmness of the reference bread increased to 33.4 N as a result of the staling process and loss of moisture. Breads with improvers (except bread baked with LC alone, Table 3) maintained a significantly softer crumb in comparison with the reference bread after 4 days of storage. Our results that hydrocolloids have a crumb softening effect on both fresh and stored composite bread is in agreement with Guarda et al. (2004) and Correa et al. (2012), who found that HPMC and pectin decreased the crumb

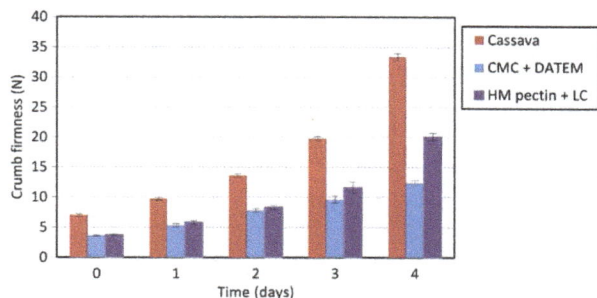

Figure 1. Effects of hydrocolloids (CMC and HM pectin) and their interactive effect with emulsifiers (DATEM and LC) on the crumb firmness of composite cassava-maize-wheat breads during storage (23°C and 50% r.h.). Error bars indicate the standard deviation. Lc, lecithin.

hardness in a wheat bread. The effect was explained by a decreased loss of water during storage (high water retention capacity of the hydrocolloids), and crumb hardening was retarded as a result. This effect could be related to the inhibition of the amylopectin retrogradation due to the water retention capacity of hydrocolloids, that is, recrystallization of amylopectin is retarded at lower water availability (Zeleznak and Hoseney 1986; Guarda et al. 2004). However, our results differ from those obtained by Lazaridou et al. (2007), who reported that pectin and CMC did not affect the crumb firmness of gluten-free breads. The contrasting results can probably be explained by the different bread formulations (gluten-free and composite cassava-maize-wheat).

The only bread with a firmness higher than the reference bread was the bread with LC. Some studies indicate that LC only has a small delaying effect on the firming of wheat starch bread (Forssell et al. 1998), or no delaying effect on the crumb firming of wheat bread (Stampfli and Nersten 1995) or gluten-free bread (Nunes et al. 2009). The increase in bread firmness with LC was explained by its inability to form complexes with the starch (Stampfli and Nersten 1995; Forssell et al. 1998).

With respect to the antistaling effect of emulsifiers, Pisesookbunterng and D'Appolonia (1983) suggested that the adsorption of emulsifier to the starch granule, as well as the formation of a starch-emulsifier complex, restrained the starch from taking up water released from gluten during the aging of the bread. Moreover, monoglycerides, which form strong complexes with amylose, will reduce granule swelling and solubilization (Gray and Schoch 1962; Gómez et al. 2004). The reduction in starch swelling and the degree of granule swelling are inversely related to crumb firmness. Lecithin, with a higher content of lysophospholipids, has been reported to retard bread staling (Forssell et al. 1998; Gray and Bemiller 2003) by complexing with starch amylose (Forssell et al. 1998). DATEM, however,

initially produces lower crumb firmness and then retards the rate of staling through its interaction with not only the amylose but also with amylopectin (Kamel and Ponte 1993; Gray and Bemiller 2003).

The combined effect of emulsifiers and hydrocolloids in reducing bread hardness was thus more pronounced than when either was added separately.

Retrogradation of starch

The amylose is responsible for setting the initial network structure but is not involved in long-term staling (Eliasson and Larsson 1993). The long-term change after 4 days of storage is therefore attributed to the amylopectin fraction, which in our composite bread is assumed to constitute about 75% of the total starch.

After storage for 4 days, an endothermal staling peak appears on the DSC thermogram in the range of 35–70°C (Schiraldi et al. 1996) as a result of the melting enthalpy of recrystallized amylopectin (ΔH_{retro}). The hydrocolloids, emulsifiers and the combinations of both exhibited different results in their effect on starch retrogradation after 4 days of storage (Table 4). However, in fresh bread no retrogradation peak was observed as previously observed by Purhagen et al., (2008).

As Table 4 shows, the onset temperature (T_o) and melting enthalpy (ΔH_{retro}) of recrystallized amylopectin of composite bread varied from 50.6–53.4°C and 6.7–23.2 J/g dry crumb, respectively, depending on the hydrocolloids (CMC and HM pectin) and/or the emulsifier types (DATEM, LC, and MG). In the reference composite bread, the retrogradation peak temperature appeared at 52°C. The addition of hydrocolloids and/or emulsifiers had a peak temperature that was 1.9–4.8°C above that of the reference bread, which indicates that the melting of recrystallized amylopectin enthalpy was delayed. The addition of improvers, except LC, in the composite bread loaves significantly reduced the melting enthalpy values compared to that of the reference bread (20.0 J/g). The lowest value was observed for CMC/LC (6.7 J/g), followed by HM pectin/DATEM (10.3 J/g) and CMC/DATEM (10.6 J/g), whereas LC showed the highest value (23.2 J/g).

Schiraldi et al. (1996) and Gujral et al. (2004) also found decreased starch retrogradation with hydrocolloids, which confirms the findings obtained in this study. The effect of hydrocolloids on starch retrogradation seems to be due to their interaction with water by limiting moisture transfer and loss and also with starch chains in the mixture (Davidou et al. 1996; Gavilighi et al. 2006). Purhagen et al. (2012) also observed that an addition of DATEM gave less retrograded amylopectin in gluten-free bread, which was attributed to the formation of amylose-emulsifier complex,

Table 4. Thermal properties of composite cassava-maize-wheat bread stored for 4 days as affected by hydrocolloids, emulsifiers, and a combination of both improvers.

Composite bread samples	T_o (°C)	T_p (°C)	ΔH_{retro} (J/g dry crumb)
No emulsifier or hydrocolloid	51.1 ± 0.4	52.1 ± 0.8	20.0 ± 0.2[g]
Emulsifiers (0.3%):			
DATEM	51.2 ± 1.2	54.7 ± 1.4	14.0 ± 0.2[def]
LC	51.9 ± 1.0	54.7 ± 2.1	23.2 ± 0.4[h]
MG	52.1 ± 0.6	52.9 ± 0.6	15.6 ± 0.8[f]
Hydrocolloids (3%):			
CMC	51.5 ± 2.2	54.8 ± 0.6	12.8 ± 0.5[cd]
HM pectin	51.6 ± 0.3	56.4 ± 1.6	14.8 ± 0.8[ef]
Hydrocolloids (3%) + emulsifiers (0.3%):			
CMC/DATEM	50.6 ± 0.7	55.2 ± 1.5	10.6 ± 0.3[b]
HM pectin/DATEM	52.8 ± 0.5	55.6 ± 1.1	10.3 ± 0.1[b]
CMC/LC	52.8 ± 1.4	55.2 ± 2.6	6.7 ± 0.4[a]
HM pectin/LC	52.9 ± 2.0	55.6 ± 1.3	12.8 ± 0.6[cde]
CMC/MG	50.8 ± 0.7	55.1 ± 0.5	11.1 ± 0.9[bc]
HM pectin/MG	53.4 ± 1.5	56.9 ± 1.4	15.8 ± 0.6[f]

Values in the fourth column followed by different letters are significantly different ($P < 0.05$). T_o, onset temperature; T_p, peak temperature; ΔH_{retro}, enthalpy of melting of the amylopectin recrystallization; CMC, carboxymethyl cellulose; HM pectin, high methoxyl pectin; DATEM, diacetyl tartaric acid esters of monoglycerides; LC, lecithin; MG, monoglycerides.

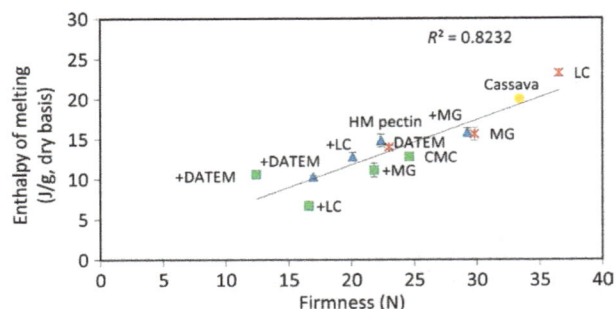

Figure 2. Relationship between amylopectin recrystallization, measured as enthalpy of melting, and firmness of composite cassava-maize-wheat bread crumb samples with hydrocolloids (CMC and HM pectin), emulsifiers (DATEM, lecithin (LC), and MG) and a combination of both improvers stored at 23°C and 50% for 4 days of storage.

thereby preventing amylopectin from re-crystallizing (retrograde) (Gudmundsson and Eliasson 1990). The effect of the addition of both hydrocolloids and emulsifiers results from an increase in the starchy lipids due to preferential binding to the gluten and displacement of nonstarchy bound lipids to the starch, and an increase in the free lipids, respectively, thus lowering the firming rate of the bread (Collar et al. 2001).

Figure 2 shows the enthalpy of melting recrystallized amylopectin as a function of the firmness of composite cassava-maize-wheat bread crumb samples after 4 days of storage. In general, the firmness of the bread crumbs increased with increased enthalpy of melting for all composite bread samples (with or without improvers) ($R^2 = 0.82$). Bread baked with LC had a significantly higher firmness and enthalpy of melting values compared with the reference bread (Tables 3 and 4). However, the composite bread prepared with the combination of CMC/DATEM had the lowest crumb firmness ($P < 0.05$), which corresponds to the lowest enthalpy of melting recrystallized amylopectin.

In general, composite bread with CMC had a lower enthalpy of melting recrystallized amylopectin and firmness compared with the corresponding bread with HM pectin. The explanation may be that HM pectin more strongly binds the water content in the bread and thereby

gives rise to a more rapid firming rate (Rogers et al. 1988). However, the retrogradation rate has been shown to be lower at low moisture starch gels (Zeleznak and Hoseney 1986).

The different types of improvers added to composite dough formulations seems to influence the rate of amylopectin recrystallization and consequently retards starch retrogradation in composite cassava-maize-wheat breads.

Conclusions

This study has shown that an addition of hydrocolloids (CMC and HM pectin) and/or emulsifiers (DATEM, LC, and MG) to composite cassava-maize-wheat bread has varying effects on quality parameters of breads during storage.

After 4 days of storage, the density and firmness of the stored bread loaves increased while the weight and moisture content was reduced in comparison with the fresh bread. Addition of emulsifiers (DATEM and MG) reduced crumb firmness but did not show a significant effect on the weight, density, or crumb moisture compared to the reference bread (with no improver). The main effect of the hydrocolloids was reduced crumb firmness, and the combination of DATEM with CMC showed the lowest crumb firmness after storage.

We found that the hydrocolloids and emulsifiers delayed the melting peak temperature for retrogradation and that the combination of both improvers further reduced the retrogradation peak temperature. CMC/LC, HM pectin/DATEM, and CMC/DATEM were especially effective in retarding starch recrystallization in composite cassava-maize-wheat bread. This suggests that emulsifiers in combination with hydrocolloids have a significant effect on retarding starch retrogradation in composite cassava-maize-wheat bread.

Effect of hydrocolloids and emulsifiers on the shelf-life of composite cassava-maize-wheat bread...

177

Acknowledgments

The authors gratefully acknowledge the financial support of the Swedish International Development Agency (SIDA) program under the project "Energy Science and Technology Research Program".

Conflict of Interest

None declared.

References

AACC. 1995. Approved methods of the American Association of Cereal Chemists. AACC, 9th ed. Minn, USA, St. Paul.

AOAC. 1990. Official Method of Analysis of the AOAC. 15th ed. Association of Official Analytical Chemists, DC Washington, USA.

Bárcenas, M. E., and C. M. Rosell. 2005. Effect of HPMC addition on microstructure, quality and aging of wheat bread. Food Hydrocolloids 19:1037–1043.

Collar, C., J. C. Martinez, and C. M. Rosell. 2001. Lipid binding of fresh and stored formulated wheat breads. Relationships with dough and bread technological performance. Food Sci. Technol. Intl. 7:501–510.

Correa, M. J., G. T. Pérez, and C. Ferrero. 2012. Pectins as breadmaking additives: effect on dough rheology and bread quality. Food Bioprocess Technol. 5:2889–2898.

Davidou, S., M. L. Meste, E. Debever, and D. Bekaert. 1996. A contribution to the study of staling of white bread: effect of water and hydrocolloid. Food Hydrocolloids 10:375–383.

Demirkesen, I., B. Mert, G. Sumnu, and S. Sahin. 2010. Rheological properties of gluten-free bread formulations. J. Food Eng. 96:295–303.

Eduardo, M., U. Svanberg, J. Oliveira, and L. Ahrné. 2013. Effect of cassava flour characteristics on properties of cassava-wheat-maize composite bread types. Int. J. Food Sci. 2013:1–10.

Eduardo, M., U. Svanberg, and L. Ahrné. 2014a. Effect of hydrocolloids and emulsifiers on baking quality of composite cassava-maize-wheat breads. Int. J. Food Sci. 2014:1–9.

Eduardo, M., U. Svanberg, and L. Ahrné. 2014b. Consumers' acceptance of composite cassava-maize-wheat breads using baking improvers. Afr. J. Food Sci. 8:390–401.

Eliasson, A.-C., and K. Larsson. 1993. Cereals in breadmaking: a molecular colloidal approach. Marcel Dekker Inc, New York (Chapter 7).

Forssell, P., S. Shamekh, H. Härkönen, and K. Poutanen. 1998. Effects of native and enzymatically hydrolysed soya and oat lecithins in starch phase transitions and bread baking. J. Sci. Food Agric. 76:31–38.

Gavilighi, H. A., M. H. Azizi, M. Barzegar, and M. A. Ameri. 2006. Effect of selected hydrocolloids on bread staling as evaluated by DSC and XRD. J. Food Technol. 4:185–188.

Gómez, M., S. del Real, C. M. Rosell, F. Ronda, C. A. Blanco, and P. A. Caballero. 2004. Functionality of different emulsifiers on the performance of breadmaking and wheat bread quality. Eur. Food Res. Technol. 219:145–150.

Gray, J. A., and J. N. Bemiller. 2003. Bread Staling: molecular basis and control. Compr. Rev. Food Sci. F. 2:1–21.

Gray, V. M., and T. J. Schoch. 1962. Effect of surfactants and fatty adjuncts on the swelling and solubilization of granular starch. Starch 14:239–246.

Guarda, A., C. M. Rosell, C. Benedito, and M. J. Galotto. 2004. Different hydrocolloids as bread improvers and antistaling agents. Food Hydrocolloids 18:241–247.

Gudmundsson, M., and A.-C. Eliasson. 1990. Retrogradation of amylopectin and the effects of amylose and added surfactants/emulsifiers. Carbohydr. Polym. 13:295–315.

Gujral, H. S., M. Haros, and M. C. Rosell. 2004. Improving the texture and delaying staling in rice flour chapati with hydrocolloids and α-amylase. J. Food Eng. 65:89–94.

He, H., and R. C. Hoseney. 1990. Changes in bread firmness and moisture during long-term storage. Cereal Chem. 67:603–607.

Kamel, B. S., and J. G. Jr Ponte. 1993. Emulsifiers in baking. Pp. 179–222 in B. S. Kamel and C. E. Stauffer, eds. Advances in baking technology. Springer Science+Business Media Dordrecht, UK.

Lagrain, B., E. Wilderjans, C. Glorieux, and J. A. Delcour. 2012. Importance of gluten and starch for structural and textural properties of crumb from fresh and stored bread. Food Biophys. 7:173–181.

Lazaridou, A., D. Duta, M. Papageorgiou, N. Belc, and C. G. Biliaderis. 2007. Effects of hydrocolloids on dough rheology and bread quality parameters in gluten-free formulations. J. Food Eng. 79:1033–1047.

Martin, M. L., K. J. Zeleznak, and R. C. Hoseney. 1991. A mechanism of bread firming. I. Role of starch swelling. Cereal Chem. 68:498–503.

Maskan, M. 2001. Kinetics of colour change of kiwifruits during hot air and microwave drying. J. Food Eng. 48:169–175.

Miles, M. J., V. J. Morris, P. D. Orford, and S. G. Ring. 1985. The roles of amylose and amylopectin in the gelation and retrogradation of starch. Carbohydr. Res. 135:271–281.

Nunes, M. H. B., M. M. Moore, L. A. M. Ryan, and E. K. Arendt. 2009. Impact of emulsifiers on the quality and rheological properties of gluten-free breads and batters. Eur. Food Res. Technol. 228:633–642.

Pisesookbunterng, W., and B. L. D'Appolonia. 1983. Bread staling studies. I. effect of surfactants on moisture migration from crumb to crust and firmness values of bread crumb. Cereal Chem. 60:298–300.

Purhagen, J. K., M. E. Sjöö, and A.-C. Eliasson. 2008. Staling effects when adding low amounts of normal and heat-treated barley flour to a wheat bread. Cereal Chem. 85(2):109–114.

Purhagen, J. K., M. E. Sjöö, and A.-C. Eliasson. 2012. The anti-staling effect of pre-gelatinized flour and emulsifier in gluten-free bread. Eur. Food Res. Technol. 235:265–276.

Ring, S. G., P. Colonna, K. J. I'Anson, M. T. Kalichevsky, M. J. Miles, V. J. Morris, et al. 1987. The gelation and crystallization of amylopectin. Carbohydr. Res. 162:277–293.

Rogers, D. E., K. J. Zeleznak, C. S. Lai, and R. C. Hoseney. 1988. Effect of native lipids, shortening, and bread moisture on bread staling. Cereal Chem. 65:398–401.

Rosell, C. M., J. A. Rojas, and C. Benedito de Barber. 2001. Influence of hydrocolloids on dough rheology and bread quality. Food Hydrocolloids 15:75–81.

Sahlström, S., A. B. Bævre, and E. Bråthen. 2003. Impact of starches properties on hearth bread characteristics. II. Purified A- and B-granule fractions. J. Cereal Sci. 37:285–293.

Schiraldi, A., L. Piazza, O. Brenna, and E. Vittadini. 1996. Structure and properties of bread dough and crumb: calorimetric, rheological and mechanical investigations on the effects produced by hydrocolloids, pentosans and soluble proteins. J. Therm. Anal. 47:1339–1360.

Sciarini, L. S., P. D. Ribotta, A. E. León, and G. T. Pérez. 2010. Effect of hydrocolloids on gluten free batter properties and bread quality. Int. J. Food Sci. Technol. 45:2306–2312.

Sciarini, L. S., P. D. Ribotta, A. E. León, and G. T. Pérez. 2012. Incorporation of several additives into gluten free breads: effect on dough properties and bread quality. J. Food Eng. 111:590–597.

Selmair, P. L., and P. Koehler. 2010. Role of glycolipids in breadmaking. Lipid Technology 22:7–10.

Seyhun, N., G. Sumnu, and S. Sahin. 2005. Effects of different starch types on retardation of staling of microwave-baked cakes. Food Bioprod. Process. 83:1–5.

Stampfli, L., and B. Nersten. 1995. Emulsifiers in bread making. Food Chem. 52:353–360.

Zeleznak, K. K., and R. C. Hoseney. 1986. The role of water in the retrogradation of wheat starch gels and bread crumb. Cereal Chem. 63:401–407.

20

Individualistic impact of unit operations of production, at household level, on some antinutritional factors in selected cowpea-based food products

Mathew K. Bolade

Department of Food Science and Technology, Federal University of Technology, P.M.B. 704, Akure, Ondo State, Nigeria

Keywords

akara, antinutrient, *gbegiri*, leguminous grains, *moin-moin*

Correspondence

Mathew K. Bolade, Department of Food Science and Technology, Federal University of Technology, P.M.B. 704, Akure, Ondo State, Nigeria.
E-mail: mkbolade@futa.edu.ng

Funding Information

No funding information provided.

Abstract

The individualistic effect of unit operations of production, at household level, on some antinutritional factors in selected cowpea-based food products (moin-moin, akara, and gbegiri) was investigated. Four cowpea types (IT93K-452-1, IT95K-499s-35, IT97K-568-18, and market sample) were used for the study, whereas the three traditional food products were produced from each of the cowpea types, respectively. The results revealed that every unit operation involved in the production of moin-moin, akara or gbegiri contributed to the overall reduction of trypsin inhibitor activity (TIA), phytic acid (PA), and tannin; though at varying degrees. In the production of moin-moin, the major contributions to the overall reduction in TIA were from steaming (64.2–72.0%), second-stage soaking (9.7–11.9%), and dehulling (9.4–10.2%). The contributions to the overall reduction in PA were from dehulling (34.0–40.4%), preliminary soaking (15.4–21.0%), and steaming (7.8–14.0%), whereas that of tannin were from dehulling (39.7–47.6%), steaming (19.6–24.7%), and preliminary soaking (9.8–15.9%). For akara production, the major contributions to TIA reduction were from deep frying (64.2–72.0%), second-stage soaking (9.7–11.9%), and dehulling (9.4–10.2%). The PA reduction was from dehulling (34.0–40.4%), preliminary soaking (15.4–21.0%), and deep frying (9.6–15.9%), whereas that of tannin reduction was from dehulling (39.7–47.6%), deep frying (20.7–25.3%), and preliminary soaking (9.8–15.9%). In the production of gbegiri, the overall reduction in TIA was contributed from pressure cooking (79.0–84.8%), preliminary soaking (5.8–11.3%), and dehulling (9.4–10.2%). The reduction in PA was contributed by dehulling (34.0–40.4%), pressure cooking (24.7–35.0%), and preliminary soaking (15.4–21.0%), whereas the overall reduction in tannin content was similarly contributed by dehulling (39.7–47.6%), pressure cooking (29.8–34.4%), and preliminary soaking (9.8–15.9%).

Introduction

Cowpea (*Vigna unguiculata* L.) is an important grain legume in East and West African countries (Hung et al. 1990). Indeed, it is the most widely consumed legume seeds in Nigeria (Onigbinde and Akinyele 1983). The grain legume serves as the largest single contributor to the total protein intake of many rural and urban families while efforts are continually being made to increase cowpea production as a means of providing a cheaper source of

protein to the teeming consumers (Ogun et al. 1989). The benefit of low-cost dietary proteins from the traditional cowpea-based food products is considered enormous, particularly in the developing countries, due to the high cost and limited availability of animal proteins (Sathe and Salunkhe 1981). The chemical and nutritional compositions of cowpea, as well as its cooking properties, vary considerably according to environmental and genetic factors (Giami 2005). Many traditional food products are derivable from cowpea, particularly in Nigeria, which include *akara* (fried cowpea paste containing seasonings), *moin-moin* (steamed cowpea paste containing seasonings), *ekuru* (steamed cowpea paste with no seasonings), *gbegiri* (cowpea soup), *ewa ibeji* (cooked and softened cowpea), etc. The processing technologies for producing these cowpea-based traditional food products are generally at artisanal level and so efforts have been made toward the development of improved technologies for producing ready-to-use cowpea meal and flour specifically for use in *akara* and *moin-moin* preparation (Ngoddy et al. 1986; Phillips and McWatters 1991).

Apart from the nutritional benefits derivable from cowpea, the grain legume also contains certain antinutritional factors which include trypsin inhibitors (Kochhar et al. 1988; Onwuka 2006; Kalpanadevi and Mohan 2013), tannin (Akinyele 1989; Ogun et al. 1989; Ghavidel and Prakash 2007), and phytic acid (Ologhobo and Fetuga 1984; Uzogara et al. 1990; Khattab and Arntfield 2009), among others. These antinutrients serve as limiting factors in the utilization of cowpea for both human and animal as they make bioavailability of certain nutrients impossible. Trypsin inhibitor is a substance that has the ability to inhibit proteolytic activity of certain enzymes especially trypsin (Liener 2001). The negative effect of tannin has to do with its interference with protein digestion by binding dietary protein into an indigestible form (Bressani et al. 1982). The phytate has been implicated to decrease the bioavailability of essential minerals (Ca, Mg, Mn, Zn, Fe, and Cu) and can also form a phytate-protein complex thereby interfering in protein utilization (Oberleas and Harland 1981).

Various attempts have been made by researchers to increase the utilization of cowpea through a wide range of appropriate processing techniques. Wang et al. (1997) investigated the combined effects of soaking, water, and steam blanching on the antinutritional factors in cowpea. It was found that a combination of soaking and steam blanching resulted in higher reduction of trypsin inhibitor activity than a combination of soaking and water blanching. Preet and Punia (2000) studied the role of soaking, dehulling, and germination on the antinutritional content of cowpea. The finding here was that each of the treatments contributed significantly in reducing the phytic acid and polyphenol content of cowpea with dehulling being the most effective in the reduction of polyphenolic content. Onwuka (2006) assessed the effect of soaking, boiling, and combination of these treatments on the antinutritional factors in cowpea. It was found that the combination of soaking and boiling was more potent than individual soaking or boiling in the reduction of trypsin inhibitor, cyanogenic glycoside, hemagglutinin, alkaloids, and tannin.

Ghavidel and Prakash (2007) also investigated the impact of germination and dehulling on the antinutrient component of cowpea. The finding here was that phytic acid and tannin were fairly reduced by germination while the impact of dehulling was more effective. Khattab and Arntfield (2009) examined the influence of physical treatments (water soaking, boiling, roasting, microwave cooking, autoclaving, fermentation, and micronization) on the antinutritional component of cowpea. It was found that all treatments evaluated caused significant decreases in tannin, phytic acid, trypsin inhibitor activity, and oligosaccharides. However, boiling caused the highest reduction in tannin followed by autoclaving and microwave cooking, whereas autoclaving and germination were the most effective in reducing phytic acid content. All the heat treatments brought a total removal of trypsin inhibitor activity. Kalpanadevi and Mohan (2013) also studied the effect of hydration, cooking, autoclaving, germination and their combination on the reduction/elimination of antinutrients in *V. unguiculata*. It was found that hydration and germination processes were less effective in reducing trypsin inhibitor activity, whereas cooking and autoclaving of presoaked seeds were very effective for reducing the content of total free phenolics, tannin, phytic acid, hydrogen cyanide, trypsin inhibitors, and oligosaccharides.

Virtually all these investigations were focused toward the role of unit operations/processes, as might be encountered in the handling and processing of cowpea, on the reduction/removal of antinutritional factors but not targeted toward the production of specific cowpea-based food products. Even where specific cowpea-based food products were targeted (Akinyele 1989; Ogun et al. 1989), it was the cumulative role of the processing methods on the antinutritional factors that was evaluated rather than the individualistic impact of the unit operations of production. However, this study examined the contributions of individual unit operations of production to such cumulative effect.

The objective of this study therefore was to evaluate the sequential impact of unit operations of production, at household level, on some antinutritional factors in selected cowpea-based food products.

Materials and Methods

Source of materials

Four cowpea types were used for this study. Three varieties, namely IT93K-452-1, IT95K-499s-35, and IT97K-568-18 were obtained from the International Institute of Tropical Agriculture (IITA), Ibadan, Nigeria; whereas the fourth one, a market sample (*Ewa Oloyin*), was obtained from Bodija market, Ibadan, Nigeria. The cowpea varieties (IT93K-452-1, IT95K-499s-35, and IT97K-568-18) were white-coated, whereas the market sample (*Ewa Oloyin*) was brown-coated.

Production of *moin-moin, akara,* and *gbegiri* using household methods

Moin-moin (steamed cowpea paste containing seasonings) and *akara* (fried cowpea paste containing seasonings) were, respectively, produced as illustrated in Figure 1. The various points where samples were taken for analysis are numbered accordingly.

Gbegiri (cowpea soup) was also produced as illustrated in Figure 2. The various points where samples were taken for analysis are also numbered accordingly.

Three different households were used to prepare the products from where samples were collected for analysis, whereas each cowpea variety was used to prepare the food products, respectively.

Sample collection and preparation

Samples for the analysis were taken from 10 different points as indicated in Figures 1 and 2. Sample preparation was done following the method of Khattab and Arntfield (2009). Each sample was dried overnight using

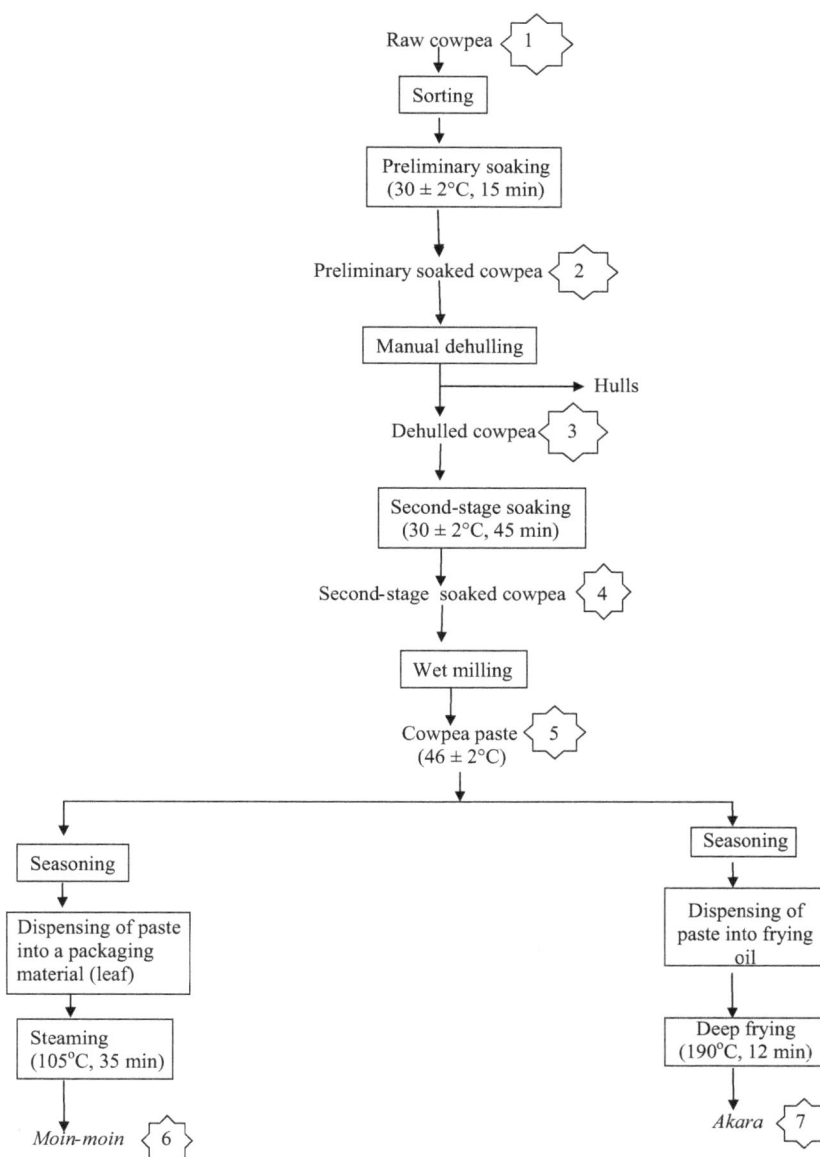

Figure 1. Flowchart illustrating the household production of *moin-moin* and *akara* from cowpea. (The indicated numbers are points of sample collection for analysis).

Raw cowpea ⟨1⟩

↓

Sorting

↓

Preliminary soaking
(30 ± 2°C, 15 min)

↓

Preliminary soaked cowpea ⟨2⟩

↓

Manual dehulling

↓ → Hulls

Dehulled cowpea ⟨3⟩

↓

Pressure cooking
(105°C, 45 min)

↓

Pressure-cooked dehulled cowpea ⟨8⟩

↓

Broom micronization

↓

Hot cowpea slurry ⟨9⟩

↓

Addition of seasonings

↓

Further cooking
(100°C, 5 min)

↓

Gbegiri (Cowpea soup) ⟨10⟩

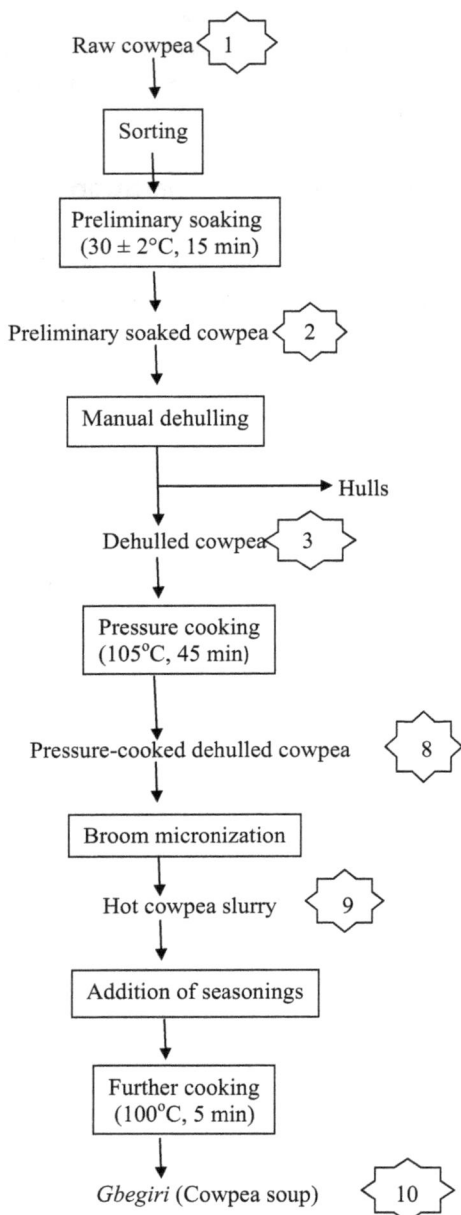

Figure 2. Flowchart illustrating the household production of *gbegiri* from cowpea. (The indicated numbers are points of sample collection for analysis).

air draught oven (Model No. DHG-910.1SA, Sanfa) at 55 ± 2°C, ground to pass through an 841-μm screen and kept frozen in polyethylene bags until required for analyzed.

Determination of trypsin inhibitor activity

The method of Kakade et al. (1974) was used to determine the trypsin inhibitor activity (TIA) of each sample using benzoyl-DL-arginine-*p*-nitroanilide (BAPNA) as substrate. A 4.0 g sample was treated with 40 mL of 0.05 mol/L sodium phosphate buffer, pH 7.5 and 40 mL of distilled water. The sample was agitated for 3 h using a magnetic stirrer and centrifuged at 700 g for 30 min at 15°C. Supernatant was diluted to obtain inhibition between 40 and 60% of enzyme activity. Incubation mixture consisted of 0.5 mL trypsin solution (5 mg/mL), 2 mL 2% (w/v) BAPNA, 1.0 mL sodium phosphate buffer (pH 7.5, 0.1 mol/L), 0.4 mL HCl (0.001 mol/L), and sample extract (0.1 mL). Total volume of incubation mixture was maintained at 4.0 mL. Incubation was carried out in a water bath at 37°C for 20 min after which 6.0 mL of 5% TCA (trichloroacetic acid) solution was added to stop the reaction. Blank sample was treated similarly through the entire determination. Absorbance (A) was read at 410 nm wavelength in a spectrophotometer (UV-160A; Shimadzu, Osaka, Japan). Results were expressed as trypsin inhibitor units (TIU). One TIU was defined as an increase of 0.01 in absorbance units under conditions of assay. Trypsin inhibitory activity was defined as the number of TIU.

Determination of phytic acid content

The method of Wheeler and Ferrel (1971) was used to determine the phytic acid content of each sample. A 2.0 g sample measurement was used for the extraction. A standard curve was prepared expressing the results as $Fe(NO_3)_3$ equivalent. Phytate phosphorus was calculated from the standard curve assuming a 4:6 iron to phosphorus molar ratio. The phytic acid content was also calculated by multiplying the amount of phytate phosphorous by the factor 3.55 based on the empirical formula $C_6P_6O_2H_{18}$ and result expressed as mg/100 g sample.

Determination of tannin content

Tannin was determined according to the method of Price and Butler (1977) with a minor modification. Sixty milligrams (60 mg) of ground sample were shaken manually for 1 min in 3.0 mL methanol. The mixture was filtered followed by mixing the filtrate with 50 mL distilled water and analyzed within an hour. About 3.0 mL of 0.1 mol/L $FeCl_3$ in 0.1 mol/L HCl were added to 1 mL filtrate, followed immediately by the addition of 3.0 mL freshly prepared $K_3Fe(CN)_6$. The absorbance was read on a spectrophotometer (Shimadzu UV-1700, Tokyo, Japan) at 720 nm after 10 min from the addition of 3.0 mL of 0.1 M $FeCl_3$ and 3.0 mL of 0.008 mol/L $K_3Fe(CN)_6$. Similar treatments were also carried on the blank. Results were expressed as tannic acid equivalent (mg/100 g sample), calculated from a calibration curve using tannic acid.

Temperature measurement

The temperature of the environment and sample was measured using different types of thermometer (0–110°C and 0–360°C).

Statistical analysis

All determinations reported in this study were carried out in triplicates. In each case, a mean value and standard deviation were calculated. Analysis of variance (ANOVA) was also performed and separation of the mean values was by Duncan's Multiple Range Test at $P < 0.05$ using Statistical Package for Social Scientists (SPSS) software, version 16.0 (SPSS Inc., Chicago, IL); on a personal computer.

Results and Discussion

Elimination of trypsin inhibitor activity in *moin-moin*, *akara*, and *gbegiri* as influenced by the unit operations of production

Table 1 summarizes the effect of unit operation of production and cowpea variety on the trypsin inhibitor activity (TIA) in *moin-moin* (steamed cowpea paste) and *akara* (fried cowpea paste), whereas Table 2 summarizes that in *gbegiri* (cowpea soup). The cowpea varieties used for preparing the food products contained different concentrations of TIA ranging between 2349.7 and 2844.2 TIU/g with significant difference ($P \leq 0.05$). Sample IT95K-499s-35 and the market sample gave the lowest and highest TIA values, respectively. Genetic and environmental factors have been implicated to be responsible for variability in the composition of antinutrients in cowpea (Terryn and Montagu 2008; Carvalho et al. 2012; Owolabi et al. 2012). The contributory reduction of TIA by the preliminary soaking, as a common unit operation in the production of *moin-moin*, *akara*, and *gbegiri*, respectively, ranged between 6.2 and 11.3%. It had earlier been observed that soaking plays a significant role in the reduction of trypsin inhibitor activity as the inhibitor is a low molecular weight protein (Clemente and Domoney 2006) capable of leaching during soaking.

The contributory reduction capacity of dehulling with respect to TIA ranged between 9.4 and 10.2%. The lowest and highest reduction capacities were observed in the market sample and IT95K-499s-35, respectively. The implication of this observation is that a modest amount of TIA is present in the seed coat while the variations in the contributory reduction of TIA, through dehulling, in the cowpea varieties are indications that varietal differences do play a role in TIA reduction

(Ologhobo and Fetuga 1984). The second-stage soaking contributed between 9.7 and 11.9% to the overall reduction of TIA in the course of *moin-moin* and *akara* production, respectively. The lowest reduction capacity was observed in IT93K-452-1, whereas the highest reduction capacity was observed in both IT95K-499s-35 and IT97K-568-18. The reduction in TIA during the second-stage soaking was higher than that of the preliminary soaking in all the cowpea varieties and this observation may be attributed to the elongated second-stage soaking period (45 min) which might have allowed greater leaching of trypsin inhibitor into the soaking water. In addition, the absence of the seed coat (hulls) during the second-stage soaking period might have allowed trypsin inhibitor to leach more easily.

The wet milling of cowpea to obtain the paste contributed between 1.6 and 4.0% to the overall reduction of TIA in the production of *moin-moin* and *akara*, respectively. This modest contributory reduction in the TIA may be attributed to an enlarged surface area of the paste with higher temperature (46 ± 2°C). The steaming (*moin-moin*) and deep frying (*akara*) operations contributed between 64.2 and 72.0% for both products. This is the highest contributory reduction in TIA for these two food products as total elimination was observed. Both IT97K-568-18 and the market sample exhibited the lowest and highest reduction capacities, respectively, for the TIA. It had earlier been observed that high-temperature processing operations such as boiling, steaming, microwave cooking, roasting, or autoclaving are capable of causing total elimination of TIA (Burns 1986; Liener 1986; Vidal-Valverde et al. 1994).

In the case of *gbegiri* production (Table 2), the TIA was equally totally eliminated by the pressure cooking giving a contributory reduction capacity of between 79.0 and 84.8%. Both IT97K-568-18 and the market sample exhibited the lowest and highest reduction capacities, respectively. The seeming higher contributory reduction capacity of pressure cooking for TIA than that of steaming and deep frying may be attributed to the higher TIA concentration in the dehulled cowpea prior to the heat treatment.

The two subsequent unit operations (broom micronization and further cooking) involved in *gbegiri* production did not contribute in any way to TIA elimination in the food product. Broom micronization is essentially a unit operation in *gbegiri* production which involves the use of a short-length broom to manually beat already cooked and softened cowpea in order to obtain free flow cowpea slurry. The zero-level TIA found in *gbegiri* as observed in this study seems to contradict the findings of a previous worker (Akinyele 1989) who reported a residual TIA in *gbegiri*.

Table 1. Effect of unit operation of production on the trypsin inhibitor activity (TIA) in *moin-moin* and *akara*[1].

Material	Corresponding unit operation	Source of *moin-moin* and *akara*							
		IT93K-452-1		IT95K-499s-35		IT97K-568-18		Market sample (*Ewa-oloyin*)	
		Trypsin inhibitor activity (TIU/g sample)	Contributory reduction capacity[2] (%)	Trypsin inhibitor activity (TIU/g sample)	Contributory reduction capacity (%)	Trypsin inhibitor activity (TIU/g sample)	Contributory reduction capacity (%)	Trypsin inhibitor activity (TIU/g sample)	Contributory reduction capacity (%)
Raw cowpea		2523.4 ± 13.4^c	–	2349.7 ± 9.8^d	–	2638.5 ± 12.6^b	–	2844.2 ± 11.7^a	–
Preliminary soaked cowpea	Preliminary soaking	2366.1 ± 6.2^b	6.2	2152.4 ± 10.2^d	8.4	2341.8 ± 8.5^c	11.3	2679.1 ± 9.7^a	5.8
Dehulled cowpea	Dehulling	2111.5 ± 11.6^b	10.1	1913.8 ± 6.6^d	10.2	$2084.2 \pm 7.6c$	9.8	2412.3 ± 5.5^a	9.4
Second-stage soaked cowpea	Second-stage soaking	1866.7 ± 5.7^b	9.7	1633.2 ± 5.1^d	11.9	1771.3 ± 4.5^c	11.9	2114.6 ± 7.2^a	10.5
Paste	Wet milling	1792.2 ± 8.3^b	4.0	1595.6 ± 9.1^d	1.6	1692.5 ± 8.1^c	3.8	2047.9 ± 6.3^a	2.3
Moin-moin	Steaming	0	71.0	0	67.9	0	64.2	0	72.0
Akara	Deep frying	0	71.0	0	67.9	0	64.2	0	72.0
Overall reduction in trypsin inhibitor activity (%)		Moinmoin = 100 / Akara = 100		Moinmoin = 100 / Akara = 100		Moinmoin = 100 / Akara = 100		Moinmoin = 100 / Akara = 100	

[1]Results are mean values of data from three different households ± standard deviation. Mean value within the same row having the same letter are not significantly different at $P \leq 0.05$.

[2]Contributory reduction capacity (%) was calculated with respect to the initial total trypsin inhibitor activity (TIA) in the raw cowpea.

Table 2. Effect of unit operation of production on the trypsin inhibitor activity (TIA) in gbegiri[1].

Material	Corresponding unit operation	Source of gbegiri							
		IT93K-452-1		IT95K-499s-35		IT97K-568-18		Market sample (Ewa-oloyin)	
		Trypsin inhibitor activity (TIU/g sample)	Contributory reduction capacity[2] (%)	Trypsin inhibitor activity (TIU/g sample)	Contributory reduction capacity (%)	Trypsin inhibitor activity (TIU/g sample)	Contributory reduction capacity (%)	Trypsin inhibitor activity (TIU/g sample)	Contributory reduction capacity (%)
Raw cowpea		2523.4 ± 13.4^c	–	2349.7 ± 9.8^d	–	2638.5 ± 12.6^b	–	2844.2 ± 11.7^a	–
Preliminary soaked cowpea	Preliminary soaking	2366.1 ± 6.2^b	6.2	2152.4 ± 10.2^d	8.4	2341.8 ± 8.5^c	11.3	2679.1 ± 9.7^a	5.8
Dehulled cowpea	Dehulling	2111.5 ± 11.6^b	10.1	1913.8 ± 6.6^d	10.2	2084.2 ± 7.6^c	9.8	2412.3 ± 5.5^a	9.4
Pressure-cooked dehulled cowpea	Pressure cooking	0	83.7	0	81.4	0	79.0	0	84.8
Hot cowpea slurry	Broom micronization	0	0	0	0	0	0	0	0
Gbegiri	Further cooking	0	0	0	0	0	0	0	0
Overall reduction in trypsin inhibitor activity of gbegiri (%)			100		100		100		100

[1]Results are mean values of data from three different households ± standard deviation. Mean value within the same row having the same letter are not significantly different at $P \leq 0.05$.
[2]Contributory reduction capacity (%) was calculated with respect to the initial total trypsin inhibitor activity (TIA) in the raw cowpea.

Table 3. Effect of unit operation of production on the phytic acid (PA) content in *moin-moin* and *akara*[1].

| | | Source of *moin-moin* and *akara* | | | | | | Market sample (*Ewa-oloyin*) | |
| | | IT93K-452-1 | | IT95K-499s-35 | | IT97K-568-18 | | | |
Material	Corresponding unit operation	Phytic acid content (mg/100 g sample)	Contributory reduction capacity[2] (%)	Phytic acid content (mg/100 g sample)	Contributory reduction capacity (%)	Phytic acid content (mg/100 g sample)	Contributory reduction capacity (%)	Phytic acid content (mg/100 g sample)	Contributory reduction capacity (%)
Raw cowpea		723.9 ± 5.6^c	–	784.2 ± 6.8^b	–	680.5 ± 7.1^d	–	984.3 ± 6.3^a	–
Preliminary soaked cowpea	Preliminary soaking	612.4 ± 4.2^b	15.4	619.6 ± 4.4^b	21.0	551.8 ± 5.5^c	18.9	792.4 ± 7.1^a	19.5
Dehulled cowpea	Dehulling	358.2 ± 6.8^b	35.0	302.8 ± 6.1^c	40.4	307.2 ± 4.9^c	35.9	458.2 ± 4.8^a	34.0
Second-stage soaked cowpea	Second-stage soaking	292.4 ± 3.7^b	9.1	238.3 ± 3.5^d	8.2	253.4 ± 3.7^c	7.9	392.3 ± 5.7^a	6.7
Paste	Wet milling	204.8 ± 4.2^b	12.1	163.7 ± 3.1^d	9.5	184.7 ± 4.6^c	10.1	304.9 ± 4.1^a	8.9
Moin-moin	Steaming	134.7 ± 2.5^b	9.7	102.4 ± 2.8^c	7.8	93.6 ± 2.5^d	13.4	166.7 ± 3.4^a	14.0
Akara	Deep frying	111.3 ± 2.9^b	12.9	88.1 ± 2.2^c	9.6	77.9 ± 2.2^d	15.7	148.2 ± 2.8^a	15.9
Overall reduction in phytic acid content (%)		Moinmoin = 81.3 Akara = 84.5		Moinmoin = 86.9 Akara = 88.7		Moinmoin = 86.2 Akara = 88.5		Moinmoin = 83.1 Akara = 85.0	

[1]Results are mean values of data from three different households ± standard deviation. Mean value within the same row having the same letter are not significantly different at $P \leq 0.05$.

[2]Contributory reduction capacity (%) was calculated with respect to the initial phytic acid (PA) in the raw cowpea.

Table 4. Effect of unit operation of production on the phytic acid (PA) content in *gbegiri*[1].

Material	Corresponding unit operation	Source of gbegiri							
		IT93K-452-1		IT95K-499s-35		IT97K-568-18		Market sample (Ewa-oloyin)	
		Phytic acid content (mg/100 g sample)	Contributory reduction capacity[2] (%)	Phytic acid content (mg/100 g sample)	Contributory reduction capacity (%)	Phytic acid content (mg/100 g sample)	Contributory reduction capacity (%)	Phytic acid content (mg/100 g sample)	Contributory reduction capacity (%)
Raw cowpea	–	723.9 ± 5.6^c	–	784.2 ± 6.8^b	–	680.5 ± 7.1^d	–	984.3 ± 6.3^a	–
Preliminary soaked cowpea	Preliminary soaking	612.4 ± 4.2^b	15.4	619.6 ± 4.4^b	21.0	551.8 ± 5.5^c	18.9	792.4 ± 7.1^a	19.5
Dehulled cowpea	Dehulling	358.2 ± 6.8^b	35.0	302.8 ± 6.1^c	40.4	307.2 ± 4.9^c	35.9	458.2 ± 4.8^a	34.0
Pressure-cooked dehulled cowpea	Pressure cooking	112.9 ± 3.1^c	33.9	109.4 ± 5.2^c	24.7	123.7 ± 3.3^b	27.0	203.7 ± 4.4^a	25.9
Hot cowpea slurry	Broom micronization	103.4 ± 2.5^{bc}	1.3	98.5 ± 4.1^c	1.4	108.6 ± 4.1^b	2.2	184.9 ± 5.2^a	1.9
Gbegiri	Further cooking	42.8 ± 2.2^c	8.4	45.7 ± 2.3^c	6.7	52.9 ± 2.7^b	8.2	63.4 ± 3.2^a	12.3
Overall reduction in phytic acid content of gbegiri (%)			94.1		94.2		92.2		93.6

[1]Results are mean values of data from three different households ± standard deviation. Mean value within the same row having the same letter are not significantly different at $P \leq 0.05$.

[2]Contributory reduction capacity (%) was calculated with respect to the initial phytic acid (PA) in the raw cowpea.

Table 5. Effect of unit operation of production on the tannin content in *moin-moin* and *akara*[1].

Material	Corresponding unit operation	Source of *moin-moin* and *akara*						Market sample (*Ewa-oloyin*)	
		IT93K-452-1		IT95K-499s-35		IT97K-568-18			
		Tannin content (mg/100 g sample)	Contributory reduction capacity[2] (%)	Tannin content (mg/100 g sample)	Contributory reduction capacity (%)	Tannin content (mg/100 g sample)	Contributory reduction capacity (%)	Tannin content (mg/100 g sample)	Contributory reduction capacity (%)
Raw cowpea	–	2219.4 ± 13.1c	–	2009.3 ± 10.5d	–	2342.8 ± 14.3b	–	2411.4 ± 12.7a	–
Preliminary soaked cowpea	Preliminary soaking	1866.2 ± 8.6c	15.9	1812.7 ± 11.2d	9.8	2053.7 ± 12.6b	12.3	2114.8 ± 11.2a	12.3
Dehulled cowpea	Dehulling	984.7 ± 6.4a	39.7	884.5 ± 7.8c	46.2	980.3 ± 7.8ab	45.8	968.2 ± 9.8b	47.6
Second-stage soaked cowpea	Second-stage soaking	804.5 ± 8.2b	8.1	780.3 ± 5.5c	5.2	788.9 ± 6.3c	8.2	857.3 ± 6.1a	4.6
Paste	Wet milling	513.6 ± 4.4c	13.1	562.8 ± 3.2b	10.8	567.4 ± 6.9b	9.5	624.7 ± 7.3a	9.7
Moin-moin	Steaming	78.2 ± 3.1a	19.6	66.9 ± 2.9bc	24.7	71.7 ± 4.8ab	21.2	62.4 ± 3.9c	23.3
Akara	Deep frying	53.4 ± 2.5a	20.7	54.8 ± 3.6a	25.3	49.3 ± 3.2a	22.1	41.5 ± 2.8b	24.2
Overall reduction in tannin content (%)		Moinmoin = 96.4 Akara = 97.5		Moin-moin = 96.7 Akara = 97.3		Moin-moin = 97.0 Akara = 97.9		Moin-moin = 97.5 Akara = 98.4	

[1]Results are mean values of data from three different households ± standard deviation. Mean value within the same row having the same letter are not significantly different at $P \leq 0.05$.
[2]Contributory reduction capacity (%) was calculated with respect to the initial tannin content in the raw cowpea.

Table 6. Effect of unit operation of production on the tannin content in *gbegiri*[1].

Material	Corresponding unit operation	Source of *gbegiri*							
		IT93K-452-1		IT95K-499s-35		IT97K-568-18		Market sample (Ewa-oloyin)	
		Tannin content (mg/100 g sample)	Contributory reduction capacity[2] (%)	Tannin content (mg/100 g sample)	Contributory reduction capacity (%)	Tannin content (mg/100 g sample)	Contributory reduction capacity (%)	Tannin content (mg/100 g sample)	Contributory reduction capacity (%)
Raw cowpea	–	2219.4 ± 13.1[c]	–	2009.3 ± 10.5[d]	–	2342.8 ± 14.3[b]	–	2411.4 ± 12.7[a]	–
Preliminary soaked cowpea	Preliminary soaking	1866.2 ± 8.6[c]	15.9	1812.7 ± 11.2[d]	9.8	2053.7 ± 12.6[b]	12.3	2114.8 ± 11.2[a]	12.3
Dehulled cowpea	Dehulling	984.7 ± 6.4[a]	39.7	884.5 ± 7.8[c]	46.2	980.3 ± 7.8[ab]	45.8	968.2 ± 9.8[b]	47.6
Pressure-cooked dehulled cowpea	Pressure cooking	221.9 ± 3.5[c]	34.4	243.8 ± 4.1[b]	31.9	259.2 ± 5.2[a]	30.8	248.7 ± 6.3[b]	29.8
Hot cowpea slurry	Broom micronization	202.5 ± 3.8[c]	0.9	230.3 ± 4.5[ab]	0.7	237.7 ± 3.9[a]	0.9	229.2 ± 4.7[b]	0.8
Gbegiri	Further cooking	62.4 ± 2.3[d]	6.3	71.6 ± 3.2[c]	7.9	80.5 ± 2.8[b]	6.7	86.6 ± 2.6[a]	5.9
Overall reduction in tannin content of *gbegiri* (%)			97.2		96.5		96.5		96.4

[1]Results are mean values of data from three different households ± standard deviation. Mean value within the same row having the same letter are not significantly different at $P \leq 0.05$.

[2]Contributory reduction capacity (%) was calculated with respect to the initial tannin content in the raw cowpea.

Effect of unit operations of production of *moin-moin, akara,* and *gbegiri* on the phytic acid

The effect of unit operations of production on the phytic acid (PA) content of *moin-moin* and *akara* is shown in Table 3 whereas that of *gbegiri* is in Table 4. There were variations in the PA content of the raw cowpea ranging between 680.5 and 984.3 mg/100 g with significant difference ($P \leq 0.05$). Both IT97K-568-18 and the market sample had the lowest and highest PA values, respectively. The preliminary soaking was found to contribute between 15.4 and 21.0% to the overall PA reduction in the production of *moin-moin, akara,* and *gbegiri*. Both IT93K-452-1 and IT95K-499s-35 exhibited the lowest and highest reduction capacities for the PA, respectively. Soaking had earlier been observed to be capable of reducing PA content in cowpea due to its leaching tendency into the soaking water (Preet and Punia 2000; Onwuka 2006) in addition to the hydrolytic tendency of PA by the endogenous phytase (Reddy and Pierson 1994; Sandberg and Andlid 2002).

The dehulling operation contributed between 34.0 and 40.4% to the overall reduction of PA in the production of *moin-moin, akara,* and *gbegiri,* respectively. The lowest and highest reduction capacities for PA were exhibited by the market sample and IT95K-499s-35, respectively. Most PA had been observed to be present in the outer aleurone layers of leguminous seeds (Deshpande et al. 1982) with the implication that dehulling could substantially remove it, hence the relative high PA reduction as observed in this study.

The contributory reduction capacity of the second-stage soaking to the overall PA reduction ranged between 6.7 and 9.1%. The reduction levels of PA by the second-stage soaking were generally observed to be lower than that of the preliminary soaking in spite of the longer soaking period. This observation may be due to the previous removal of the hulls which, most probably, had contributed to the reduced leaching of PA into the soaking water. The wet milling of the dehulled cowpea contributed between 8.9 and 12.1% to the overall PA reduction in the course of *moin-moin* and *akara* production, respectively. Wet milling essentially resulted in paste with a larger surface area and higher temperature ($46 \pm 2°C$) which could facilitate an effective enzyme-substrate interaction. Grenier and Konietzny (1999) had earlier observed that the optimal temperatures for the intrinsic plant phytases were between 45°C and 65°C. This might have influenced the modest contribution of wet milling to the overall reduction of the PA.

The steaming operation in *moin-moin* production was observed to contribute between 7.8 and 14.0% to the overall PA reduction, whereas deep frying in *akara* production exhibited a contributory reduction capacity of between 9.6 and 15.9%. Both IT95K-499s-35 and the market sample contributed the lowest and highest reduction levels, respectively, for both products. Heat treatment generally had been observed to reduce PA concentration partly due to the heat-labile nature of the acid coupled with possible formation of insoluble complexes such as calcium and magnesium phytates (Crean and Haisman 1963; Udensi et al. 2007). Nevertheless, the contributory reduction capacity of deep frying to the overall PA reduction was generally observed to be higher than that of steaming due to its higher temperature (190°C) as against that of 105°C for steaming. The overall reduction in PA content in *moin-moin* was observed to range between 81.3 and 86.9%, whereas that in *akara* was between 84.5 and 88.7%. This implies that the two food products, as consumed, could still contain residual PA.

In the production of *gbegiri* (Table 4), the contributory reduction capacity of pressure cooking in the overall reduction of PA was between 24.7 and 33.9%. Broom micronization, as a unit operation of production, accounted for 1.3–2.2% reduction capacity level in the overall PA reduction, whereas further cooking also accounted for 6.7–12.3%. The contributory reduction capacities of pressure cooking and further cooking may be attributed to thermal destruction, whereas that of broom micronization may be attributed to surface area enlargement which might have facilitated somewhat mechanical destruction. The overall reduction in PA content in *gbegiri* was observed to range between 92.2 and 94.2%; which implies that *gbegiri,* as consumed, could still contain residual PA.

Reduction of tannin content in *moin-moin, akara,* and *gbegiri* as influenced by the unit operations of production

The effect of unit operation of production on the tannin content of *moin-moin* and *akara* is presented in Table 5, whereas that of *gbegiri* is in Table 6. The concentration of tannin in the raw cowpea ranged between 2009.3 and 2411.4 mg/100 g with significant difference ($P \leq 0.05$). Both IT95K-499s-35 and the market sample gave the lowest and highest values, respectively. The preliminary soaking, as a common unit operation of production for *moin-moin, akara,* and *gbegiri,* exhibited a contributory reduction capacity of between 9.8 and 15.9%. Samples IT93K-452-1 and IT95K-499s-35 gave the lowest and highest reduction levels, respectively. The tannin reduction through preliminary soaking may be attributed to water solubility property of tannin (Kumar et al. 1979) which predisposed it toward solubilizing into the soaking water.

The contributory reduction capacity of dehulling in the overall tannin reduction during *moin-moin, akara,* and *gbegiri* production was observed to range between 39.7 and 47.6%. IT93K-452-1 and the market sample gave the lowest and highest reduction levels, respectively. It had earlier been observed that tannin is predominantly located in the leguminous seed coats (Reddy and Pierson 1994) and therefore dehulling has the capability of its substantial removal. The second-stage soaking contributed between 4.6 and 8.2% to the overall tannin reduction; lower than that of the preliminary soaking. This may be due to the initial removal of the hulls before soaking which, most probably, had contributed to a reduced solubilization of tannin into the soaking water.

The contribution of wet milling, as a unit operation, to the overall reduction of tannin during *moin-moin* and *akara* production ranged between 9.5 and 13.1%. IT97K-568-18 and IT93K-452-1 gave the lowest and highest reduction levels, respectively. During wet milling, paste is formed and the temperature of the paste usually increases to 46 ± 2°C, whereas the overall surface area of the paste is enlarged. All these factors could facilitate an efficient enzyme-substrate interaction in the paste thereby leading to tannin reduction. It had earlier been observed that tannin could be oxidized by polyphenol oxidase (endogenous enzyme) which might lead to reduction in its concentration (Reddy and Pierson 1994).

The contributory reduction capacity of steaming in the overall tannin reduction ranged between 19.6 and 24.7%. However, the contributory reduction capacity of deep frying ranged between 20.7 and 25.3%. The high temperature of steaming (105°C) and deep frying (190°C) may be implicated for this modest tannin reduction. It had earlier been observed that tannin could be degraded upon heat treatment such as boiling, roasting, and microwave cooking (Rakic et al. 2007; Udensi et al. 2007).

In the production of *gbegiri* (Table 6), the contributory reduction capacity of pressure cooking for tannin ranged between 29.8 and 34.4%. The market sample and IT93K-452-1 contributed the lowest and highest reduction levels, respectively. Broom micronization accounted for 0.7–0.9% reduction levels in tannin while further cooking also accounted for 5.9–7.9% reduction level. Pressure cooking and further cooking had the capacity for thermal destruction of tannin, whereas broom micronization might have caused somewhat marginal mechanical destruction. Nevertheless, the overall reduction of tannin in *moin-moin* (96.4–97.5%), *akara* (97.3–98.4%), and *gbegiri* (96.4–97.2%); all indicated that the food products, as consumed, could still contain residual tannin content. This observation seems to contradict the findings of previous workers (Ogun et al. 1989) who reported zero-level tannin in *moin-moin*.

Conclusion

The conclusion that can be drawn from this study is that the various unit operations through which cowpea is processed, in the course of preparing *moin-moin, akara,* and *gbegiri*, contributed individually to the overall reduction in TIA, PA, and tannin. In the preparation of *moin-moin*, the highest contributory reduction capacity for TIA was obtained from the steaming operation (64.2–72.0%); dehulling and preliminary soaking operations exhibited contributions of 34.0–40.4% and 15.4–21.0% for PA, respectively, whereas dehulling and steaming operations showed contributions of 39.7–47.6% and 19.6–24.7% for tannin, respectively.

For *akara* production, the deep frying operation had the highest reduction capacity (64.2–72.0%) for TIA; dehulling and preliminary soaking operations similarly showed contributions of 34.0–40.4% and 15.4–21.05 for PA, respectively, whereas dehulling and deep frying operations revealed contributions of 39.7–47.6% and 20.7–25.3% for tannin, respectively.

In the production of *gbegiri*, it was the pressure cooking that exhibited the highest contributory reduction capacity (79.0–84.8%) for TIA, whereas dehulling and pressure cooking operations exhibited contributions of 34.0–40.4% and 24.7–33.9% for PA as well as contributions of 39.7–47.6% and 29.8–34.4% for tannin, respectively.

Acknowledgments

The author is grateful to the families of Mrs. E.A. Majiyagbe, Mrs. C.O. Fabiyi, and Mrs. F.F.A. Ibironke whose households were used for sample preparation. The author also wishes to thank Messrs. B.D. Ayodeji and O.E. Adetiba for their involvement in sample collection and logistics.

Conflict of Interest

None declared.

References

Akinyele, I. O. 1989. Effects of traditional methods of processing on the nutrient content and some antinutrient in cowpeas (*Vigna unguiculata*). Food Chem. 33:291–299.

Bressani, R., L. G. Elias, and J. E. Braham. 1982. Reduction of digestibility of legume proteins by tannins. J. Plant Food 4:43–55.

Burns, R. A. 1986. Protease inhibitors in processed plant foods. J. Food Prot. 50:161–166.

Carvalho, A. F. U., N. M. DeSousa, D. F. Farias, L. C. D. Rocha-Bezerra, R. M. P. Silva, M. P. Viana, et al. 2012.

Nutritional ranking of 30 Brazilian genotypes of cowpeas including determination of antioxidant capacity and vitamins. J. Food Compos. Anal. 26:81–88.

Clemente, A., and C. Domoney. 2006. Biological significance of polymorphism in legume protease inhibitors from the Bowmane Birk family. Curr. Prot. Pept. Sci. 7:210–216.

Crean, D. E. C., and D. R. Haisman. 1963. The interaction between phytic acid and divalent cations during the cooking of dried peas. J. Sci. Food Agric. 14:824–833.

Deshpande, S. S., S. K. Sathe, D. K. Salunkhe, and D. P. Cornforth. 1982. Effects of dehulling on phytic acid, polyphenols and enzyme inhibitors of dry bean (Phaseolus vulgaris L.). J. Food Sci. 47:1846–1850.

Ghavidel, R. A., and J. Prakash. 2007. The impact of germination and dehulling on nutrients, antinutrients, in vitro iron and calcium bioavailability and in vitro starch and protein digestibility of some legume seeds. LWT-Food Sci. Technol. 40:1292–1299.

Giami, S. Y. 2005. Compositional and nutritional properties of selected newly developed lines of cowpea (Vigna unguiculata L.Walp). J. Food Compos. Anal. 18:665–673.

Grenier, R., and U. Konietzny. 1999. Improving enzymatic reduction of myo-inositol phosphates with inhibitory effects on mineral absorption in black beans (Phaseolus vulgaris var. preto). J. Food Process. Preserv. 23:249–261.

Hung, Y. C., K. H. McWatters, R. D. Phillips, and M. S. Chinnan. 1990. Effect of pre-decortication drying treatment on the microstructure of cowpea products. J. Food Sci. 55:774–776.

Kakade, M. L., J. J. Rackis, J. E. McGhee, and G. Puski. 1974. Determination of trypsin inhibitor activity of soy bean products: a collaborative analysis of an improved procedure. Cereal Chem. 51:376–382.

Kalpanadevi, V., and V. R. Mohan. 2013. Effect of processing on antinutrients and in vitro protein digestibility of the underutilized legume, Vigna unguiculata (L.) Walp subsp. nguiculata. LWT-Food Sci. Technol. 51:455–461.

Khattab, R. Y., and S. D. Arntfield. 2009. Nutritional quality of legume seeds as affected by some physical treatments. 2. Antinutritional factors. LWT-Food Sci. Technol. 42:1113–1118.

Kochhar, N., A. F. Walker, and D. J. Pike. 1988. Effect of variety on protein content, amino acid composition and trypsin inhibitor activity of cowpeas. Food Chem. 29:65–78.

Kumar, N. R., A. N. Reedy, and K. N. Rao. 1979. Levels of phenolic substances in the leachates of cicer seeds. Indian J. Exp. Biol. 17:114–116.

Liener, I. E. 1986. Trypsin inhibitors: concern for human nutrition or not? J. Nutr. 116:920–923.

Liener, I. E. 2001. Antinutritional factors related to proteins and amino acids. Pp. 257–298 in Y. H. Hui, R. A. Smith and D. G. Spoerke, eds. Foodborne disease handbook. 2nd ed. Marcel Dekker, New York.

Ngoddy, P. O., N. J. Enwere, and V. I. Onuorah. 1986. Cowpea flour performance in akara and moin-moin preparation. Trop. Sci. 26:101–119.

Oberleas, D., and B. F. Harland. 1981. Phytate content of foods: effect on dietary zinc bioavailablility. J. Am. Diet. Assoc. 79:433–436.

Ogun, P. O., P. Markakis, and W. Chenoweth. 1989. Effect of processing on certain antinutrients in cowpeas (Vigna unguiculata). J. Food Sci. 54:1084–1085.

Ologhobo, A. D., and B. L. Fetuga. 1984. Distribution of phosphorus and phytate in some Nigerian varieties of legumes and some effects of processing. J. Food Sci. 49:199–201.

Onigbinde, A. O., and I. O. Akinyele. 1983. Oligosaccharide content of 20 varieties of cowpeas in Nigeria. J. Food Sci. 48:1250–1251.

Onwuka, G. I. 2006. Soaking, boiling and antinutritional factors in pigeon peas (Cajanus cajan) and cowpeas (Vigna unguiculata). J. Food Process. Preserv. 30:616–630.

Owolabi, A. O., U. S. Ndidi, B. D. James, and F. A. Amune. 2012. Proximate, antinutrient and mineral composition of five varieties (improved and local) of cowpea, Vigna unguiculata, commonly consumed in Samaru community, Zaria-Nigeria. Asian J. Food Sci. Technol. 4:70–72.

Phillips, R. D., and K. H. McWatters. 1991. Contribution of cowpea to nutrition and health. Food Technol. 45:127–130.

Preet, K., and D. Punia. 2000. Antinutrients and digestibility (in vitro) of soaked, dehulled and germinated cowpeas. Nutr. Health 14:109–117.

Price, M. L., and L. G. Butler. 1977. Rapid visual estimation and spectrophotometric determination of tannin content of sorghum grain. J. Agric. Food Chem. 25:1268–1273.

Rakic, S., S. Petrovic, J. Kukic, M. Jadranin, V. Tesevic, D. Povrenovic, et al. 2007. Influence of thermal treatment on phenolic compounds and antioxidant properties of oak acorns from Serbia. Food Chem. 104:830–834.

Reddy, N. R., and M. D. Pierson. 1994. Reduction in antinutritional and toxic components in plant foods by fermentation. Food Res. Int. 27:281–290.

Sandberg, A. S., and T. Andlid. 2002. Phytogenic and microbial phytases in human nutrition. Int. J. Food Sci. Technol. 37:823–833.

Sathe, S. K., and D. K. Salunkhe. 1981. Preparation and utilization of protein concentrates and isolates for nutritional and functional improvement of foods. J. Food Qual. 4:145–233.

Terryn, N., and M. V. Montagu. 2008. Genetic transformation of Phaseolus and Vigna species. Pp. 159–168 *in* P. B. Kirti, ed. Handbook of new technologies for genetic improvement of legumes. CRC Press, Boca Raton.

Udensi, E. A., F. C. Ekwu, and J. N. Isinguzo. 2007. Antinutrient factors of vegetable cowpea (Sesquipedalis) seeds during thermal processing. Pak. J. Nutr. 6:194–197.

Uzogara, S. G., L. Morton, and J. W. Daniel. 1990. Changes in some antinutrients of cowpea (*Vigna unguiculata*) processed with "kanwa" alkaline salt. Plant Food Hum. Nutr. 40:249–258.

Vidal-Valverde, C., J. Frias, I. Estrella, M. J. Gorospe, R. Ruiz, and J. Bacon. 1994. Effect of processing on some antinutritional factors of lentils. J. Agric. Food Chem. 42:2291–2295.

Wang, N., M. J. Lewis, J. G. Brennan, and A. Westby. 1997. Effect of processing methods on nutrients and anti-nutritional factors in cowpea. Food Chem. 58:59–68.

Wheeler, E. L., and R. E. Ferrel. 1971. A method for phytic acid determination in wheat and wheat fractions. Cereal Chem. 48:312–320.

21

Effect of water yam (*Dioscoreaalata*) flour fortified with distiller's spent grain on nutritional, chemical, and functional properties

Wasiu Awoyale[1,2], Busie Maziya-Dixon[1], Lateef Oladimeji Sanni[3] & Taofik Akinyemi Shittu[3]

[1]International Institute of Tropical Agriculture, PMB 5320 Oyo Road, Ibadan, Oyo State, Nigeria
[2]Department of Food, Agriculture and Bioengineering, Kwara State University, PMB 1530, Malete, Kwara State, Nigeria
[3]Department of Food Science and Technology, Federal University of Agriculture Abeokuta, PMB 2240, Abeokuta, Ogun State, Nigeria

Keywords
Amino acids, dietary fiber, functional properties, nutrition, protein, Yam flour

Correspondence
Busie Maziya-Dixon, International Institute of Tropical Agriculture (IITA), Carolyn House 26 Dingwall Road, Croydon CR9 3EE, U.K.

E-mail: b.maziya-dixon@cgiar.org

Funding Information
This research was supported and carried out at the International Institute of Tropical Agriculture (IITA) Ibadan, Oyo State, Nigeria.

Abstract

It was envisaged that the inclusion of treated distiller's spent grain (DSG) to yam flour might increase its nutritional value, with the aim of reducing nutritional diseases in communities consuming yam as a staple. Hence, yam flour was fortified with DSG at 5–35%. The effects of this fortification on the nutritional, chemical, and functional properties of yam flour were investigated. The result showed a significant increase ($P \leq 0.001$) in fat, ash, protein, total amino acids, total dietary fiber, and insoluble dietary fiber contents of the blends as DSG increased except for starch and soluble dietary fiber contents, which decreased. The functional properties showed a significant ($P \leq 0.001$) reduction with DSG inclusion. The inclusion of DSG increased both the tryptophan and methionine contents of the blends. Therefore, the DSG fortified yam flour could contribute to quality protein intake in populations consuming yam as a staple, due to its indispensible amino acid content.

Introduction

Yam is a staple crop in growing areas (Asiedu et al. 1992) with over 90% of the global production coming from West Africa with Nigeria as the leading producer (FAO, 2003). Yam is consumed in different ways, such as boiled, fried, or baked. Tubers are often dried and milled into flour for reconstituting into a stiff paste (*amala*), which is eaten with preferred vegetable soup (Awoyale et al. 2010). It is an elite crop and preferred over other crops in growing regions. It can be stored longer than other root and tuber crops, ensuring a food supply even at times of general scarcity. Yam is of major importance in the diet and economic life of people in West Africa, the Caribbean islands, Asia, and Oceania (Ravindran and

Wanasundera 1992; Girardin et al. 1998). Information on the nutritive value of yam has been highlighted (Bradbury and Holloway 1988; Opara 1999; Alves 2000; Afoakwa and Sefa-Dedeh 2001).

Yam tubers consist of about 21% dietary fiber and are rich in carbohydrates, vitamin C, and essential minerals. Worldwide annual consumption of yam is 18 million tons, with 15 million in West Africa. Annual consumption in West Africa is 61 kg per capita. Because of its perishability and bulkiness, it is processed into yam flour (Onwuka and Ihuma 2007). Yam flour is either in the fermented or unfermented form produced from white yam (*Dioscorea rotundata*) or water yam (*Dioscorea alata*). However, because of the low protein content of *Dioscorea spp.* (Onayemi and Potter 1974), protein-energy malnutrition is prevalent in rural populations where yam is a staple especially among women and children (Adamson 1989). Therefore, improving the nutritional quality of yam flour through fortification using a protein rich source will contribute in improving the nutrition status of women and children.

Fortification refers to the practice of deliberately increasing the content of essential nutrients in a food irrespective of whether the nutrients were originally in the food before processing or not, so as to improve the nutritional quality of the food supply and to provide a public health benefit with minimal risk to health and enrichment; which is defined as "synonymous with fortification refers to the addition of nutrients which are lost during processing to a food (FAO/WHO, 1994). The levels of food fortification depend on the nutritional needs of the population, amount consumed, and regulations in the country (Awoyale et al. 2010). Some research has been reported on the fortification of yam flour with different fortificants (Akingbala et al. 1996; Babajide et al. 2004; Abulude and Ojediran 2006). At present, there islimited information on the use of distillers' spent grain (DSG) to fortify yam flour.

Distiller's spent grain is a by-product of cereal fermentation in the production of alcoholic beverages. It contains a high amount of proteins ranging from 23 to 35% and dietary fiber ranging from 27 to 55% (Rasco and McBurney 1989). It also contains yeast cell, vitamin B-complexes, and other nutrients formed during the fermentation–distillation process (Kaiser 2006). The potential to utilize DSG products from wheat and other cereal grains as a high protein and fiber ingredient in formulated foods has received increasing attention (Morad et al. 1984; Wu et al. 1985; Rasco et al. 1987a,b). For instance, DSG has been added up to 35% by mass in brownies, chocolate chip, and spice and lemon molasses cookies, about 30 to 50% of yeast bread, 30% of quick breads to produce highly acceptable products (Rasco and McBurney 1989). In addition, Tsen et al. (1982) reported that chocolate chip cookies containing 15% dried distillers' grain residues were as acceptable as chocolate chip cookies containing no distillers' grain. However, no work has been reported on the effect of using DSG as a fortificant on staples such as yam flour commonly consumed in developing countries.

Therefore, this study was aimed at evaluating the effect of DSG fortification on the nutritional composition and functional properties of yam flour produced from water yam (*Dioscorea alata*).

Methods

The water yam variety TDd98/01166 was grown at the Research Farm of the International Institute of Tropical Agriculture (IITA), Ibadan, Nigeria and its tubers were used for the investigation. The DSG was obtained from United State Department of Agriculture-Agricultural Research Service (USDA-ARS), North Dakota, USA.

Treatment of DSG

A suspension was made from 100 g of DSG containing alcohol and residual sugar in 400 mL of water and fermented with yeast (0.8 g) for 1 h to convert residual sugar to alcohol, which was then removed by distillation. The pH of the suspension was then adjusted to about 6.0 to 7.0 by adding sodium hydroxide (7 mL). The resulting suspension was dried to about 5–10% moisture content as reported by Awoyale et al. (2010).

Production of yam flour

The yam tubers were peeled, washed, sliced into cubes, and dried in a hot air oven at 65°C for 48 h. The dried yam chips were milled into flour using an attrition mill, packaged in polyethylene bags, and stored until needed (Udensi et al. 2008).

Formulation of DSG-yam flour blends

Yam flour and DSG were weighed and mixed in ratios 100:0, 95:5, 90:10, 85:15, 80:20, 75:25, 70:30 and 65:35. Each mixture was thoroughly blended with a laboratory blender, packed and sealed in a low-density polythene bag until required (Adelakun et al. 2004).

Determination of chemical composition

The moisture, ash, and fat contents were determined using standard laboratory procedures (AOAC, 1990). The protein content was determined by Kjeldahl method using KjeltecTM model 2300 protein analyzer (Foss Analytical Manual, 2003). A conversion factor of 6.25 was used to convert total nitrogen to percent crude protein. The total sugar and starch contents were analyzed in duplicate, according to the method described by Dubois et al. (1956).

Samples were analyzed for soluble (SDF) and insoluble dietary fiber (IDF) fractions using the enzymatic-gravimetric procedure (Prosky et al. 1988). Total dietary fiber (TDF) was calculated as the sum of SDF and IDF.

Amino acids were determined by reverse phase liquid chromatography (Waters 1500 Series HPLC; Milford, MA) with UV detection at 254 nm as reported by Cohen et al. (1988) at the South African Grain Laboratory, Pretoria, South Africa.

Bulk density

Bulk density was determined using a standard laboratory method (AOAC, 1990). Flour samples were weighed (7 g) into a 50-mL graduated measuring cylinder. The cylinder was tapped gently against the palm of the hand until a constant volume was obtained. Bulk density was calculated as weight of sample / volume of sample after tapping. All analyses were carried out in duplicates.

Water absorption capacity and oil absorption capacity

Water absorption capacity (WAC) and oil absorption capacity (OAC) were determined using the method as described by Beuchat (1977). Flour sample (1 g) was mixed with 10 mL of distilled water for WAC and 10 mL of oil for OAC and blended for 30 sec. Each sample was allowed to stand for 30 min after which it was centrifuged at 1303 g for 30 min at room temperature. The supernatant was decanted. The weight of water or oil absorbed by the flour was calculated and expressed as percentage WAC or OAC.

Swelling capacity

The method described by Ukpabi and Ndimele (1990) was used. Flour samples (10 g) were placed in a washed, dried, and weighed graduated measuring cylinder and 100 mL of distilled water was added. The suspension was stirred and allowed to stand for 1 h. The supernatant was discarded and the cylinder with its content weighed to obtain the weight of the net sample. Swelling capacity on volume basis was calculated as difference in final to initial volume of the sample.

Least gelation concentration

The least gelation concentration was determined by the method described by Coffman and Garcia (1977). Ten suspensions (2, 4, 6, 8, 10, 12, 14, 16, 18, and 20% w/v) of the samples in 5-mL distilled water were prepared in test tubes. The test tubes containing the suspensions were heated in a boiling water bath (Thelco, model 83, Missouri City, Texas, United States) for 1 h. The tubes and contents were cooled rapidly under running water and then cooled further for 2 h at 4°C. The tubes were then inverted to see if the contents would fall or slip off. The least gelation concentration is that concentration when the sample from the inverted test tube does not fall or slip off.

Pasting properties

Pasting characteristics was determined using a Rapid Visco Analyzer (Model RVA-4C, Newport Scientific, Warriewood,

Australia) interfaced with a personal computer equipped with the Thermocline software supplied by the same manufacturer (Deffenbaugh and Walker 1989). A sample of 3 g (moisture content less than 12%) was weighed into a canister and made into slurry by adding 25 mL of distilled water. The canister (covered with a stirrer) was inserted into the RVA. The slurry was held at 50°C for 1 min, heated to 95°C within 3 min, and then held at 95°C for 2 min., cooled to 50°C within 3 min and then held at 50°C for 2 min, while maintaining a rotation speed of 160 rpm. The viscosity is expressed as Rapid Viscosity Units (RVU). Records of peak viscosity (the maximum viscosity during pasting), breakdown viscosity (the difference between the peak viscosity and the minimum viscosity during pasting), setback viscosity (the difference between the maximum viscosity during cooling and the minimum viscosity during pasting), final viscosity (the viscosity at the end of the RVA run), pasting temperature (°C) (the temperature at which there is a sharp increase in the viscosity of the flour suspension after the commencement of heating), and peak time (min) (time taken for the paste to reach the peak viscosity) were taken.

Statistical analysis

Data were subjected to analysis of variance (ANOVA) using Statistical Analysis System (SAS) package (version 9.1, SAS Institute, Inc., Cary, NC) (SAS, 2003). Means were separated using Fischer's protected Least Significant Difference (LSD) test.

Results

Chemical composition

Presented in Table 1 is the chemical composition of the yam flour: DSG blends. Mean total ash content was 3.2% and ranged from 2.4% for 100% yam flour to 3.5% for the 90% yam: 10% DSG blends. There was a significant difference ($P \leq 0.001$) in the ash content of the fortified yam flour. The fortified yam flour moisture content was also significantly ($P \leq 0.001$) higher in 65% yam flour: 35% DSG blend (4.6%) and lower in 100% yam flour (3.6%) (Table 1). Furthermore, there was a significant ($P \leq 0.001$) decrease in the starch content of the blends from 62% for 100% yam flour to 48% for 65% yam flour: 35% DSG blend (Table 1), showing that as the quantity of DSG increased, the starch content of the blends decreased. Additionally, there was a significant ($P \leq 0.001$) increase in the fat content from 0.43% (100% yam flour) to 4.30% (65% yam flour: 35% DSG blend) (Table 1). Similarly, the protein content increased significantly ($P \leq 0.001$) from 7.2% for 100% yam flour to 15.10%

Table 1. Chemical composition of yam flour (Y) and distillers' spent grain (DSG) blends.

Samples	Ash (%)	Moisture (%)	Protein (%)	Sugar (%)	Starch (%)	Fat (%)	pH
100% DSG	2.37 ± 0.02^b	5.06 ± 0.04^a	29.80 ± 0.13^a	3.37 ± 0.06^d	23.86 ± 0.40^g	9.63 ± 0.02^a	5.70 ± 0.01^i
100% Y	3.29 ± 0.02^c	3.56 ± 0.08 ^{g}	7.15 ± 0.31^i	2.71 ± 0.10^e	62.02 ± 0.11^a	0.43 ± 0.00^i	6.82 ± 0.00^a
95% Y:5% DSG	3.33 ± 0.01^a	3.62 ± 0.02 ^{fg}	7.92 ± 0.10^h	2.41 ± 0.03^f	61.76 ± 0.63^a	0.90 ± 0.02^h	6.73 ± 0.00^b
90% Y:10% DSG	3.35 ± 0.01^a	3.68 ± 0.01^f	9.15 ± 0.03^g	2.82 ± 0.10^e	61.40 ± 0.94^{ab}	1.58 ± 0.07^g	6.64 ± 0.01^c
85% Y:15% DSG	3.30 ± 0.01^{bc}	3.79 ± 0.04^e	10.50 ± 0.07^f	3.50 ± 0.18^d	60.80 ± 0.12^b	1.92 ± 0.05^f	6.56 ± 0.00^d
80% Y:20% DSG	3.34 ± 0.016^a	3.94 ± 0.06^d	11.55 ± 0.03^e	3.58 ± 0.07^{cd}	57.81 ± 0.22^c	2.53 ± 0.12^e	6.38 ± 0.01^e
75% Y:25% DSG	3.33 ± 0.03^a	4.20 ± 0.02^c	12.76 ± 0.03^d	3.71 ± 0.05^c	53.03 ± 0.05^d	3.79 ± 0.01^d	6.33 ± 0.00^f
70% Y:30% DSG	3.34 ± 0.02^a	4.46 ± 0.04^b	13.83 ± 0.01^c	4.04 ± 0.06^b	49.29 ± 0.51^e	4.18 ± 0.00^c	6.27 ± 0.01^g
65% Y:35% DSG	3.32 ± 0.02^{ab}	4.63 ± 0.01^a	15.08 ± 0.10^b	4.50 ± 0.18^a	48.02 ± 0.42^f	4.30 ± 0.08^b	6.23 ± 0.00^h
Mean	3.22	4.10	13.08	3.40	53.11	3.25	6.41
P-level	***	***	***	***	***	***	***

***$P \leq 0.001$. Means with different superscript along the same column are significantly different at $P \leq 0.05$.

Table 2. Amino acid composition (g/100 g sample) of yam flour (Y) and distillers' spent grains (DSG) blends.

Amino Acids	100% DSG	100% Y	95% Y: 5% DSG	90% Y: 10% DSG	85% Y: 15% DSG	80% Y: 20% DSG	75% Y: 25% DSG	70% Y: 30% DSG	65% Y: 35% DSG
Methionine	0.49	0.12	0.16	0.18	0.18	0.24	0.23	0.23	0.24
Aspartic acid	2.15	0.53	0.74	0.81	0.88	0.94	1.01	1.06	1.13
Glutamic acid	6.22	0.89	1.26	1.51	1.74	1.94	2.17	2.43	2.61
Serine	1.74	0.42	0.48	0.56	0.62	0.67	0.73	0.81	0.84
Glycine	1.24	0.25	0.31	0.36	0.41	0.45	0.50	0.56	0.59
Histidine	0.87	0.18	0.23	0.26	0.29	0.31	0.37	0.41	0.42
Arginine	1.34	0.47	0.50	0.54	0.58	0.63	0.67	0.79	0.88
Threonine	1.24	0.24	0.30	0.35	0.40	0.43	0.50	0.54	0.60
Alanine	2.45	0.30	0.41	0.53	0.62	0.71	0.80	0.92	0.99
Proline	2.79	0.27	0.40	0.56	0.66	0.76	0.89	1.02	1.13
Tyrosine	1.07	0.23	0.23	0.23	0.26	0.37	0.38	0.41	0.49
Valine	1.45	0.27	0.34	0.41	0.45	0.51	0.57	0.63	0.68
Isoleucine	1.09	0.23	0.29	0.34	0.36	0.41	0.45	0.51	0.53
Leucine	3.87	0.43	0.60	0.77	0.90	1.07	1.21	1.37	1.51
Phenylalanine	1.66	0.34	0.41	0.47	0.50	0.58	0.64	0.72	0.74
Lysine	0.82	0.24	0.25	0.26	0.27	0.33	0.35	0.41	0.41
Trypthophan	0.23	0.09	0.09	0.10	0.15	0.15	0.14	0.12	0.14
TAa	30.72	5.48	6.98	8.22	9.28	10.49	11.6	12.93	13.91
TIAa With Histidine	11.73	2.14	2.66	3.12	3.50	4.03	4.46	4.93	5.26
TIAa Without Histidine	10.85	1.96	2.43	2.86	3.20	3.71	4.09	4.53	4.83
TCIAa	2.42	0.69	0.73	0.77	0.84	0.99	1.05	1.20	1.37
TDAa	16.58	2.65	3.60	4.33	4.95	5.47	6.09	6.79	7.29

TIAa, total indispensable amino acids; TCIAa, total conditional indispensable amino acids; TDAa, total dispensable amino acids; TAa, total amino acid.

for 65% yam flour: 35% DSG blend (Table 1), revealing that for every increase in the quantity of DSG, there is a corresponding increase in the protein content of the blends.

Amino acids composition

The total amino acid composition of the fortified yam flour increased from 5.48 g/100 g for 100% yam flour to 13.91 g/100 g for the 65% yam flour: 35% DSG blend, with glutamic and aspartic acids contributing the highest percentage while tryptophan and methionine contributes the least (Table 2). It was observed that the amount of

each amino acid in the blends increased as the quantity of DSG increased. Additionally, the inclusion of DSG increased both the quantity and quality of the protein in the yam flour (Table 2). The result also revealed that methionine and tryptophan were the limiting amino acids in the 100% yam flour. These amino acids were compensated for with the inclusion of DSG to the yam flour (Table 2).

Dietary fiber content

The total dietary fiber (TDF) content of the fortified yam flour increased from 6.02% for 100% yam flour to 11.7%

for 65% yam flour: 35% DSG blend, out of which the insoluble dietary fiber (IDF) had the highest value (23.50%) and the soluble dietary fiber (SDF) the least (0.90%) (Table 3).

Functional properties

The functional properties of yam flour and DSG blend are presented in Table 4. It was observed that blending yam flour with DSG significantly ($P \leq 0.001$) decreased oil absorption capacity (OAC) of 100% yam flour from 213 to 183% for 90% yam flour: 10% DSG blend (Table 4). There was a decrease in the water absorption capacity (WAC) from 260% for 100% yam flour to 241% for the 75% yam flour: 25% DSG blend (Table 4). Swelling capacity of the fortified yam flour similarly decreased from 3.40 for 100% yam flour to 2.40 for 65% yam flour: 35% DSG blend (Table 4). Significant differences ($P \leq 0.001$) were observed for bulk density of the

fortified yam flour; which decreased as the amount of added DSG increased. It ranged from 57% for 80% yam flour: 20% DSG blend to 70% for 100% yam flour (Table 4).

Pasting properties

Table 5 showed the pasting properties of the yam flour and DSG blends. The peak viscosity, which is the highest viscosity of the range from 44RVU for 65% yam flour: 35% DSG blend to 198RVU for 100% yam flour. The values of the fortified yam flour final viscosity ranged between 63RVU for 65% yam flour: 35% DSG blend and 235RVU for 100% yam flour (Table 5). The setback viscosity also ranged from 26RVU for 65% yam flour: 35% DSG blend to 41RVU for 100% yam flour (Table 5). The result also revealed that the peak time ranged between 6 and 7 min and the pasting temperature ranged from 61.90°C for the 80% yam flour: 20% DSG blend to 62.30°C for the 75% yam flour: 25% DSG blend, respectively (Table 5).

Discussion

Chemical composition

Yam flour is a fermented or unfermented flour produced from either white yam (*Dioscorea rotundata*) or water yam (*Dioscorea alata*). However, because of its low protein content (Onayemi and Potter 1974), protein-energy malnutrition is prevalent in rural populations where yam is consumed as a staple, especially among women and children (Adamson 1989). Hence, the reason for the fortification of yam flour with DSG.

The result of this investigation showed that the protein content of the flour blends increased as the quantity of DSG increased. This could be attributed to the high

Table 3. Dietary fiber content of yam flour (Y) and distillers' spent grain (DSG) blends.

Samples	Dietary fibers (%)		
	Insoluble	Soluble	Total
100% DSG	23.53	0.67	24.20
100% Y	5.10	0.92	6.02
95% Y:5% DSG	5.47	0.91	6.37
90% Y: 10% DSG	6.38	0.87	7.25
85% Y:15% DSG	8.27	0.85	9.13
80% Y:20% DSG	8.67	0.78	9.45
75% Y:25% DSG	9.59	0.76	10.34
70% Y:30% DSG	10.05	0.75	10.80
65% Y:35% DSG	11.01	0.72	11.73
Mean	9.79	0.80	10.59
Minimum	5.10	0.67	6.02
Maximum	23.53	0.92	24.20

Table 4. Functional characteristics of yam flour (Y) and distillers' spent grain (DSG) blends.

Samples	OAC (%)	WAC (%)	SWC	BD (%)	Amylose (%)	LGC (%)
100% DSG	216.00 ± 0.02[a]	291.00 ± 0.03[a]	2.27 ± 0.03[h]	58.00 ± 0.00[e]	3.18 ± 0.24[e]	20.10
100% Y	213.00 ± 0.13[a]	260.00 ± 0.03[b]	3.42 ± 0.02[a]	70.00 ± 0.00[a]	22.34 ± 0.15[a]	20.20
95% Y:5% DSG	191.00 ± 0.03[bc]	261.00 ± 0.02[b]	3.23 ± 0.05[b]	60.00 ± 0.01[d]	22.07 ± 0.24[a]	20.06
90% Y:10% DSG	183.00 ± 0.01[c]	243.00 ± 0.02[de]	3.09 ± 0.03[c]	63.00 ± 0.01[c]	19.83 ± 0.88[b]	20.02
85% Y:15% DSG	189.00 ± 0.03[bc]	245.00 ± 0.00[d]	2.72 ± 0.02[e]	66.00 ± 0.02[b]	17.77 ± 0.35[c]	20.04
80% Y:20% DSG	194.00 ± 0.01[b]	243.00 ± 0.01[de]	2.83 ± 0.06[d]	57.00 ± 0.01[e]	17.63 ± 0.38[c]	20.02
75% Y:25% DSG	193.00 ± 0.02[bc]	241.00 ± 0.01[d]	2.75 ± 0.02[e]	60.00 ± 0.01[d]	17.24 ± 0.65[c]	20.06
70% Y:30% DSG	191.00 ± 0.08[bc]	252.00 ± 0.02[c]	2.55 ± 0.02[f]	60.00 ± 0.01[d]	16.36 ± 0.12[d]	20.06
65% Y:35% DSG	186.00 ± 0.01[bc]	253.00 ± 0.01[c]	2.35 ± 0.02[g]	63.00 ± 0.01[c]	15.92 ± 0.03[d]	20.06
Mean	195.11	254.33	2.80	61.89	16.85	20.07
P-level	***	***	***	***	***	

OAC-Oil absorption capacity, WAC-Water absorption capacity, SWC-Swelling capacity, BD-Bulk density, LGC-Least gelation concentration.
***$P \leq 0.001$. Means with different superscript along the same column are significantly different at $P \leq 0.05$.

Table 5. Pasting properties of yam flour (Y) and distillers' spent grain (DSG) blends.

Samples	Peak (RVU)	Trough (RVU)	Break down (RVU)	FinalVisc. (RVU)	Set back (RVU)	Peak Time (min)	Pasting Temp (°C)
100% DSG	6.54 ± 0.54[i]	−11.50 ± 0.58[i]	18.04 ± 1.13[a]	−6.21 ± 3.71[i]	5.29 ± 4.29[e]	0.03 ± 0.00[d]	61.37 ± 0.39[b]
100% Y	198.46 ± 1.88[a]	194.38 ± 0.13[a]	4.08 ± 2.00[d]	235.25 ± 2.67[a]	40.88 ± 2.79[b]	6.53 ± 0.07[b]	61.90 ± 0.00[a]
95% Y:5% DSG	157.54 ± 2.04[b]	147.13 ± 1.96[b]	10.42 ± 0.08[b]	195.08 ± 2.58[b]	47.96 ± 0.63[a]	6.23 ± 0.10[c]	61.93 ± 0.03[a]
90% Y:10% DSG	139.46 ± 5.88[c]	135.67 ± 4.92[c]	3.79 ± 0.96[d]	176.92 ± 7.42[c]	41.25 ± 2.50[b]	6.23 ± 0.10[c]	61.93 ± 0.03[a]
85% Y:15% DSG	115.46 ± 2.88[d]	114.33 ± 3.17[d]	1.13 ± 0.29[f]	154.79 ± 2.46[d]	40.46 ± 0.71[b]	6.47 ± 0.13[b]	61.90 ± 0.05[a]
80% Y:20% DSG	97.67 ± 2.42[e]	95.67 ± 1.83[e]	2.00 ± 0.58[ef]	134.96 ± 3.71[e]	39.29 ± 1.88[bc]	6.90 ± 0.03[a]	61.88 ± 0.03[a]
75% Y:25% DSG	75.50 ± 0.75[f]	72.04 ± 0.54[f]	3.46 ± 1.29[de]	108.38 ± 0.63[f]	36.33 ± 0.08[c]	6.93 ± 0.07[a]	62.25 ± 0.50[a]
70% Y:30% DSG	57.88 ± 1.29[g]	50.67 ± 1.50[g]	7.21 ± 0.21[c]	79.42 ± 2.00[g]	28.75 ± 0.50[d]	7.00 ± 0.00[a]	61.90 ± 0.05[a]
65% Y:35% DSG	43.71 ± 1.46[h]	36.83 ± 1.75[h]	6.88 ± 0.29[c]	63.08 ± 1.08[h]	26.25 ± 0.67[d]	7.00 ± 0.00[a]	61.95 ± 0.05[a]
Mean	99.14	92.80	6.33	126.85	34.05	5.92	61.89
P-level	***	***	***	***	***	***	NS

***$P \leq 0.001$. Means with different superscript along the same column are significantly different at $P \leq 0.05$.

protein content of DSG (29.80%) compared to that of the 100% yam flour (7.15%). Similar results for the protein content of the 100% yam flour had been reported for the flour produced from different yam varieties (Bokanga 2000; Udensi et al. 2008; Alozie et al. 2009; Baah et al. 2009). Furthermore, Abulude and Ojediran (2006) reported comparable results with some of yam blends, for the protein contents of yam–cassava and yam–plantain blends.

The increase in fat content of the flour blends with DSG inclusion could be attributed to the high fat content of DSG. Comparable value (0.42%) with the fat content of the 100% yam flour was reported by Bradbury and Holloway (1988) and Souci et al. (1994). Additionally, the fat contents (0.04–2.00%) reported by Osagie (1992) for yam tubers were also comparable with that of the present study.

Ash content is a reflection of the mineral status, even though contamination can indicate a high concentration in a sample (Baah et al. 2009). The ash content of different varieties of *Dioscorea alata* flour reported by Lebot et al. (2005) and Baah et al. (2009) was in range with that of the 100% yam flour, while that of Osagie (1992) and Udensi et al. (2008) were lower compared to the results of this investigation.

The decrease in the starch content of the flour blends with corresponding increase in DSG inclusion may be attributed to the low starch content of DSG. This might have a negative effect on the acceptability of the reconstituted paste (*amala*), as starch is a very important factor for gel formation. However, similar observations with the starch content of the 100% water yam flour were made by other researchers (Maziya-Dixon and Asiedu 2003; Lebot et al. 2005; Baah et al. 2009).

The moisture content of all the yam flour blends is still below the recommended safe level (12–13%) for storage of flour (FAO, 1992). This implied that all the blends

might be stored for a long period before being used for the preparation of *amala*, without microbial contamination (Pierre 1989), if properly packaged in an airtight packaging material.

Amino acid composition

Amino acid assay of foods is an important quality index, from which useful information on the nutritional quality and authenticity of food products and sources of raw materials used in food manufacture could be revealed (Alozie et al. 2009). The results obtained for the amino acids composition of the 100% yam flour was comparable with the observation made by Ekpeyong (1984) on yam tuber. However, it was observed that the inclusion of DSG to yam flour improves the amount of its limiting amino acids (methionine and tryptophan). This could be due to the level of these amino acids in DSG. In addition, FAO/WHO/UNN (1985) reported that the total indispensable amino acid requirements (g/100 g protein) are between 24.10 and 12.70 (with histidine) and between 22.20 and 11.10 (without histidine) for school children (10–12 years) and adults, respectively. Consequently, the consumption of the reconstituted paste (*amala*) produced from all the yam blends with total indispensable amino acid content (g/100 g protein) range of 28.09–34.34 (with histidine) and 25.89–31.72 (without histidine) by the target group (children and adults) might increase their total indispensible amino acids level required for growth and repair of worn-out tissue.

Dietary fiber content

Fiber is a type of carbohydrate that the body cannot digest. Though most carbohydrates are broken down into sugar molecules, fiber cannot be broken down into sugar molecules, and instead it passes through the body undigested. Fiber helps regulate the body's use of sugars, helping to keep hunger and blood sugar in check. Children and adults need at least 20 to 30 grams of fiber per day for good health. Great sources are whole fruits and vegetables, whole grains, and beans (Bauer and Turler-Inderbitzin 2008). The values for the TDF content of the 100% yam flour in this study was similar to those reported by Baah et al. (2009) for different varieties of *Dioscorea alata*. Furthermore, Bauer and Türler-Inderbitzin (2008), reported that decreased risk of coronary heart disease is correlated with increase in consumption of SDF and that, high water-binding capacity of IDF results in the formation of softer stools which reduces the pressure necessary for the elimination of stools through the system faster, thus, less constipation and low incidence of maladies. Hence, *amala* produced from all the yam blends of this investigation might be able to suite this purpose when consumed by the target groups.

Functional properties

As the functional properties of foods is known to affect the end use of any food and how such a food behaves during preparation for consumption, the functional properties of the fortified yam flour such as oil absorption capacity (OAC), water absorption capacity (WAC), swelling capacity (SWC), bulk density (BD), amylose content, and pasting properties would be of importance to the end users.

The OAC is important as oil acts as a flavor retainer and improves the mouth feel of foods (Kinsella 1976). The corresponding decrease in the OAC of the flour blends with DSG inclusion may be attributed to the low OAC of DSG. However, the OAC of this study was higher compared with that reported by Abulude and Ojediran (2006) on yam flour fortified with cassava and plantain flour. The WAC on the other hand, is the amount of water that an insoluble starch is able to hold in relation to its weight. High WAC is attributed to lose association of amylose-amylopectin ratio in the native starch granule (Ayermor 1976). Similarly, the reduction in WAC of the flour blends may be due to the low WAC of the DSG, and reduction in amylose content of the blends as the proportion of DSG increased (Ayermor 1976). This result was comparable to the values (240–301%) reported by Abulude and Ojediran (2006) for the WAC of yam flour fortified with cassava and plantain flour.

Furthermore, the decrease in the SWC of the flour blends with increase in DSG may be due to the low starch content of DSG (Houssou and Ayernor 2002). Contrary to the observations made by Tester and Morrison (1990), the SWC of the flour blends increased as the amylose content increased.

Brennan et al. (1976) observed that high BD increases the rate of dispersion and as a result it is important in the reconstitution of yam *fufu* dough. This implied that 100% yam flour reconstituted into *amala* might be finer in texture compared to that of 80% yam flour: 20% DSG blend. In addition, 80% yam flour: 20% DSG blend might be easily packed for storage compared to that of 100% yam flour due to its low BD (Ikujenlola 2008).

The amylose content of the 100% yam flour of the present investigation was comparable to the work on *Dioscorea alata* and *Dioscorea rotundata* found in literature (Rasper and Coursey 1976; Bokanga 2000). The low starch content of DSG may be attributed to the low amylose content of the flour blends.

Pasting properties

As the flour blends would be reconstituted to a thick paste (*amala*) before consumption, the pasting properties become important in predicting its behavior during and after cooking. Final viscosity is the most commonly used parameter to determine a particular starch-based samples quality as it indicates the ability of the material to form gel after cooking (Sanni et al. 2006). The results showed that, the higher the proportion of DSG in the flour blend, the lower the final viscosity. This implied that, the 100% yam flour might quickly form a paste (*amala*) compared to the others due to its high final viscosity. Additionally, the higher the setback viscosity, the higher the rate of syneresis or weeping, and the easier it is for the food to be digested (Shittu et al. 2001). This means that *amala* produced from 65% yam flour: 35% DSG might keep longer before "weeping" and as well digest fast when consumed due to its low setback value compared with that of 100% yam flour (Shittu et al. 2001). However, all the flour blends might be cooked to *amala* at a temperature of <63°C and time of <8 min., this implied low energy cost (Fasasi et al. 2007).

Conclusion

The fat, ash, protein, sugar, total amino acids composition, total dietary fiber, and insoluble dietary fiber contents of the fortified yam flour increased while the starch and the soluble dietary fiber contents decreased as the DSG increased. In addition, aspertic and glutamic acids contributed the highest percentage of amino acids while tryptophan and methionine contributed the least in the flour blends. However, the inclusion of DSG to the yam flour increased both the tryptophan and methionine contents. There was also a decrease in the peak, final, and setback viscosities of the fortified yam flour. The WAC, OAC, swelling capacity, bulk density, and amylose content of the flour blends decreased with DSG inclusion. However, the DSG fortified yam flour could serve as a quality protein source when prepared to *amala* and consumed with preferred soup due to its high indispensible amino acid content.

Acknowledgments

This research was supported and carried out at the International Institute of Tropical Agriculture (IITA) Ibadan, Oyo State, Nigeria. We acknowledge Dr K. Dashiell, formally of United State Department of Agriculture-Agricultural Research Service (USDA-ARS), North Dakota, United States America for supplying DSG.

Conflict of Interest

None declared.

References

Abulude, F. O., and V. A. Ojediran. 2006. Development and quality evaluation of fortified *amala*. Acta Sci. Pol. Technol. Aliment. 5:127–134.

Adamson, I. 1985. Dietary fibre of yam and cassava. In: Advances in Food research, (ed.) Osuji, Pp. 321–340.

Adelakun, O. E., J. A. Adejuyitan, J. O. Olajide, and B. K. Alabi. 2004. Effect of soybean substitution on some physical, compositional and sensory properties of *kokoro* (a local maize snack). European Food Res. Technol. 220:79–82.

Afoakwa, E. O., and S. Sefa-Dedeh. 2001. Chemical composition and quality changes in *Dioscorea dumetorum pax* tubers after harvest. Food Chem. 75:85–91.

Akingbala, J. O., T. B. Oguntimein, and A. O. Sobande. 1996. Physicochemical properties and acceptability of yam flour replaced by soy flour. Plant Foods Hum. Nut. 48:73–80.

Alozie, Y., M. I. Akpanabiatu, E. U. Eyong, B. I. Umoh, and G. Alizie. 2009. Amino acid composition of *D. dumentorum* varieties. Pakistan J. Nut. 8:103–105.

Alves, R. M. L. 2000. Caracterizacao de ingredientes obtidos de cara (*Dioscorea alata*) potencial aplicao industrial. Ph.D. thesis, Universidade Estadual de Londrina, Brazil.

AOAC. 1990. Official Methods of Analysis of the Association of Official Analytical Chemists.

Asiedu, R., S. Y. C. Ng, D. Vuylsteke, R. Terauchi, and S. K. Hahn. 1992. Analysis of the need for biotechnology research on cassava, yam and plantain. Pp. 70–74 in G. Thottappilly, L. M. Monti, D. R. Mohan Raj, and A. W. Moore, eds. Biotechnology: Enhancing research on tropical crops in Africa. CTA/IITA, Ibadan, Nigeria.

Awoyale, W., B. Maziya-Dixon, L. O. Sanni, and T. A. Shittu. 2010. Nutritional and sensory properties of *amala* supplemented with distiller's spent grain (DSG). J. Food Agric. Environ. 8:66–70.

Ayermor, G. S. 1976. Particulate properties and rheology of pre gelled yam (*Dioscorea rotundata*) products. J. Food Sci. 41:180–182.

Baah, F. D., B. Maziya-Dixon, R. Asiedu, I. Oduro, and W. O. Ellis. 2009. Nutritional and biochemical composition of *D. alata* (*Dioscorea* spp.) tubers. J. Food Agric. Environ. 7:373–378.

Babajide, J. M., S. O. Babajide, and A. K. Kadri. 2004. Pasting viscosity of yam flour "elubo" enriched with defatted soybean flour and vegetable "ugu" powder. Pp. 295–296 in G. O., Adegoke, L. O., Sanni, K. O. Falade, and P. I. Uzo-Peters, eds. Proceedings of the 28th Annual Conference/Annual General Meeting of Nigerian Institutes of Food Science and Technology (NIFST), held at University of Ibadan, Ibadan, Nigeria. 12–14 October 2004.

Bauer, W., and S. Turler-Inderbitzin. 2008. Nestlé professional nutrition magazine dietary fibre and its various health benefits.

Beuchat, L. R. 1977. Functional and electrophoretic characteristic of succinylated peanut flour. J. Agricult. Food Chem. 25:258–261.

Bokanga, M. 2000. Raw material characteristics of root and tubers. Pp. 45 in Root and tubers in the global food system. A vision statement to the year 2020. A report to the Technical Advisory Committee (TAC) of the Consultative Group on International Agriculture (CGIAR) by the committee on Inter-Centre Root and Tubers Crops Research (CICRTCR).

Bradbury, J. H., and W. D. Holloway. 1988. Chemistry of tropical root crops. Australian Center for International Agricultural Research, Canberra. Pp. 101–119.

Brennan, J. G., J. R. Butters, N. D. Cowell, and A. E. V. Lielly. 1976. Effect of agglomeration on the properties of spray-dried roaibos tea. Int. J. Food Sci. Technol. 23:43–48.

Coffman, C. W., and V. V. Garcia. 1977. Functional properties and amino acid content of protein isolate from mung bean flour. J. Food Technol. 12:473–484.

Cohen, S. A., A. Meys, and T. L. Tarvin. 1988. The pico-tag method: a manual of advanced techniques for amino acid analysis. Waters Chromatography Division, Milford, MA.

Deffenbaugh, L. B., and C. E. Walker. 1989. Comparison of starch pasting properties in the brabender Viscoamylograph and the Rapid Visco-Analyzer. Cereal Chem. 66:499.

Dubois, M., K. A. Gilles, J. K. Hamilton, P. A. Rebers, and F. Smith. 1956. Colorimetric method for determination of sugars and related substances. Anal. Chem. 28:350–356.

Ekpeyong, T. E. 1984. Composition of some tropical tuberous foods. Food Chem. 15:31–36.

FAO. 1992. Maize in human nutrition. FAO food and nutrition series 25. FAO, Rome, Italy.

FAO. 2003. Federal Office of Statistic (FAOSTAT) data. FAO, Rome, Italy.

FAO/WHO. 1994. Methods of analysis and sampling. Joint FAO/WHO Food Standards Programme Codex Alimentarius Commission, Vol. 13, 2nd edition.

FAO/WHO/UNN. 1985. Expert consultation. Energy and protein requirements. Technical Report Series 724. World Health Organization, Geneva, Switzerland.

Fasasi, O. S., I. A. Adeyemi, and O. A. Fagbenro. 2007. Functional and pasting characteristics of fermented maize and Nile Tilapia (Oreochromis niloticus) flour diet. Pakistan J. Nut. 6: 304–309.

Foss Analytical Manual, A. B. 2003. Manual for Kjeltec System 2300 Distilling and Titration Unit. Kjeldahl method of protein determination.

Girardin, O., C. Nindjin, Z. Farah, F. Eshcher, P. Stamp, and D. Otokore. 1998. Effect of storage on system and sprout removal on post-harvest yam (Dioscorea spp.) fresh weight losses. J. Agric. Sci. 130:329–336.

Houssou, P., and G. S. Ayernor. 2002. Appropriate processing and food functional properties of maize flour. African Journal of Science and Technology (AJST). Sci. Eng. Series 3:126–131.

Ikujenlola, A. V. 2008. Chemical and functional properties of complementary food from malted and unmalted Acha (Digitaria exilis), Soybean (Glycine max) and Defatted sesame seeds (Sesamun indicum L). J. Eng. Appl. Sci. 39:471–475.

Kaiser, R. M. 2006. Utilizing the Growing Supply of Distiller's Grains. University of Wisconsin – Extension, USA. Diary Agent. Pp. 1–6. http://dysci.wisc.edu/uwex/brochures/brochures/ADDProceedings02.pdf.

Kinsella, J. E. 1976. Functional properties of proteins in foods. Crit. Rev. Food Sci. Nut. 1:219–280.

Lebot, V., R. Malapa, T. Molisale, and J. L. Marchand. 2005. Physicochemical characterization of yam (Dioscorea alata L.) tubers from Vanuatu. Genet. Resour. Crop Evol. 53:1199–1208.

Maziya-Dixon, B., and R. Asiedu. 2003. Characterisation of physicochemical and pasting properties of yam varieties. 2003 Project Annual Report, International Institute of Tropical Agriculture, Ibadan, Nigeria.

Morad, M. M., C. A. Doherty, and L. W. Rooney. 1984. Utilization of dried distillers' grain from sorghum in baked food systems. Am. Assoc. Cereal Chemis. 61:409–414.

Onayemi, O., and N. N. Potter. 1974. Preparation and storage properties of drum dried white yam (Dioscorea rotundata Poir) flakes. J. Food Sci. 39:539–541.

Onwuka, G. I., and C. Ihuma. 2007. Physicochemical composition and product behaviour of flour and chips from two yam spp. (Dioscorea rotundata and Dioscorea alata). Res. J. Appl. Sci. 2:35–38.

Opara, L. U. 1999. Yam storage. In: Bakker-Arkema, FW (eds.), CIGR Handbook of agricultural engineering, Vol. IV, Agro Processing. The American Society of Agricultural Engineers, MI, Pp. 182–214.

Osagie, A. U. 1992. Yam tubers in storage. Pp. 33–84. Post harvest research bulletin Biochemistry. University of Benin, Nigeria.

Pierre, S. 1989. Cassava. Macmillan Education Ltd. London. In cooperation with the Technical Centre for Agriculture and Rural Cooperation, Wageningen, The Netherlands, Pp. 1–2, Pp. 30–76.

Prosky, L., N. G. Asp, T. E. Schweizer, J. W. Devries, and I. Furda. 1988. Determination of insoluble, soluble and total dietary fiber in foods and food products: inter-laboratory study. J. Assoc. Analyt. Chem. 71:1017–1024.

Rasco, B. A., and W. J. McBurney. 1989. Human food product produced from dried distillers' spent cereal grains and soluble. US Patent No. 4,828,846.

Rasco, B. A., F. M. Dong, A. E. Hashisaka, S. S. Gazzaz, S. E. Downey, and M. L. San Buenaventura. 1987a. Chemical composition of distiller's dried grains with soluble (DDGS) from soft white wheat, hard red wheat and corn. J. Food Sci. 52:236–237.

Rasco, B. A., S. E. Downey, and F. M. Dong. 1987b. Consumer acceptability of baked goods containing distiller's dried grains with solubles from soft white winter wheat. Cereal Chem. 64:139–143.

Rasper, V., and D. G. Coursey. 1976. Properties of starches of some West African yams. J. Sci. Food Agricult. 18:240–244.

Ravindran, G., and J. P. D. Wanasundera. 1992. Chemical changes in Yam (Dioscorea alata and D. esculenta) tubers during storage. Trop. Sci. 33:57–62.

Sanni, L. O., A. A. Adebowale, T. A. Filani, O. B. Oyewole, and A. Westby. 2006. Quality of flash and rotary dried fufu flour. J. Food Agric. Environ. 4:74–78.

SAS. 2003. SAS Institute Inc. SAS® 9.1 Qualification tools user's guide. SAS Institute Inc., Cary, NC.

Shittu, T. A., O. O. Lasekan, L. O. Sanni, and M. O. Oladosu. 2001. The effect of drying methods on the functional and sensory characteristics of pupuru—a fermented cassava product. Asset Series 1:9–16.

Souci, S. W., W. Fachmann, and H. Kraut. 1994. Food composition and nutrition tables. Medpharm, Scientific Publ., Stuttgart – CRC Press, Boca Raton, Ann Arbor, London. Pp. 1091.

Tester, R. F., and W. R. Morrison. 1990. Swelling and gelatinization of cereal starches. Cereal Chem. 67:558–563.

Effect of water yam (Dioscoreaalata) flour fortified with distiller's spent grain on nutritional, chemical...

203

Tsen, C. C., W. Eyestone, and J. Weber. 1982. Evaluation of the quality of biscuits supplemented with distillers dried grain flours. J. Food Sci. 47:684–685.

Udensi, E. A., H. O. Oselebe, and O. O. Iweala. 2008. The Investigation of chemical composition and functional properties of water yam (D.alata): effect of varietal differences. Pakist. J. Nut. 7:342–344.

Ukpabi, U., and C. Ndimele. 1990. Evaluation of the quality of Gari produced in Imo State Nigeria. Nigeria Food J. 8:106–108.

Wu, Y. V., K. R. Sexson, and A. A. Lagoda. 1985. Protein rich alcohol fermentation residues from corn dry milled fractions. Cereal Chem. 62:470–473.

22

Comparison of nutritional properties of Stinging nettle (*Urtica dioica*) flour with wheat and barley flours

Bhaskar Mani Adhikari[1], Alina Bajracharya[1] & Ashok K. Shrestha[2]

[1]Department of Food Technology, National College of Food Science and Technology, Kathmandu, Nepal
[2]Nutrition and Food Science, School of Science and Health, Hawkesbury Campus, University of Western Sydney, Penrith NSW 2751, Australia

Keywords
Antioxidant activity, barley flour, nettle powder, polyphenol

Correspondence
Bhaskar Mani Adhikari, Department of Food Technology, National College of Food Science and Technology, Kathmandu, Nepal.

E-mails: vaskarmani@gmail.com; bm.adhikari@nist.edu.np

Funding Information
No funding information provided.

Abstract

Stinging nettle (Urtica dioica. L) is a wild, unique herbaceous perennial flowering plant with Stinging hairs. It has a long history of use as a food sources as a soup or curries, and also used as a fiber as well as a medicinal herb. The current aim was to analyze the composition and bioactive compounds in Nepalese Stinging nettle. Chemical analysis showed the relatively higher level of crude protein (33.8%), crude fiber (9.1%), crude fat (3.6%), total ash (16.2%), carbohydrate (37.4%), and relatively lower energy value (307 kcal/100 g) as compared to wheat and barley flours. Analysis of nettle powder showed significantly higher level of bioactive compounds: phenolic compounds as 129 mg Gallic acid equivalent/g; carotenoid level 3497 µg/g; tannin 0.93 mg/100 g; anti-oxidant activity 66.3 DPPH inhibition (%), as compared to wheat and barley. This study further established that nettle plants as very good source of energy, proteins, high fiber, and a range of health benefitting bioactive compounds.

Introduction

Stinging nettle (*Urtica dioica, L. Urticaceae*) is a ubiquitous herb which is available in large part of the world. *Urtica dioica* is a moderately shade-tolerant species, which occurs on most moist or damp, weakly acid or weakly basic, richly fertile soils. Its stems and leaves are densely covered with Stinging hairs, which release potentially pain-inducing toxins, is rarely eaten by castles and rabbits (Taylor, 2009). This species is known as tenacious weeds, able to live in the toughest conditions, and notoriously known for inflicting pain. From ancient times, the fresh Stinging nettle is used for flailing arthritic or paralytic limbs with fresh Stinging nettle to stimulate circulation and bring warmth to joints and extremities in a treatment known as "urtication" (Green, 1820). Ancient Egyptians also reportedly used the infusion for the relief of arthritis and lumbago (Harrison 1966). Above mentioned practice of urtication or rubefaction became a standard in folk medicine as a remedy for arthritis, rheumatism, and muscular paralysis and is perhaps the most ancient medicinal use of Stinging nettle (Upton 2013). Nettle can be used to foster health and vitality of the people. Due to the nutritional and functional qualities of nettle, it has been utilized to alleviate symptoms associated with allergic rhinitis and improve oxidative stability in brine anchovies. It is also rich in fatty acids, carotenoid, and phenolic compounds, while its extracts have been reported to improve oxidative stability in brined vegetables (Rutto et al. 2013).

A comprehensive proximate analysis showed the shoots harvested from Stinging nettle (Shoot) showed close to 90% moisture and rests are proteins (3.7%), fat (0.6%), ash (2.1%), dietary fiber (6.4%), total carbohydrate (7.1%), and total calories (45.7 kcal/100 g) (wet basis). Besides, Stinging nettle (shoot) contains vitamin A, vitamin C, calcium, iron, sodium, and rich fatty acid profile (Rutto et al. 2013). Farag et al. (2013) studied the geographic, taxonomical and morphological diversity, genetics, etc., under control conditions. *Urtica dioica* is the only species of Urtica to be cultivated commercially for pharmaceutical purposes, and the commercial extraction of chlorophyll and stem fibers. He further reported *U. dioica* as a good source of flavonoids, phenylpropanoids, and caffeic acid analogues. Besides, the use of Nettle extract for rheumatism, eczema, allergic rhinitis, and arthritis is well studied (Harrison 1966; Upton 2013).

Stinging nettle is rarely domesticated due to its sting but the species remains popular as food and medicines in poor countries like Nepal (Uprety et al. 2012). From the centuries, in the foothills of Nepal's Himalayas, the Himalayan Stinging nettle has naturally grown in the wild. Recently, founder of Himalayan Wild Fibers, is in the process of developing the nettle fiber industry with the local community. It is expected to help in the development of strong fiber that would create work and income to many Nepalese and bring a durable and sustainable textile to the market (Tree hugger).

In Georgia, a meal of boiled Stinging nettle seasoned with walnut is common. Romanians use sour soup made from fermented wheat bran vegetables and green nettle leaves harvested from young plants (Costa et al. 2013). However, one of the most underutilized and neglected crops are now getting attention on their commercial utilization due to its nutrition and functional properties. Production and processing of different products from it will also support to uplift the economic status of the local people from third world countries (Palikhe 2012). Stinging nettle is very popular as a vegetable in a range of countries, particularly among the lower socioeconomic people. More work is needed to learn the nutritive value of Nepalese cultivars. Stinging nettle is consumed primarily as a boiled or cooked fresh vegetable whereby it is added to soups, cooked as a pot herb, or used as a vegetable complement in dishes. Although popular in Nepal, there is almost no study on Stinging nettle (Palikhe 2012).

In Nepal, wheat and barley are two most consumed cereal grains, after rice. These are the major source of starch, fiber, proteins, lipid, minerals, etc. Barley grain is reported to be effective in lowering blood cholesterol because of its high β-glucan content, 2–9% (Hassan et al. 2012). The recent approval of soluble barley beta-glucan health claims by the Food and Drug Administration of the USA for lowering blood cholesterol level could further boost food product development from barley and consumer interest in eating these food products (Yamlahi and Ouhssine 2013). Wheat is the staple diet for a majority

of global population, containing carbohydrates, protein, minerals, B-group vitamins, and dietary fiber, etc. Starch is the major component of wheat, providing calorie as well as the inner bran coats, phosphates, and other mineral salts; the outer bran, the much-needed roughage the indigestible portion that helps easy movement of bowels; the germ, vitamins B and E. Kumar et al. (2011) have comprehensively reviewed the nutritional content and medicinal properties of wheat.

Nepal is one of the developing countries in the world but it is rich in flora and fauna. Stinging nettle is one of the most popular wild edible plants (WEP) that provide staple and supplement foods. Often these WEPs become the top cash income to the local communities which contribute to food security to the region. However, there is hardly any work on the composition and nutritional properties of Stinging nettle in Nepal. Therefore, this work studied the nutritional and functional properties of stinging nettle dried powder. Besides, this work also compared the properties of Stinging nettle powder with wheat as well as barley flour.

Materials and Methods

Raw material collection and preparation

Stinging nettle (*Urtica dioica*) leaves were collected from Kirtipur, Nepal. The collected samples were carefully carried afresh to the laboratory for chemical analysis. The Stinging nettle leaves were cleaned (washed) so that the foreign particles are removed. The leaves were blanched for 1 min at 80°C in wet condition. The leaves were then drained, placed on the trays and so as to remove excess water. Then the trays were put inside the cabinet drier for drying at 60°C for 2 days, till the crispy texture was observed. Dried leaves were ground in a coffee grinder and sieved through the 80 size mesh making into a fine Nettle powder as done previously (Palikhe 2012). Barley and wheat flours were purchased from the local market. All dried flours were analyzed for moisture content and transferred into the hermitically sealed container.

Chemical analysis

All three flour samples *viz.*, wheat, barley and Stinging nettle were immediately analyzed in triplicates. Moisture content was measured by oven drying at 100°C until the constant weight was reached. Total crude oil of all three samples was extracted using hexane in Soxhlet System HT2 Texator, (Sweden). Total ash values were obtained by incineration of flour samples for overnight at 550°C at minimum 6 h (AOAC, 2005). Calcium content was measured by precipitation as calcium oxalate, dissolving in concentrated sulfuric acid and titration with standard potassium permanganate (Ranganna 2001). Iron was determined by converting iron present in foods into ferric form and treating thereafter with potassium thiocyanate to form the red ferric thiocyanate which is measured by colorimetry at 480 nm (Ranganna 2001). Total carbohydrate content was measured by the difference method. All chemical analyses were conducted by the methods as recommended by Ranganna (2001).

A number of functional properties of above samples were also analyzed. The antioxidant activity (AA, DPPH inhibition %) was determined by the method described by Nuengchamnong et al. (2009). Total polyphenol (TP, mg GAE/g) and carotenoids were determined as per the method described by Ranganna (2001). Tannin as an anti-nutritional factor was determined according to AOAC (2005).

Data analysis

Data were statistically processed by Gen Stat for analysis of variance (ANOVA), Microsoft Excel-2007 for analysis. Means of data were separated whether they are significant or not by using LSD (least square difference) method at 5% level of significance.

Results and Discussion

Proximate analysis of raw materials

Wheat and barley flours were purchased from the local market but nettle was processed into fine powders as mentioned in the methodology section. The chemical analyses of wheat flour, barley flour and nettle powder were carried out and results are presented in dry basis. The mean values of chemical composition of wheat flour, barley flour, and nettle powder are presented in Table 1. The initial moisture content of the leaves was not measured. However, previous studies have shown that the Nettle

plant contains relatively high level of moisture of, for example, 89% (Rutto et al. 2013) and 84.4% (Mishra 2007). The moisture content of wheat flour was 12.4% which is common in commercial wheat flour as previously reported by Kent and Evers (2004). Barley flour had similar moisture content, 12.2%. After cabinet drying of leaves followed by grinding, the moisture content of the Nettle powder was reduced significantly to 7.0% (Table 1).

Protein content of the ground wheat, barley and Stinging nettle were 10.6%, 11.8%, and 33.8% (db), respectively. Analytical data of the nettle powder exhibits about 3 times protein level as compared to the traditional source of cereal proteins, that is, rice, wheat, and barley. Previous study also showed a relatively higher amount of protein content, 33.6% (dry basis) in the Nettle powder. Considering a higher level of protein in Nettle powder, this species expected to supply higher concentrations of essential amino acids. Besides, it has a better amino acid profile than most of the other leafy vegetables (Rutto et al. 2013).

Rutto et al. (2013) has reported relatively higher amounts of all essential amino acid content in Stinging nettle, except leucine and lysine. Nettle leaf flour has been incorporated in many recipes, for example, bread, pasta, and noodles dough that suggest it could be used as a protein-rich supplement in starchy diets associated with poor and undernourished population like Nepal. As compared to the conventional source of proteins, Nettle powder contains 3.2 and 2.9 times greater amount of proteins as compared to wheat and barley flours, respectively. Nettle powder has one of the richest sources of crude fiber (9.1%, db) (Table 1). The amount of crude fiber in the nettle powder is significantly higher than most of the cereals and other plant foods, more than 9 times higher as compared to wheat and barley flour. Published literatures showed the Nettle powder has 6.4% (db) crude fiber (Rutto et al. 2013). The level of crude fat is relatively low at 3.6% (db), but this value is still higher than wheat (1.7%) and barley flour (1.7%).

Table 1. Chemical composition of wheat, barley, and nettle powders[1].

Parameters	Wheat flour	Barley flour	Nettle powder
Moisture (%)	12.37 ± 0.25	12.2 ± 0.19	7.04 ± 0.77
Crude protein (%,db)	10.6 ± 0.23	11.84 ± 0.09	33.77 ± 0.35
Crude fiber (%,db)	0.65 ± 0.13	1.03 ± 0.08	9.08 ± 0.14
Crude fat (%,db)	1.68 ± 0.23	1.73 ± 0.67	3.55 ± 0.06
Total ash (%,db)	0.56 ± 0.07	3.6 ± 0.08	16.21 ± 0.54
Carbohydrate (%,db)	86.51 ± 0.27	81.8 ± 0.08	37.39 ± 0.72
Calcium (mg/100 g)	18.94 ± 0.08	17.51 ± 0.26	168.77 ± 1.47
Iron (mg/100 g)	3.37 ± 0.29	3.63 ± 0.11	227.89 ± 0.21

db, dry basis.
[1]Values are mean ± Standard deviation of triplicates.

Stinging Nettle is rich in minerals. Current study showed the Nettle powder contained 16.2% (db) ash content which is much higher than conventional cereals (Table 1). Rutto et al. (2013) reported that the total ash content of Nettle powder is 19.1% (db) or 2.1% in wet basis. The higher level of minerals in Nettle is also demonstrated by higher level of calcium (169 mg/100 g) and iron (277 mg/100 g) (Table 2). Once again, these values are much higher than those from wheat and barley flours. USDA data showed the Nettle powder contains 4% calcium (db), 2.8% (db) potassium followed by phosphorus, magnesium and traces of iron, sodium and zinc (USDA, 2008). Based on this data, Nettle powder probably is one of the richest sources of minerals among the plant foods. In comparison, wheat flour and barley flour have much lower total ash content, 0.6% and 3.6%, respectively.

Nettle powder contained the lowest amount of carbohydrate (37.4%, db) as compared to regular cereals, for example, wheat (86.5%) and barley (81.8%). It shows the Nettle powder is much less glycemic as compared to the conventional sources of plant foods such as cereals and tuber in particular. Table 1 shows the crude fiber (9.1% db) forms a significant component of Nettle powder. Rutto et al. (2013) shows the carbohydrate content of Nettle powder is close to the currently analyzed value, 7.1% db. The calorific value of wheat flour was higher, that is, 381.9 kcal/100 g than barley flour and nettle powder which were 369.7 kcal/100 g and 307.2 kcal/100 g, respectively. This also shows the Nettle powder is low in calorie as compared to conventional cereals.

Carbohydrate levels in the nettle powder decreases with increase in the protein content, fiber, ash, and fat as shown in Table 1. The results agree to the report given by Palikhe (2012) Thapaliya (2010). Therefore, the current finding showed that the use of nettle powder and barley in bakery products likely to increase the protein, ash and fiber whereas decrease in calorific value as well as increase in bioactive compound (discussed later). The incorporation of barley flour in the cereal based food products such as biscuits, breakfast cereals, noodles, etc., potentially lower the blood cholesterol, cardiovascular and other diet related because of soluble β-glucan (Hassan et al. 2012).

Functional properties of raw materials

Nettle plant and its associated products are reported to be rich in a number of bioactive compounds (Knipping et al. 2012; Johnson et al. 2013; Rutto et al. 2013). It has been reported that the natural phenolic compounds play an important role in cancer prevention and treatment. Phenolic compounds from medicinal herbs and dietary plants include phenolic acids, flavonoids, tannins, stilbenes, curcuminoids, coumarins, lignans, quinones, and others. Various bioactivities of phenolic compounds are responsible for their chemopreventive properties (e.g., antioxidant, anticarcinogenic, or antimutagenic and anti-inflammatory effects) and also contribute to their inducing apoptosis by arresting cell cycle, regulating carcinogen metabolism and ontogenesis expression, inhibiting DNA binding and cell adhesion, migration, proliferation or differentiation, and blocking signaling pathways (Knipping et al. 2012; Johnson et al. 2013).

Table 2 shows the nettle powder contained relatively higher level of bioactive compounds, for example, tannin, total polyphenol (TP), antioxidant activity (AA), carotenoid, and total caloric value as compared to wheat and barley flours. The total phenolic content of nettle powder was 129 mg GAE (Gaelic Acid Equivalent)/g, which is much higher than the wheat flour (1.3 GAE/g) and barley flour (1.7 GAE/g). One of the quantitative analysis of plant phenolics in Nettle plant showed only 21 of the 45 compounds in levels above the reliable quantification limit (Orcic et al. 2014). Natural phenolic compounds reported to play important role in cancer prevention and treatment. A comprehensive review showed the compounds from medicinal herbs such as Nettle plant and dietary plants include phenolic acids, flavonoids, tannins, curcuminoids, coumarins, lignans, etc. (Huang et al. 2010).

The carotenoid content of wheat flour, barley flour and nettle powder were 320, 382.3, and 3496.7 $\mu g/g$, respectively. Nettle powder appeared to have almost ten times higher amount of carotenoid as compared to wheat flour and barley flour (Table 2). According to Rutto et al. (2013), the blanched nettle at 98°C for 1 min contained 4689 $\mu g/g$ amount of carotenoids. It seems that in both

Table 2. Analysis of wheat flour, barley flour, and nettle powder[1].

Parameters	Wheat flour	Barley flour	Nettle powder
Tannin content (% as is)	ND	0.53 ± 0.03	0.93 ± 0.01
Total polyphenol (mg GAE/g, db)	1.31 ± 0.01	1.76 ± 0.01	128.75 ± 0.21
Antioxidant activity (DPPH inhibition, % as is)	23.72 ± 0.45	28.64 ± 0.03	66.3 ± 0.12
Carotenoids ($\mu g/g$, db)	320.05 ± 0.08	382.3 ± 0.56	3496.67 ± 0.56
Calorific value (kcal/100 g)	381.93 ± 0.05	369.68 ± 0.84	307.24 ± 0.13

db, dry basis; ND, not detected.
[1]Values are mean ± standard deviation of triplicates.

cases, the most significant reductions might have occurred due to the longer exposure to heat during the drying process. It has been reported that the amount of vitamin A, iron, and calcium are significantly affected by the heat (Rutto et al. 2013).

Carotenoids are the precursors of vitamin A and similar compounds. β-carotene is one of most commonly known carotenoids which is a potent antioxidant as well as a dietary factor for growth. It is a precursor of vitamin A that has important role in vision, as the prosthetic group of the light sensitive proteins in retina, and a major role in the regulation of gene expression and tissue differentiation (Bender, 2003). Deficiency of vitamin A is a major public health problem around the world. The prevention of vitamin A deficiency is one of the three micronutrient priorities of the World Health Organization (WHO), others are iron and iodine.

Tannins (Polyphenols) occur in cereals, especially in the seed coat (Reilly et al. 2009). The tannin content of barley flour and nettle powder was 0.53 and 0.93 mg/100 g, respectively, whereas, no tannin was observed in the wheat flour. Polyphenols occur in cereals and these form complexes with proteins and inhibition of digestive enzymes. Nettle powder had higher level of anti-oxidant activity of 66.3 DPPH inhibition (%) as compared to wheat flour 23.72 DPPH inhibition (%) and barley flour (28.64 DPPH inhibition (%). Higher level of antioxidant activity (AA) was also observed in Nepalese nettle powder. Current data showed similar amino acid value in Nepalese nettle, as reported by Thapaliya (2010).

A prospective, randomized, double-blind, placebo-controlled, crossover study showed the *Urtica dioica* reported to have beneficial effects in the treatment of symptomatic benign prostatic hyperplasia (BPH). Further clinical trials should be conducted to confirm these results before concluding that *Urtica dioica* is effective.

Conclusions

Stinging nettle (*Urtica dioica*) is a common herb and its stem and leaves are densely covered with Stinging hairs that inflict pain. It is eaten as a curry, sour soup, vegetable complement in dish, etc. Stinging nettle has a great medicinal value such as relieve of arthritis, rheumatism, muscular pain, etc. Nettle powder contains high amount of protein (38%), crude fiber (9%), total ash (16.2%), calcium (0.17%), iron (0.23%), and relatively low in carbohydrate (37%). As compared to barley and wheat flour, it has much higher protein, crude fiber, fat, ash, calcium and iron, and low in glycemic index. Besides, it has excellent health enhancing functional properties as compared to conventional grains. As compared to barley and wheat, it has much higher level of tannin content, total polyphenol,

antioxidant activity, carotenoids, and lower calorific value. Bioactivities of these functional components may play important role in arthritis, rheumatism, muscular paralysis, potentially cancer prevention, etc.

Conflict of Interest

None declared.

References

AOAC. 2005. Official methods of analysis of AOAC International, 18th ed. AOAC International, Washington, DC.

Bender, D. A. (2003). Nutritional biochemistry of the vitamins. Cambridge, UK.

Costa, H. S., T. G. Albuquerque, S. S. Silve, and P. Finglas. 2013. New nutritional composition data on selected traditional foods consumed in Black Sea Area countries. *Journal of the Science of Food and Agriculture* 93:3524–3534.

Farag, M. A., S. H. El-Ahmady, F. S. Elian, and L. A. Wessjohann. 2013. Metabolomics driven analysis of artichoke leaf and its commercial products via UHPLC-q-TOF-MS. Phytochemistry 95:177–187.

Green, T. 1820. The universal herbal or botanical, medical, and agricultural dictionary containing an account all the known plants in the world. Vol. 2, Henry Fisher (Caxton Press.), Liverpool.

Harrison, R. K. 1966. Healing herbs of the Bible. Publisher Leiden, B., 58 p.

Hassan, A. A., N. M. Rasmy, M. I. Foda, and W. K. Bahgaat. 2012. Production of functional biscuits for lowering blood lipids. *World Journal of Dairy & Food Sciences* 7:1–20.

Huang, W. Y., Y. Z. Cai, and Y. Zhang. 2010. Natural phenolic compounds from medicinal herbs and dietary plants: potential use for cancer prevention. *Nutrition and Cancer* 62:1–20.

Johnson, T. A., J. Sohn, W. D. Inman, L. F. Bjeldanes, and K. Rayburn. 2013. Lipophilic stinging nettle extracts possess potent anti-inflammatory activity, are not cytotoxic and may be superior to traditional tinctures for treating inflammatory disorders. Phytomedicine 15, 20, 143–147.

Kent, N. L., and A. D. Evers. 2004. Technology of bread making. In: Kents technology of cereals. 4th ed. Oxford: Pergmon Press.

Knipping, K., J. Garssen, and B. Van't Land. 2012. An evaluation of the inhibitory effects against rotavirus infection of edible plant extracts. Virol. J. 9:135–137.

Kumar, P., R. K. Yadava, G. Gollen, S. Kumar, R. K. Verma, and S. Yadav. 2011. Nutritional contents and medicinal properties of wheat: a review. Life Sciences and Medicine Research 11:1–10.

Mishra, A. 2007. Preservation of *Sisnu* (nettle buds) by drying and fermentation process and its quality evaluation, M. Tech (food) Thesis, Tribhuvan University., Nepal.

Nuengchamnong, N., K. Krittasilp, and K. Ingkaninan. 2009. Rapid screening and identification of antioxidants in aqueous extracts of Houttuyniacordata using LC–ESI–MS coupled with DPPH assay. Food Chemistry. 117:750–756.

Orčić, D., M. Francišković, K. Bekvalac, E. Svirčev, I. Beara, M. Lesjak, et al. 2014. Quantitative determination of plant phenolics in *Urtica dioica* extracts by high-performance liquid chromatography coupled with tandem mass spectrometric detection. Food Chem. 143:48–53.

Palikhe, M. 2012. Preparation of biscuit incorporating nettle powder and its quality evaluation. B. Tech. (Food) Thesis. College of applied food and dairy technology, Purwanchal University, Nepal.

Ranganna, S. 2001. Handbook of analysis and quality control for fruit and vegetable products. Tata McGraw-Hill publishing Co. Ltd, New Delhi.

Reilly, A., C. Tlustos, L. O'Connor. 2009. Food Safety: A public health issue of growing importance. *in* M. J. Gibney, et al., eds. Introduction to Human Nutrition, 2nd edn. Wiley-Blackwell, West Sussex, UK.

Rutto, L. K., Y. Xu, E. Ramirez, and M. Brandt. 2013. Mineral properties and dietary values raw and processed Stinging nettle (*Urtica dioica* L.). Int. J. Food Sc. 2013. doi: 10.1155/2013/857120.

Taylor, K. 2009. Biological flora of the British Isles: *Urtica dioca* L. J. Ecol. 97:1436–1458.

Thapaliya, P. 2010. Effect of different blanching conditions, drying methods and temperatures and retention of vitamin C on *sisnu* powder. M. Tech (Food) Thesis, Tribhuvan University, Nepal.

Tree hugger. In Nepal's Himalayas Grows a Sustainable Textile Fiber. http://www.treehugger.com/sustainable-fashion/in-nepals-himalayas-grows-a-sustainable-textile-fiber-photos.html.

Uprety, Y., R. C. Poudel, K. K. Shrestha, S. Rajbhandary, N. N. Tiwari, U. B. Shrestha, et al. 2012. Diversity of use and local knowledge of wild edible plant resources in Nepal. J. Ethnobiol. Ethnomed. 8:16.

Upton, R. 2013. Stinging nettles leaf (*Urtica dioca* L.): extraordinary vegetable medicine. Journal of Herbal Medicine. 3:9–38.

USDA National Nutrient Database for Standard Reference Release 27. 2008. Basic Report 35205, Stinging Nettles, blanched (Northern Plains Indians).

Yamlahi, A. E., and M. Ouhssine. 2013. Utilization of barley (Hordeum vulgare L.) flour with common wheat (Triticum aestivum L.) flour in bread-making. Annals of Biological Research. 2013:119–129.

Permissions

The contributors of this book come from diverse backgrounds, making this book a truly international effort. This book will bring forth new frontiers with its revolutionizing research information and detailed analysis of the nascent developments around the world.

We would like to thank all the contributing authors for lending their expertise to make the book truly unique. They have played a crucial role in the development of this book. Without their invaluable contributions this book wouldn't have been possible. They have made vital efforts to compile up to date information on the varied aspects of this subject to make this book a valuable addition to the collection of many professionals and students.

This book was conceptualized with the vision of imparting up-to-date information and advanced data in this field. To ensure the same, a matchless editorial board was set up. Every individual on the board went through rigorous rounds of assessment to prove their worth. After which they invested a large part of their time researching and compiling the most relevant data for our readers.

The editorial board has been involved in producing this book since its inception. They have spent rigorous hours researching and exploring the diverse topics which have resulted in the successful publishing of this book. They have passed on their knowledge of decades through this book. To expedite this challenging task, the publisher supported the team at every step. A small team of assistant editors was also appointed to further simplify the editing procedure and attain best results for the readers.

Apart from the editorial board, the designing team has also invested a significant amount of their time in understanding the subject and creating the most relevant covers. They scrutinized every image to scout for the most suitable representation of the subject and create an appropriate cover for the book.

The publishing team has been an ardent support to the editorial, designing and production team. Their endless efforts to recruit the best for this project, has resulted in the accomplishment of this book. They are a veteran in the field of academics and their pool of knowledge is as vast as their experience in printing. Their expertise and guidance has proved useful at every step. Their uncompromising quality standards have made this book an exceptional effort. Their encouragement from time to time has been an inspiration for everyone.

The publisher and the editorial board hope that this book will prove to be a valuable piece of knowledge for researchers, students, practitioners and scholars across the globe.

List of Contributors

Abiodun Omowonuola Adebayo-Oyetoro, Olade-inde Olatunde Ogundipe and Kehinde Nojeemdeen Adeeko
Department of Food Technology, Yaba College of Technology, P.M.B 2011, Yaba, Lagos, Nigeria

Genet Gebremedhin Heshe and Ashagrie Zewdu Woldegiorgis
Center for Food Science and Nutrition, College of Natural Sciences, Addis Ababa University, Addis Ababa, Ethiopia

Habtamu Fekadu Gemede
Center for Food Science and Nutrition, College of Natural Sciences, Addis Ababa University, Addis Ababa, Ethiopia
Department of Food Technology and Process Engineering, Wollega University, Nekemte, Ethiopia

Gulelat Desse Haki
Department of Food Science and Technology, Botswana College of Agriculture, Private Bag 0027, Gaborone, Botswana

Nagat S. Mahmoud, Sahar H. Awad, Rayan M. A. Madani, Fahmi A. Osman and Amro B. Hassan
Environment and Natural Resource and Desertification Research Institute (ENDRI), National Center for Research, Khartoum, Sudan

Khalid Elmamoun
Sudanese Atomic Energy Commission (SAEC), Khartoum, Sudan

Vincent Mlotha, Agnes Mbachi Mwangwela, William Kasapila, Edwin W.P. Siyame and Kingsley Masamba
Faculty of Food and Human Sciences, Lilongwe University of Agriculture and Natural Resources (LUANAR) Lilongwe, Malawi

Noha I. A. Mutwali, Abdelmoniem I. Mustafa and Isam A. Mohamed Ahmed
Department of Food Science and Technology, Faculty of Agriculture, University of Khartoum, Shambat, 13314 Khartoum, Sudan

Yasir S. A. Gorafi
Wheat Research Program, Agricultural Research Corporation, Wad Medani, Sudan

Natthakarn Nammakuna and Puntarika Ratana-triwong
Department of Agro-Industry, Faculty of Agriculture Natural Resources and Environment, Naresuan University, Phitsanulok 65000, Thailand

Sheryl A. Barringer
Department of Food Science and Technology, The Ohio State University110 Parker Hall2015 Fyffe Road, Columbus, Ohio 43210

Dulce Nhassico, Adelaide Cumbane, Luis Sitoe and Humberto Muquingue
Department of Biochemistry, Faculdade de Medicina, Universidade Eduardo Mondlane, Maputo, Mozambique

James Howard Bradbury, Ian C. Denton and Matthew P. Foster
EEG, Research School of Biology, Australian National University, Canberra, ACT 2601, Australia

Julie Cliff
Department of Community Health, Faculdade de Medicina, Unversidade Eduardo Mondlane, Maputo, Mozambique

Rita Majonda Constantino Cuambe and Joao Pedro
Instituto de Investigacao Agraria de Mocambique (IIAM), Nampula, Mozambique

Arlinda Martins
Direccao Provincial de Saude, Nampula, Mozambique

Olusegun A. Olaoye, Stella C. Ubbor and Ebere A. Uduma
Department of Food Science and Technology, Michael Okpara University of Agriculture, Umudike, Abia State, Nigeria

Ganiyat O. Olatunde, Folake O. Henshaw and Michael A. Idowu
Department of Food Science and Technology, Federal University of Agriculture, Abeokuta, Ogun State, Nigeria

Keith Tomlins
Foods and Markets Department, Natural Resources Institute, University of Greenwich, Medway, Kent, United Kingdom

Omobolanle O. Oloyede, Samaila James, Ocheme B. Ocheme, Chiemela E. Chinma and V. Eleojo Akpa
Department of Food Science and Technology, Federal University of Technology, PMB 65, Minna, Niger State, Nigeria

Adebukola T. Omidiran, Olajide P. Sobukola, Abdul-Rasaq A. Adebowale, Olusegun A. Obadina and Lateef O. Sanni
Departments of Food Science and Technology, Federal University of Agriculture, Abeokuta, Nigeria

Ajoke Sanni
Nutrition and Dietetics, Federal University of Agriculture, Abeokuta, Nigeria

Keith Tomlins
Natural Resources Institute, University of Greenwich, Greenwich, U.K.

Tosch Wolfgang
SABMiller Plc, Surrey, U.K.

Małgorzata Przygodzka, Henryk Zieliński and Grzegorz Lamparski
Division of Food Science, Institute of Animal Reproduction and Food Research of the Polish Academy of Sciences, Tuwima 10, 10-748 Olsztyn 5, Poland

Zuzana Ciesarová and Kristina Kukurová
National Agriculture and Food Centre – Food Research Institute, Priemyselná 4, 824 75 Bratislava 26, Slovak Republic

Poonam Singha and Kasiviswanathan Muthukumarappan
Department of Agricultural and Biosystems Engineering, South Dakota State University, Brookings, South Dakota 57007

Yui Sunano
Graduate School of Bioagricultural Sciences, Nagoya University, Furocho, Chikusa-ku, Nagoya, Aichi 464-8601, Japan

Virgílio Gavicho Uarrota, Bianca Coelho, Rodolfo Moresco and Marcelo Maraschin
Plant Science Center, Plant Morphogenesis and Biochemistry Laboratory, Postgraduate Program in Plant Genetic Resources, Federal University of Santa Catarina, Rodovia Admar Gonzaga 1346, CEP 88.034-001 Florianópolis, SC, Brazil

Eduardo da Costa Nunes, Luiz Augusto Martins Peruch, Enilto de Oliveira Neubert and Hernan Ceballos
Santa Catarina State Agricultural Research and Rural Extension Agency (EPAGRI), Experimental Station of Urussanga (EEUR), Rd. SC 446Km 19 S/N, Urussanga, Florianópolis, SC, CEP 88840-000, Brazil

Moralba Garcia Domínguez, Teresa Sánchez, Jorge Luis Luna Meléndez, Luis Augusto Becerra Lopez-Lavalle and Clair Hershey
International Center for Tropical Agriculture (CIAT), Apartado Aéreo 6713 Cali, Colombia

Dominique Dufour
International Center for Tropical Agriculture (CIAT), Apartado Aéreo 6713 Cali, Colombia
Centre de Coopération Internationale en Recherche Agronomique pour le Développement (CIRAD), UMR Qualisud, 73 Rue Jean-Francois Breton, TAB-95/16, 34398 Montpellier Cedex 5, France

Miguel Rocha
Centre of Biological Engineering, University of Minho, Campus de Gualtar, 4710-057 Braga, Portugal

Maria del Carmen Wacher-Rodarte and Tanya Paulina Trejo-Muñúzuri
Depto de Alimentos y Biotecnología, Facultad de Química, UNAM, Ciudad Universitaria Coyoacán, 04510 México, Distrito Federal, México

Jesús Fernando Montiel-Aguirre
Depto de Bioquímica, Facultad de Química, UNAM, Ciudad Universitaria Coyoacán, 04510 México, Distrito Federal, México

Maria Elisa Drago-Serrano, Raúl L. Gutiérrez-Lucas, Jorge Ismael Castañeda-Sánchez and Teresita Sainz-Espuñes
Depto. Sistemas Biológicos, UAM-XochimilcoCalzada del Hueso No.1100, Coyoacan, 04960 Mexico, Distrito Federal, México

Lanhong Zheng, Kangli Yang, Mi Sun, Jiancheng Zhu, Mei Lv, Daole Kang, Wei Wang, Mengxin Xing and Zhao Li
Key Laboratory for Sustainable Utilization of Marine Fisheries Resources, Ministry of Agriculture, Yellow Sea Fisheries Research Institute, Chinese Academy of Fishery Sciences, Qingdao 266071, China

Jia Liu
Medical College, Qingdao University, Qingdao 266021, China

Si Zhu and Guibing Chen
Center for Excellence in Post-Harvest Technologies, North Carolina A&T State University, The North Carolina Research Campus, 500 Laureate Way, Kannapolis, North Carolina 28081

Maria Eduardo
Department of Biology and Biological Engineering/ Food and Nutrition Science, Chalmers University of Technology, Gothenburg, Sweden
Departamento de Engenharia Química, Faculdade de Engenharia, Universidade Eduardo Mondlane, Maputo, Moçambique

Lilia Ahrné
Department of Biology and Biological Engineering/ Food and Nutrition Science, Chalmers University of Technology, Gothenburg, Sweden
Process and Technology development, SP Technical Research Institute of Sweden, Food and Bioscience, Gothenburg, Sweden

Ulf Svanberg
Department of Biology and Biological Engineering/ Food and Nutrition Science, Chalmers University of Technology, Gothenburg, Sweden

Mathew K. Bolade
Department of Food Science and Technology, Federal University of Technology, P.M.B. 704, Akure, Ondo State, Nigeria

Wasiu Awoyale
International Institute of Tropical Agriculture, PMB 5320 Oyo Road, Ibadan, Oyo State, Nigeria
Department of Food, Agriculture and Bioengineering, Kwara State University, PMB 1530, Malete, Kwara State, Nigeria

Busie Maziya-Dixon
International Institute of Tropical Agriculture, PMB 5320 Oyo Road, Ibadan, Oyo State, Nigeria

Lateef Oladimeji Sanni and Taofik Akinyemi Shittu
Department of Food Science and Technology, Federal University of Agriculture Abeokuta, PMB 2240, Abeokuta, Ogun State, Nigeria

Bhaskar Mani Adhikari and Alina Bajracharya
Department of Food Technology, National College of Food Science and Technology, Kathmandu, Nepal

Ashok K. Shrestha
Nutrition and Food Science, School of Science and Health, Hawkesbury Campus, University of Western Sydney, Penrith NSW 2751, Australia

Index